New Trends in Ionic (Co)Polymers and Hybrids

NEW TRENDS IN IONIC (CO)POLYMERS AND HYBRIDS

ECATERINA STELA DRAGAN
EDITOR

Nova Science Publishers, Inc.
New York

For permission to use material from this book please contact us:
Telephone 631-231-7269; Fax 631-231-8175
Web Site: http://www.novapublishers.com

LIBRARY OF CONGRESS CATALOGING-IN-PUBLICATION DATA

ISBN 978-1-60021-611-4

Published by Nova Science Publishers, Inc. ; New York

Contents

PREFACE

Ionic polymers are present almost everywhere in the world, biological systems being the most important reservoir of ionic polymers such as nucleic acids, and many proteins, alginic acid and so on. Artificial ionic polymers derived from natural polymers such as chitin, cellulose or starch, by adequate chemical transformations, and a multitude of synthetic ionic polymers are generated in the laboratory. They can contain ionic or ionizable groups. Ionic polymers, either artificial or synthetic, have a multitude of applications such as: stabilization or destabilization of dispersions, thickening of solutions, oil recovery aids, water purification, corrosion inhibitors, soil conditioners, antistatic agents, additives in cosmetics and foods, surface modification, and so forth. Solid ionic hybrid materials found applications such as polymer electrolytes or substrates for the organic synthesis in solid phase. Many books and article reviews were dedicated to this wide class of polymers but the permanent diversification of their structures and applications ask for new and updated overviews on this field. Therefore, this book reports updated information from literature as well as original contributions in the field of ionic (co)polymers and hybrids, mainly on the applications of ionic polymers and hybrids such as: self-assembled multilayers, ionic polymers containing azobenzene chromophore, phase separation processes, chelating ion exchangers, polymer electrolytes, functionalized solid surfaces and ionic hybrid hydrogels.

Chapter 1 - Some molecular and supramolecular systems containing azobenzene chromophores have been surveyed in this review. Three main topics have been discussed in more detail: the interaction between some strong polycations, bearing quaternary ammonium salt centers in the backbone, and a large variety of multicharged azo dyes in dilute aqueous solution, supramolecular order of the solid state polycation/azo dye complexes, and the heat-induced reactions of some polycation/methyl orange (MO) complexes. Thermochemical transformation of the electrostatically formed complexes between MO and polycations containing primary amine groups as ammonium salts afforded new polymers with a high content of covalently bound 4-N,N-dimethylaminoazobenzene groups in the side chain. Transformation of sulfonate-ammonium ion pairs into sulfonamide bonds, by heating at elevated temperature, was supported by elemental analyses, Fourier transform infrared spectroscopy (FTIR), ^{1}H-NMR and UV-vis spectroscopy, and thermogravimetric analyses (TGA). The spacer length between the backbone and azobenzene groups influenced the polymer properties before and after the heat-induced reaction.

Chapter 2 - Polyelectrolytes are charged polymers which have found important applications in the major fields of science and engineering such as chemistry, physics, biology, chemical and materials engineering. The review presents some of the more important and recent experimental and theoretical data regarding the application of PELs in processes which imply the phase separation (*solid-liquid and liquid-liquid*) in aqueous systems. Selected papers in this area are briefly discussed with particular attention given to the influence of several parameters related to the characteristics of the suspended particle (concentration, size), the polyelectrolyte (type of functional groups, structure, molar mass, charge density, dose) and the environment (ionic strength, pH, temperature) on the polymers efficiency in the flocculation process.

Chapter 3 - Organic chelating sorbents have been intensively investigated because they proved to be effective materials for the removal/recovery of polluting metal ions from industrial effluents, water supplies and mine waters. This review surveys the main contributions concerning organic chelating sorbents containing amidoxime and iminodiacetate functional groups. Also, the classification of organic chelating sorbents according to the structure of the functional group with specifications on the type of interaction between chelating groups and polymer matrix, as well as the nature and physical status of polymer matrix has been described in the paper. Two less explored approaches for the synthesis of crosslinked chelating polymers as beads containing iminodiacetic groups have been shown. The preparation of anion exchangers with primary amine groups by the aminolysis-hydrolysis reaction with 1,2-diaminoethane of the nitrile groups, contained in some porous copolymers of acrylonitrile-divinylbenzene, and by the Hofmann degradation of the amide groups contained in acrylamide-divinylbenzene copolymers, followed by the carboxymethylation of the primary amine groups to prepare chelating ion exchangers, has been presented. The influence of porogen nature, monomers dilution and crosslinking degree on the properties of the ion exchangers was examined. Structural characterization, functionalization-morphology correlations, the ion exchange properties, and potential applications for the ion exchangers thus prepared have been discussed.

Chapter 4 - Polymer electrolyte has been studied since 1973, and its ionic conductivity was still not high enough. Recently, nanotechnology provided some new insights, and some marked progress was achieved, indicating great promise for its practical application. After introduction of nano fillers, the formed nanocomposites with poly(ethylene oxide) as polymer electrolytes for lithium ions present evidently an enhancement in ionic conductivity, which is due to several factors such as decreasing the content of crystal phase and producing new pathways for the movement of lithium ions along the surface of the nanoparticles. The behavior of the interphase between polymer electrolytes and electrodes is also ameliorated. Of course, different species of nano fillers and their amount and surface groups will lead to different effects. Prior to its practical applications, there are still some challenging questions to solve.

Chapter 5 - Controlling the surface chemical composition and mastering its modification at the nanometer scale are critical issues for high-added value applications. Three main directions of this topic have been followed in the chapter: modification of membrane properties by sequential adsorption of polyelectrolytes, multicomponent stimuli-responsive systems based on chemically or physically modification of solid substrates with appropriate stimuli-responsive polymers, and generation and uses of polyelectrolyte-colloids reactive hybrids. The original contribution of the authors concerning the preparation of reactive hybrids by the adsorption of poly(vinylamine-*co*-vinylformamide) onto silica particles as a function of polycation concentration and the ionic strength, and the post-reactions of crosslinking with epichlorohydrin, and the preparation of oligopeptides on the resulted hybrid were discussed in more detail. The reactive hybrid and the peptide formation on this new substrate were followed by total carbon analysis, electrokinetic measurements and X-ray photoelectron spectroscopy.

Chapter 6 - The swelling capacity of the classical hydrogels, generally consisting in hydrophilic networks, can be controlled by co-networking with a hydrophobic phase resulting in amphiphilic co-networks. Such phase can be formed by a polysiloxane well known for its hydrophobic character. Polysiloxanes, on the other hand, can be chemically modified by introduction of ionic groups to induce hydrophilicity and other interesting properties depending on the cationic centers content and the substituent structure. Recent developments in this direction were briefly reviewed in the paper. Own results on the synthesis and characterization of some novel ionic organic/siloxane hydrogels were also presented. Ionic organic/siloxane crosslinked copolymers containing quaternary ammonium salt (QAS) groups in the backbone were obtained by a polyaddition reaction between 1,3-bis(3-aminopropyl)tetramethyldisiloxane and siloxane oligomers, on the one side, and some reactive cationic oligomers containing preformed QAS groups, on the other side. Ionic hybrid hydrogels containing QAS sequences in the side chain were obtained by the quaternization reaction of either polyethylene glycol chains bearing tertiary amine groups or oligo(N,N-dimethylaminoethylmethacrylate) with a polyfunctional chloroalkylated siloxane. Thermooxidative stability and swelling properties of the ionic hybrid hydrogels in different solvents were evaluated.

Chapter 7 - The self-assembled multilayers (SAMs) built with a large variety of building blocks such as synthetic polyelectrolytes, biopolymers, organic compounds, organic and inorganic colloidal particles and their applications, were briefly reviewed in this Chapter, on the basis of the the newest information, being well known that the number of publications in this field increases continuously with every year. The mechanism of the multilayer growth and factors which influence the stability of these nanoarchitectures are still being investigated with interest for both fundamental and practical reasons. The main tools which allow to control the properties of these nanostructured materials and to tailor nanoobjects for specific applications have been discussed in detail, with a special concern on the influence of ionic strength, pH, solvent quality, and temperature, the influence of the last parameter being less investigated so far.

In: New Trends in Ionic (Co) Polymers and Hybrids
Editor: Ecaterina Stela Dragan, pp. 1-30
ISBN 978-1-60021-611-4

Chapter 1

MOLECULAR AND SUPRAMOLECULAR SYSTEMS CONTAINING AZOBENZENE CHROMOPHORES

Ecaterina Stela Dragan* and Marcela Mihai
"Petru Poni" Institute of Macromolecular Chemistry,
Aleea Grigore Ghica Voda 41 A, RO-700487 Iasi, Romania

ABSTRACT

Some molecular and supramolecular systems containing azobenzene chromophores have been surveyed in this review. Three main topics have been discussed in more detail: the interaction between some strong polycations, bearing quaternary ammonium salt centers in the backbone, and a large variety of multicharged azo dyes in dilute aqueous solution, supramolecular order of the solid state polycation/azo dye complexes, and the heat-induced reactions of some polycation/methyl orange (MO) complexes. Thermochemical transformation of the electrostatically formed complexes between MO and polycations containing primary amine groups as ammonium salts afforded new polymers with a high content of covalently bound 4-N,N-dimethylaminoazobenzene groups in the side chain. Transformation of sulfonate-ammonium ion pairs into sulfonamide bonds, by heating at elevated temperature, was supported by elemental analyses, Fourier transform infrared spectroscopy (FTIR), ^{1}H-NMR and UV-vis spectroscopy, and thermogravimetric analyses (TGA). The spacer length between the backbone and azobenzene groups influenced the polymer properties before and after the heat-induced reaction.

Keywords: azobenzene, photoisomerization, polycations, azo dyes, metachromasy, polycation/azo dye complexes, methyl orange, UV-vis spectroscopy, heat induced reaction, FTIR spectroscopy, ^{1}H-NMR, thermogravimetric analyses (TGA).

* E-mail: sdragan@icmpp.ro

LIST OF ABBREVIATIONS

AFM	Atomic Force Microscopy
Am	Amaranth
ATRP	atom transfer radical polymerization
Az-PAA	poly(allylamine) derivative bearing a small amount of azobenzene residues as pendant active ester (N-hydroxysuccinimide ester of azobenzene carboxylate)
CR	Congo Red
CS	Crystal Scarlet
CSMOO	Crocein Scarlet MOO
DB1	Direct Blue 1
DR80	Direct Red 80
DSC	Differential Scanning Calorimetry
DTG	derivative thermogravimetry
ESAM	electrostatic self-assembled multilayers
FTIR	Fourier transform infrared spectroscopy
^{1}H-NMR	Proton Nuclear Magnetic Resonance
LbL	layer-by-layer
LCST	lower critical solution temperature
MO	methyl orange
NaPSS	poly(sodium styrenesulfonate)
PAEAH	poly(2-aminoethyleneacrylamide hydrochloride)
PAH	poly(allylamine hydrochloride)
PAI	poly(amide imide)
PAZO	poly{1-4[4-(3-carboxy-4-hydroxyphenylazo)benzensulfonamido]-1,2-ethanediyl sodium salt}
PDADMAC	poly(diallyldimethylammonium chloride)
PEI	polyethyleneimine
PVS	poly(vinylsulfonate)
PCA_5	polycation containing 95 mole % of (N,N-dimethyl-2-hydroxypropyleneammonium chloride) units in the backbone
PCMO	polycation/MO complex
PolyAZOB	polymer with azobenzene groups in the side chain
PSS	Ponceau SS
TGA	thermogravimetric analysis
WAXD	Wide Angle X-ray Diffraction

1. INTRODUCTION

From the first organic material with reversible photoinduced birefringence based on methyl orange (MO) introduced in poly(vinylalcohol) reported by Todorov et al. [1], molecular systems incorporating azobenzene chromophores constituted a very attractive research area both in practice and academia. Azobenzene chromophores are well-known to undergo reversible isomerization from the thermodynamical more stable *trans* configuration (E-form) to the *cis* configuration (Z-form) by UV-light irradiation. The reverse process takes place by thermal treatment and/or by visible light irradiation ($\lambda > 450$ nm) [1-7]. The reversible photoisomerization process can be applied to design promising materials for optical information storage (reversible optical memory) and switching applications in micro- and optoelectronics. It is known that the rates of photoinduced *trans* to *cis,* and thermal induced *cis* to *trans* izomerization of azobenzene chromophores attached to polymers depend on the polymer morphology [2,3,7]. Reversible changes of optical properties are achieved by molecular photoreactions changing the supramolecular structure of polymer films, an optical anisotropy being thus generated. When the free volume in the vicinity of an azobenzene group is above a critical limit, the *trans* to *cis* photoizomerization rapidly occurs, unlike the azobenzene molecules located in positions with a low free volume which isomerize more slowly, but are characterized by a faster reversibility attributed to the presence of *cis*-forms in stressed sites, with a lower free volume [7]. Azobenzene groups could be part of a cationic or anionic bipolar amphiphile [8-11], a polymer with azobenzene in the side chains [7,12-44] or in the backbone [45-50], block copolymers [51-53], azo dyes and their complexes with polycations in bulk and solution [54-60] or in multilayer thin films [61-78], etc.

Numerous papers still concern the synthesis, characterization and photoizomerization properties of the side chain azobenzene containing polymers having a variable content in azobenzene containing monomers [16-39]. The side azobenzene chains are contained either in a soluble polymer [16-33,35-38], in a hybrid [6] or in a network [34,40,41]. The main requirements for the azobenzene containing polymers are to exhibit a high contrast, a fast response time, a good optical quality and a long thermal stability. Erasability and rewritability on these materials are another willing properties which enhace their value. In their investigations on the photoinduced birefringence in films of amorphous copolymers containing azobenzene groups using linearly polarized light, Natansohn et al. have demonstrated a cooperative motion in these copolymers, the polarity of the involved groups being the major factor below the glass transition temperature (T_g) [18,22,24]. It has been shown that the excess birefringence could be correlated with the amount of azobenzene containing units in copolymer [18]. Stumpe et al. synthesized poly(amide imide)s (PAIs) with azobenzene units characterized by high T_g and good film forming properties, the surface properties of polymer being successfully combined with the photopatterning [19]. The molecular orientation of PAIs films induced by irradiation with linearly polarized light, determining the orientation of the azobenzene groups, has been strongly influenced by the 4' substituent of the azobenzene groups and by the polymer backbone. Kanbara et al. have reported on some new azobenzene containing polymers synthesized by palladium-catalyzed amination and polycondensation [20]. Comparison of photoinduced isomerization rate constants of the polymers and the corresponding model compound showed little differences, thus suggesting no significant influence of the polymer backbone on the azobenzene *cis-trans* isomerization.

Azobenzene chromophores have been successfully attached to natural cellulose of ultrahigh molecular weight, polymers with varying degrees of substitution ranging from 0.13 to 0.99 being obtained and characterized by Tripathy et al. [21]. These polymers showed photoinduced surface relief gratings directly formed on the thin films.

Organic-inorganic polymer hybrids prepared by the sol-gel technique from azobenzene-modified poly(2-methyl-2-oxazoline) and tetraalcoxysilanes have been studied by Imai et al.

[6]. The *trans* to *cis* isomerization and the kinetics of the thermal *cis* to *trans* isomerization of azobenzene chromophores have been compared with those in dilute polymer solution or in polymer film, as well as with model compounds doped in sol-gel glass.

Polyanions containing azobenzene chromophore in the side chain were used in the electrostatically self-assembled multilayers (ESAM) thin film construction by the interaction with either ionenes, polyethyleneimine (PEI), or poly(diallyldimethylammonium chloride) (PDADMAC) [28-30]. The influence of the polycation structure on the film formation with poly{1-4[4-(3-carboxy-4-hydroxyphenylazo)benzensulfonamido]-1,2-ethanediyl sodium salt} (PAZO) as polyanion and on its photoresponse has been studied by Advincula et al. [28-30]. The photoisomerization reaction was studied by irradiating the sample with UV light and recording the spectra over different time intervals until a photostationary state was reached. The influence of the polycation structure manifested as follows: when the photoisomerization reaction has been induced by UV light in the PDADMAC/PAZO film, a spectral change occured, with a progressive decrease in the π-π^* band maximum intensity, the maximum position slowly undergoes a blue shift, from 372 to 366 nm, at the end of the UV light irradiation; when the photoisomerization reaction has been induced in the PEI/PAZO film, the position of the π-π^* band changed but with a more complicated behavior. Thus, the initial position of the 372 nm band has been red shifted first by about 4 nm and then returned to a constant value at 366 nm. Even if no clear band assigned to the *cis* isomer appeared (a weak absorption band centered at 460 nm), the *trans-cis* isomerization reaction occured to some extent. Although, the ionic conditions were the same in both systems, the different extent of protonation or the branched nature of PEI could change the ionic conditions on the film surface and the microenvironment inside the film. The red shift observed in PDADMAC/PAZO system has been interpreted by the formation of J aggregates. A more complex behavior emerged from the PEI/PAZO interaction in which a red shift in the first part of the absorption process was followed by a blue shift as the number of the deposited bilayers increases. The authors assumed that either both types of aggregates coexist in this system or that a complex molecular arrangement gives rise to nonmonotonic changes in the spectra.

A more complex self-assembled film has been designed by Suzuki et al. by employing PAZO as polyanion and a poly(allylamine) derivative bearing a small amount of azobenzene residues as a pendant active ester (N-hydroxysuccinimide ester of azobenzene carboxylate) (Az-NHS) [31]. X-ray reflectivity measurements revealed a well-ordered layer-by-layer (LbL) structure of PAZO and polycation layers. The azobenzene chromophore in the multilayer films was found to be isomerized from *trans* to *cis* upon UV light irradiation, while a visible light irradiation did not induce reverse reaction from *cis* to *trans* isomer. Furthermore, the thermal isomerization reaction from *cis* to *trans* form was not fully reversible. A successful photoisomerization and fully reversibility was found when a new type of polycation containing a small amount of chromophore was prepared from PAH containing 0.005 mole % of azobenzene chromophore as side chain and assembled with NaPSS. The introduction of azobenzene chromophores after the multilayer has been constructed with poly(allylamine) or PEI containing reactive amino groups has been also studied by Suzuki et al. [33]. This technique has been successfully applied to incorporate new functional moieties into the multilayer films, being the first paper reporting the chemical reaction of an active ester, Az-NHS, to polyelectrolyte multilayers. It was observed that Az-

NHS reacted mainly with the outermost layer of PAH, directly exposed to the reactive ester solution. The influence of polyanion on the azobenzene aggregation has been reflected in the formation of H-aggregates in the multilayer films containing poly(vinylsulfonate) (PVS), but did not in the poly(styrenesulfonate) (NaPSS) based films. The azobenzene-modified films showed E-Z photoisomerization to some extent, though the photoresponse was suppressed when the azobenzene chromophore formed aggregates.

Bipolar amphiphiles, either anionic or cationic, containing azoaromatic groups have been also used as building blocks in the construction of photoreactive ESAM thin films with controlled architecture [8-10]. Photoswitchable compounds reported by Toutianoush et al. [8,9] comprised cationic and anionic bipolar amphiphiles containing azobenzene and ionene polyions carrying the azobenzene units in the main chain. It was observed that the aggregate formation was less pronounced in the multilayers of the azobenzene containing ionenes than in the multilayers of azobenzene bolamphiphiles, a higher photoconversion being found in the first case. Hong et al. [10] have controlled supramolecular ordering in the photochromic layers by the ionic strength in bolamphiphile solution. The absorption maximum in solution at 341 nm was slightly blue shifted to 337 nm by about 4 nm with increasing ionic strength to 0.1 M, and more shifted to 335 nm with increasing ionic strength up to 1.0 M. A strong influence of the ionic strength on the photoisomerization characteristics has been observed. The *trans-cis* photoisomerization of the azobenzene units in the sample with no salt was completed after 5 min. In contrast to that, the increased amount of the added salt caused a much slower isomerization process, by almost ten times.

An interesting study on the aggregation modes of the azobenzene-substituted amphiphiles in aqueous vesicular dispersions has been developed by Kuiper and Engberts [11] to get more insights into the impact of the vesicles preparation on the H-aggregation and into the impact of H-aggregation on membrane permeability. UV-vis spectra gave information about the aggregation state within the membrane based on phosphate amphiphiles containing azobenzene groups. A λ_{max} between 297 and 305 nm has been attributed to maximal H-aggregation. The closest interaction between the azobenzenes with a strong interchain interaction in the bilayer. *Trans-cis* isomerization of azobenzenes in this type of aggregates was not possible. *Trans-cis* isomerization was easier in the monomeric or dimeric forms of azobenzene at a λ_{max} around 350 nm.

Backbone azoaromatic chromophore functionalized polymers are very attractive because the contour length of the polymer could be selectively lengthned or shorted by switching between the *trans*- and *cis*- azo configuration, this being a significant contribution toward the molecular level machines [49,50]. Lee et al. combined the polyelectrolyte character and the optical modulation property in a water soluble azobispiridinium oligomer, which offers the possibility to construct ESAMs by the alternate interaction with a polyanion [49]. The holographic grating formation of bulk azobispyridinium films has been evidenced as a three-dimensional atomic force microscopy (AFM) image.

A special attention has been lately focused on the block copolymers containing azobenzene moieties because the photoisomerization of azobenzene may be used to influence the morphology or the nanostructures that can self-assemble in block copolymers employing an ATRP method, to understand the nanostructure-functionality correlations [51-53]. The ATRP method has been used for synthesis because this method can be applied to a large variety of monomers and leads to polymers with narrow polydispersities. A novel series of multifunctional block copolymers, which contain both a well defined length of poly(ethylene glycol) as the hydrophilic block and different length of poly(methacrylate) with an azobenzene unit as the hydrophobic, liquid crystalline, and photoisomerizable block have been synthesized by Tian et al. [51]. The azobenzene group could act as a probe to study the mutual influences of the photoisomerization on the microphase separation, and of the block copolymer morphologies on the photochromism of the azobenzene units. Cui et al. observed that the photoinduced orientation is much more difficult to develop in the diblock copolymers, namely poly{styrene-*b*-6-[4-(4-methoxyphenylazo)phenoxy]hexyl methacrylate} than in the homopolymer [52]. The azobenzene groups were constrained due to the microdomain structures, being confined either inside the lamellae or between the PS cylinders. These constraints hindered the development of photoinduced alignment of azobenzene groups. Water soluble azobenzene-containing diblock copolymers have been synthesized by Ravi et al., one block being all the time poly(2-dimethylaminoethyl methacrylate) [53]. The authors observed that the aqueous solutions of the diblock copolymers showed a lower critical solution temperature (LCST), the LCST increasing upon UV light irradiation due to the increasing the concentration in more hydrophilic *cis* conformers by *trans-cis* isomerization.

In the next sections we will summarize the main recently published articles in the field of the molecular and supramolecular systems containing azo dyes and our recent contributions focused on this topic.

2. Polycation/Azo Dyes Complexes in Aqueous Solution

The ability of polyelectrolytes to change the environment of the dyes in solution is well known [79-89] and has been intensively studied for many azo dyes, the most investigated being MO and its homologous [79-85]. Other azo dyes have been also used for this purpose [90-92]. Numerous studies have been focused on the determination of the thermodynamic parameters of the polyion-dye binding [79-85], and on the improving the dyebility of fibers [86-88]. Dye binding onto ionic polymers is usually controlled by three main types of interactions: the electrostatic interactions of ionic dye molecules with oppositely charged sites of polyions; the nonspecific interactions, the hydrophobic interactions between the dye molecules and nonpolar sites of polyions being very important in solubilization of the dyes; the interaction between the π electrons of the bound dye molecules, the main cause of the metachromatic behavior of some dyes [83,85,89-92]. As a global definition, metachromasy is the phenomenon whereby the ionic dyes change their color in aqueous solution as a result of their interaction with oppositely charged macroions, or charged surfaces [89]. This is a consequence of the interactions of the dye orbitals with each other or with the orbitals of polyion, when the dye molecules stack in close proximity on the macroion chain or on the surface [89].

The nature of the interaction between dyes and polyelectrolytes, the effects of this interaction on the absorption or fluorescence properties of the dyes, the types of dyes interacting with polyelectrolytes and undergoing a metachromatic shift, the types of polyelectrolytes suitable as substrates for metachromatic dyes, the applications of the metachromasy are the main directions concerning this phenomenon [56,73,93,94]. There are some factors which control the metachromasy: a minimum critical chain length of 7 - 20 monomer units, a low ionic strength, a high dielectric constant of the medium, all of them leading to the increase of the dye aggregation on the polymer chain. Numerous polycations with cationic centers in the side chain like quaternized polyvinylimidazole [82] and polyvinylpyridine [83,85,89], poly(vinylbenzyltrialkylammonium chloride) [79,85], or in the backbone, like in PEI [81] have been used as substrates for metachromatic dyes. Chitosan as a natural chromotrope polycation with primary amine groups has been also used as a support for anionic dyes [97]. The authors assumed that in this case the metachromatic effect was generated by the increase of the dye concentration close to the chitosan chains, the aggregation of the dye molecules being a consequence of the high concentration of the dye.

The influence of the structural characteristics of both the polycation and the dyes on the metachromatic behavior of the dyes concerned the following applications: quantitative determination of the small amounts of cationic polyelectrolytes in natural and industrial waters [89,91,98,99], to probe the ionization level of polymers in solution [79], the evidence of the energy transfer between dye molecules, and so forth.

The conformation of polyelectrolytes, i.e., their shape and coilling degree, greatly influence their performances in different applications, such as: flocculants for colloids and organic matter from natural and waste waters, control of corrosion and fouling in cooling water systems, deemulsifiers for oil-water emulsions, surface modifiers, etc. Therefore, a very interesting application of metachromasy in the study of the conformational changes of strong polyelectrolytes in aqueous solution as a function of the environment properties has been promoted [73,85,89]. By the addition of sufficiently high concentrations of low molar mass salts, a reversal of the metachromasy has been observed by Kugel for some metachromatic dyes in the presence of strong polycations [89]. This clearly demonstrates that metachromasy can serve as an indicator of the polyion conformation, being well known that polyion chains can coil up at high ionic strength.

Up to date, our attention has been mainly focused on the interaction between some strong polycations, bearing quaternary ammonium salt centers in the backbone, and a large variety of multicharged azo dyes [56-59,73,92]. The chromotropic properties of polycations and the metachromatic properties of the dyes have been deeply investigated. The structures of the azo dyes selected for the studies focused on the influence of polycations on the dye behavior, in dilute aqueous solution, are presented in Figure 1.

The stoichiometry of the polycation/azo dye complex and the metachromatic behavior of the dyes were followed by viscometry and UV-vis spectrometry [56,59,92]. As the molar ratio between polycation and dye, P/D, corresponding to the stoichiometry was close to the theoretical number of sulfonic groups of the dye, we have assumed that the main driving force in the spectral changes of the dyes was the electrostatic attraction between sulfonic groups and the oppositely charged groups along the macroion chain, and not a simple agglomeration of the dye molecules in the vicinity of polyelectrolyte chain, followed by the dye aggregation.

A metachromatic behavior has been observed, i.e., some UV-vis spectral changes accompanied the binding of the dye to the polycation chain as a consequence of the π-π interactions between the dye molecules located in a very close proximity with each other. Spectral shifts were observed to longer or shorter wavelengths depending on the mutual orientation of the dipole moments [93,94].

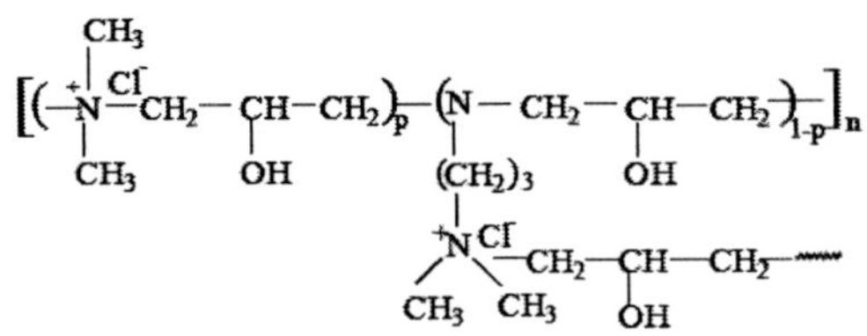

PCA$_5$, p = 0.95; PCA$_{20}$, p = 0.80

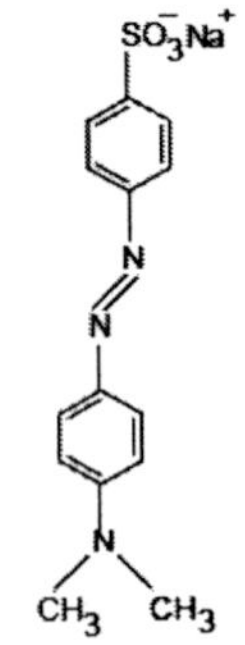

Methyl Orange (MO)

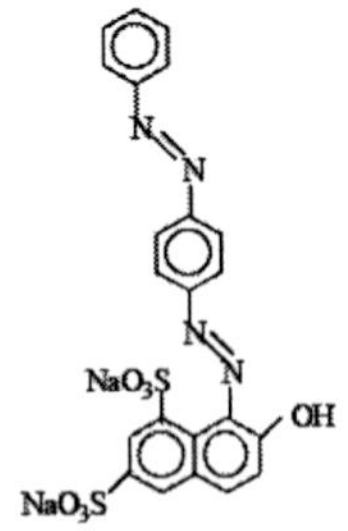

Crocein Scarlet MOO (CSMOO)

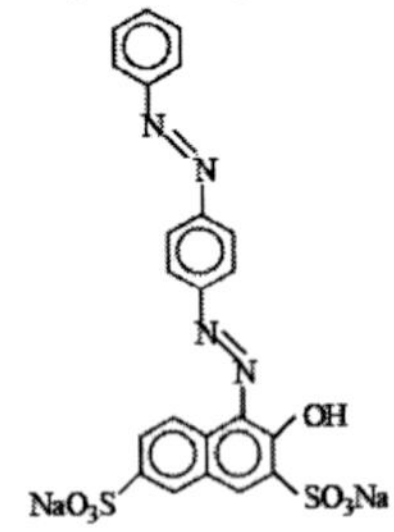

Ponceau SS (PSS)

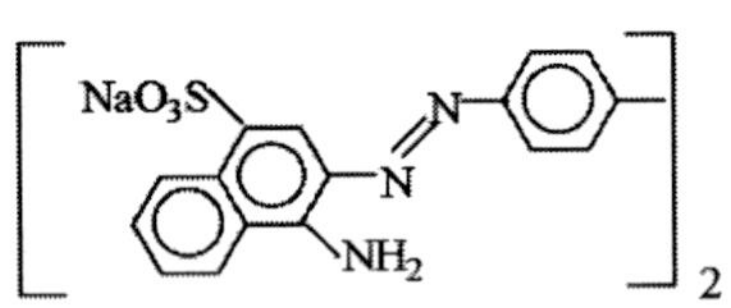

Direct Blue 1 (DB1)

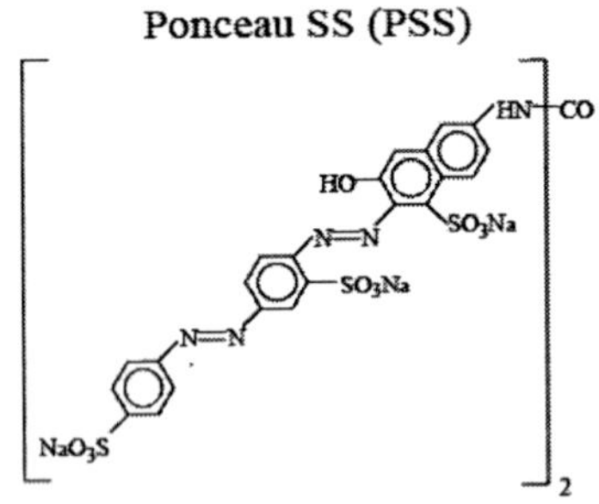

Direct Red 80 (DR80)

Figure 1. The structures of the polycations and azo dyes used in the investigations of the systems containing azobenzene chromophore.

The first azo dye employed for this study was MO, the most often used azo dye for such investigations. As Figure 2 shows, a small amount of polycation (P/D = 0.2) led only to the decrease of the intensity of the characteristic absorption band at 462 nm of the dye monomer. By further increase of the polycation content, P/D = 1.6, i.e., an excess of polycation led to the dramatic decrease of the absorbance corresponding to 462 nm and to the appearance of a new sharp band at 370 nm, assigned to the high aggregation of the dye onto PCA$_5$ chains. Such strong blue shifts have been previously explained by a high degree of stacking comprising a greater number of interacting chromophores formed when MO molecules are bound to certain polycations [79,82,85]. Increasing the P/D ratio, brought about complicated changes in MO spectra, i.e., the intensity of the 370 nm band increased and then shifted to 368 nm, this blue shifted maximum being actually inferred in literature for MO in the presence of other polycations [79]. By the increase of PCA$_5$ concentration to much higher values (P/D = 200), the maximum at 368 nm decreased and that one characteristic to the dye monomer increased without an isosbestic point.

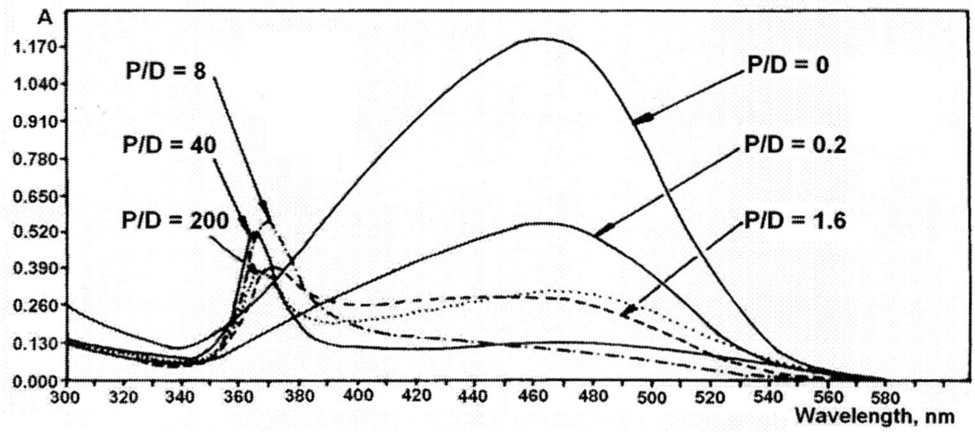

Figure 2. Absorption spectra of MO with a concentration of 5 x 10^{-5} $mol.L^{-1}$ in the presence of PCA_5 at different molar ratios P/D.

The lack of isosbestic point could show the presence of an equilibrium between different absorbing forms of the dye [79]. The appearance of the band at 368 nm, which intensity depends on the polycation concentration, can be also correlated with the polycation conformation, the charged centers available for the dye binding being dependent also on the polycation concentration.

The characteristic wavelengths of the nonaggregated molecules of CSMOO, PSS, DB1 and DR80 were found at 510, 514, 620, and 521 nm, respectively, but some spectral shifts appeared in the presence of the polycation PCA_5. For instance, CSMOO and PSS, which differ from each other only by the position of the two sulfonic groups, showed differences in the metachromatic effects in the presence of PCA_5 in both dilute aqueous solution and in the multilayers [70,73]. A blue shift of the characteristic maximum from 510 nm to about 500 nm and a small shoulder at about 550 nm have been observed in the CSMOO spectrum in the presence of PCA_5. A red shift from 514 nm to about 530 nm and a distinct shoulder at about 580 nm were found for PSS in the presence of PCA_5 (Figure 3).

The small blue shift in the case of CSMOO shows a better organization of the dye molecules in the proximity of the polycation chain. The clear red shift of the maximum in the case of PSS suggests a J-aggregation of the dye molecules by their interaction with PCA_5 macroions, in dilute aqueous solution. The intensity of the characteristic maxima decreased first with the increase of the polycation/dye molar ratio, P/D, the minimum being found at a P/D of about 2, for both dyes. The intensity of the characteristic maximum increased again with the increase of the P/D molar ratio, but the metachromatic effect was still evident at P/D = 500. It is known that the larger the polycation concentration required to restore the characteristic band of the dye monomer, the stronger is the dye binding cooperativity [85], a higher stacking of the dye being specific for the polymers with a high charge density due to the short distance between the binding sites on the polymer chain.

It is well known that the dyes aggregate as a consequence of their flat geometry, but the aggregation level and mode depend on the dye structure and the environmental characteristics. Thus, the changes in ionic strength allow to control the polyion conformation and this significantly influenced the metachromatic effects observed in the PCA_5/CSMOO and PCA_5/PSS self-assembled thin films [73]. It was observed that the UV-vis spectra of the $(PCA_5/CSMOO)_n$ multilayers with PCA_5 adsorbed from 0.2 M Na_2SO_4 aqueous solution were completely similar with those found for the polycation deposition from salt-free aqueous solution, only the intensity values were much higher (see details about LbL deposition of the PCA_5/azo dye ESAMs in the last Chapter of this book).

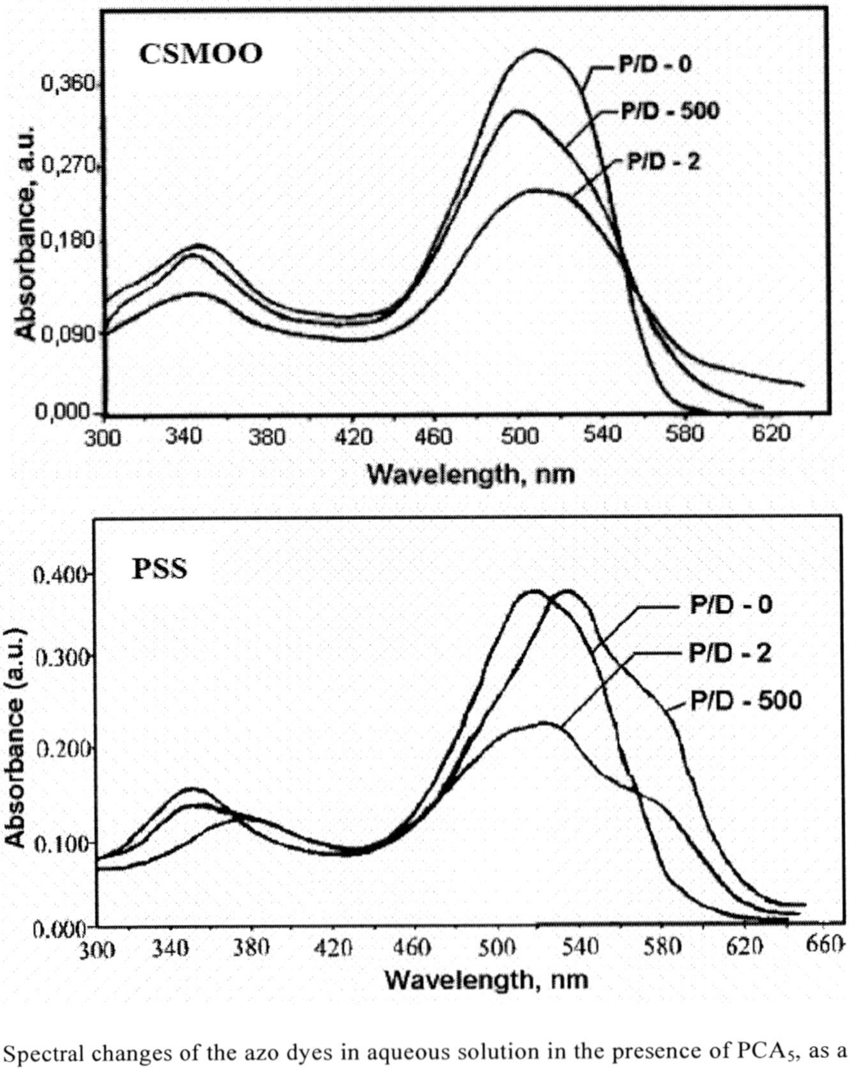

Figure 3. Spectral changes of the azo dyes in aqueous solution in the presence of PCA_5, as a function of P/D molar ratio.

With the two multicharged azo dyes, DB1 and DR80, a different behavior was observed in the presence of the polycation PCA_5. As Figure 4 shows, two new absorption maxima appeared in the UV-vis spectrum of both dyes, in dilute aqueous solutions. A hypsochromic shift at 610 nm and a bathochromic shift at 680 nm, comparative with the characteristic maximum of the DB1 monomer can be observed for DB1 in the presence of PCA_5.

Comparative with the characteristic maximum of the DR80, located at 521 nm, a hypsochromic shift at 505 nm and a bathochromic shift at 542 nm can be observed in Figure 4 for DR80 in the presence of PCA_5. The intensity of the bathochrome band was lower than that of the hypsochrome band for P/D molar ratio in the range 1 – 20, but the order was reversed for higher P/D values. The cooperativity of the dye binding was observed in a very large range of the P/D molar ratios, in the case of DR80. If dye stacking is the cause of the new bands, increasing the probability for the dye to spread out along the polymer chain by increasing the P/D ratio should result in a gradual disappearance of the high-energy band at 521 nm and of low energy band at 542 nm and appearance of the monomeric band for the bound dye molecules.

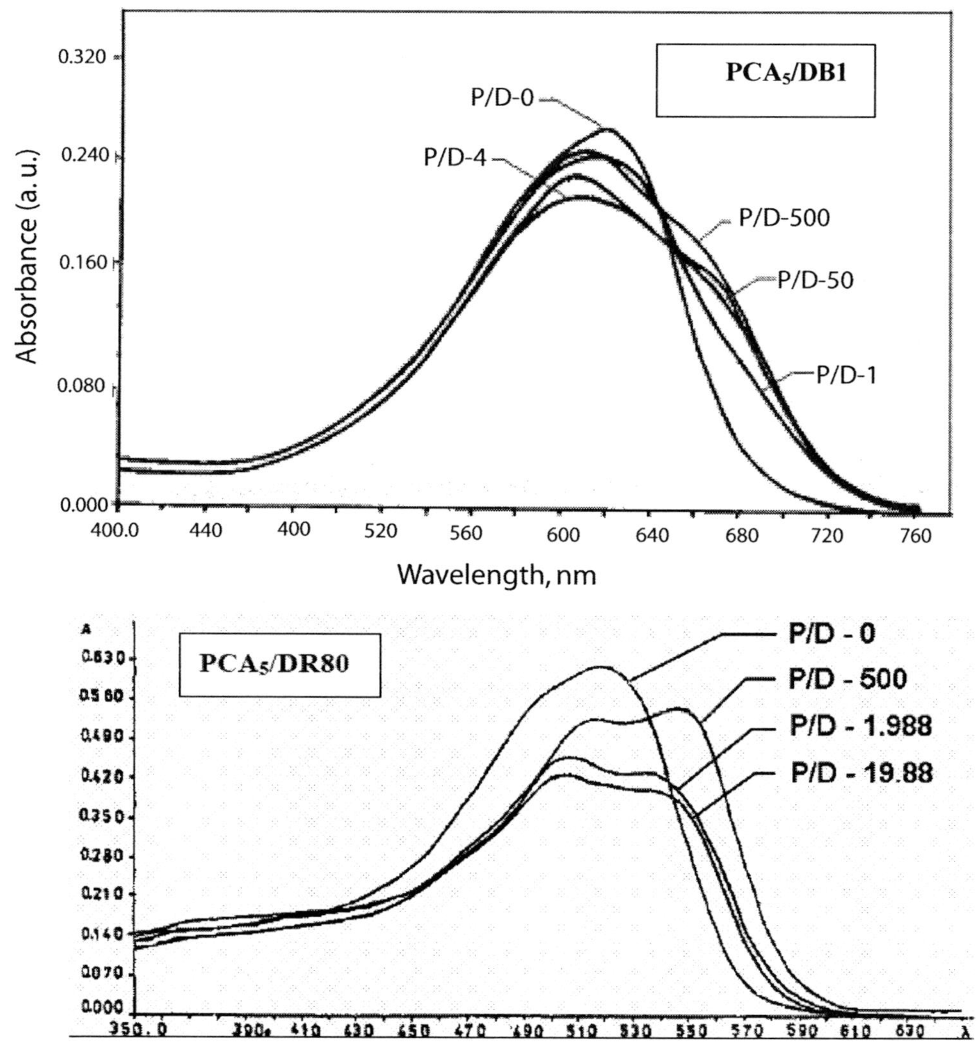

Figure 4. Absorption spectra of the dyes DB1 and DR80 in the absence and in the presence of PCA_5, in saltfree aqueous solutions.

As Figure 4 shows, both maxima increased with the increase of the P/D molar ratio, but the shape of the UV-vis spectrum remained unchanged even at P/D of 500. This shows a small influence of the polycation in excess on the metachromatic behavior of the dye DR80, and can be explained by the interaction between dye molecules bound on the same chain when the concentration of the polycation is high enough and the chains start to coil up. A much faster return of the spectrum to that of the dye monomer, i.e., a lower stability of the polycation/DR80 complex, was observed when a polycation with a higher degree of branching was used as polycation (PCA_{20}, Figure 1) [56].

3. Solid-State Properties of Some Polycation/Azo Dyes Complexes

Valuable information on the mechanisms of polycation/azo dye multilayer growth and structure formation in such supramolecular nanoarchitectures could be obtained by comparison of the structures of bulk polyelectrolyte/dye complexes and polyelectrolyte/dye self-assemblies. So far, the information concerning the solid state polyelectrolyte/dye complexes are relatively scarce [55,100] and those concerning the polycation/azo dye complexes are even less. This lack led us to the preparation and characterization of some PCA_5/azo dye complexes in bulk, in the absence and in the presence of NaCl [57,58]. The azo dyes used for this purpose were different by either the number or the position of sulfonic groups as follows:

- Crystal Scarlet (CS) and Amaranth (Am), having two and respectively three sulfonic groups, both derived from 7-hydroxy-8-(1-naphtylazo)naphtalen;
- Congo Red (CR), which is very often used in the polycation/azo dye multilayer construction;
- two azo dyes having the basic structure of 3-hydroxy-4-(4-phenylazo)phenylazonaphtalen and two sulfonic groups placed in different position, CSMOO and PSS (Figure 1);
- two multicharged azo dyes with four sulfonic groups (DB1) and six sulfonic groups (DR80), Figure 1.

Taking into account the complex stoichiometry determined by the UV-vis spectrometry in dilute aqueous solutions and by viscometric tirations [56,73,92], a theoretical structure of the PCA_5/azo dye complexes could be calculated. The solid state polycation/azo dye complexes were characterized by elemental analyses, DSC and WAXD [57]. Some results of the elemental analysis as a function of the dye structure and preparation conditions were collected in Table 1. As Table 1 shows, a good agreement between the theoretical and the experimental values of the elemental analysis was found when these complexes were obtained by mixing saltfree aqueous solutions of PCA_5 and dyes. Higher PCA_5/dye molar ratios were observed when NaCl was present. This shows a lower dye amount was bound on the polycation chains due to the screening effect of the counterions in excess on the positive charges of the polycation.

The influence of the ionic strength on the complex structure was the most evident in the case of DR80, the dye with six sulfonic groups. This different behavior could be explained by the compensation difficulties of the opposite charges when PCA_5 chains start to coil up and also by the decrease of the critical aggregation concentration (cac) of the dye in the presence of salt. DSC measurements were performed between 25 ^{o}C and at the beginning of the thermal decomposition. The presence or absence of the glass transition temperatures, T_g, gave information on the strength of the interactions between the polycation and the dye. Thus, the T_g found for the PCA_5/azo dye complexes formed with CS, CR, CSMOO and Am showed that the interaction between polycation and azo dyes was not so strong. On the other side, the absence of T_g for the complexes formed between PCA_5 and DB1, DR80 showed a very strong interaction between polycation and the corresponding multicharged azo dyes [57]. Hence, the number of sulfonic groups of the dye has a strong influence on the supramolecular structure of the corresponding polycation/azo dye complex.

Table 1. Characterization by Elemental Analyses of the PCA_5/Azo Dye Complexes[a]

Azo Dye	PCA_5/Dye Ratio at the Stoichiometry		Cl, %		N, %		S, %	
	Calc.[b]	Found	Calc.[c]	Found	Calc.[c]	Found	Calc.[c]	Found
CSMOO	2	2.2	0	<0.50	11.66	10.65	8.86	8.14
CSMOO, 0.1 M NaCl	2	2.4	0	0.90	11.66	-	8.86	8.34
PSS	2	1.9	0	<0.50	11.66	11.05	8.86	8.68
DB1	4	4.1	0	<0.50	10.60	10.02	9.69	9.57
DR80	6	5.7	0	<0.50	12.01	11.38	10.3	9.35
DR80, 0.1 M NaCl	6	7.0	0	1.15	12.01	9.22	10.3	7.63

[a] Preparation conditions: the dye solution was slowly dropped into the aqueous solution of PCA_5 with a concentration of 2 unit mM, under a vigorous stirring at room temperature (~ 20 °C), until the phase separation was achieved;

[b] Corresponding to the number of sulfonic groups of the dye;

[c] According to the theoretical structure of the complex repeat unit.

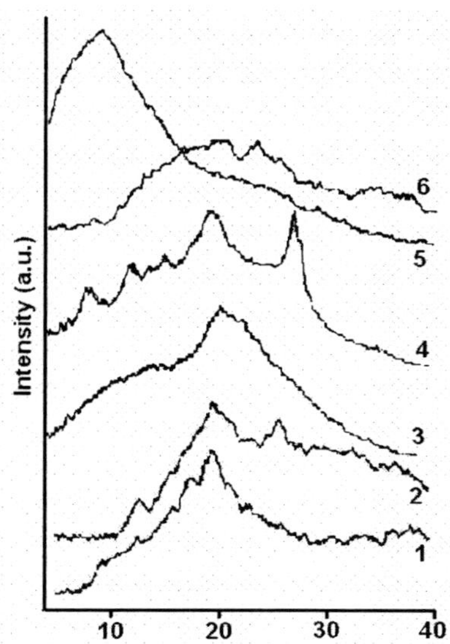

Figure 5. X-ray diffractograms obtained from: 1 = PCA_5/CSMOO, 0.1 M NaCl; 2 = PCA_5/CSMOO, saltfree aqueous solution; 3 = PCA_5/DB1, saltfree aqueous solution; 4 = PCA_5/PSS, saltfree aqueous solution; 5 = PCA_5/DR80, saltfree aqueous solution; 6 = PCA_5/DR80, 0.1 M NaCl.

The differences in the supramolecular order between the starting polycation, completely amorphous, and some PCA_5/azo dye complexes were evidenced by WAXD (Figure 5). The WAXD patterns of PCA_5/CSMOO complex obtained in water and 0.1 M NaCl were characterized by a crystallinity degree of 42 %, and 39 %, respectively. The PCA_5/PSS diffraction pattern showed several well defined sharp peaks indicating also a high degree of crystallinity of about 37%. The 2θ values of the inner diffraction maxima at about 6.4° and the derived spacing of about 9.7 Å support a supramolecular order of this complex. The diffractogram of PCA_5/DB1 complex obtained in water pointed several broad maxima and a lower crystalline phase of 26 %. X-ray diffraction of the PCA_5/DR80 revealed clearly the influence of the formation conditions on the supramolecular order of this complex. A diffraction maximum at about 9° for 2θ was observed in salt-free aqueous solution, but the formation of the complex in 0.1 M NaCl caused the decrease of the crystallinity degree to 29%.

The results presented above demonstrate the strong influence of the dye structure and of the polycation/azo dye complex formation conditions on the supramolecular structure of the corresponded complex in solid state.

4. Heat-Induced Reactions of Polycation/MO Complexes

As it was already mentioned in Section 2, MO is one of the most accessible and convenient azo dye used in the studies of the polyelectrolyte/dye interactions. The study of the polycation/MO (PCMO) complexes in solid state has attracted less attention. These observations led us to the preparation, in a first stage, of some complexes between polycations containing primary amine groups as ammonium salts and MO, which contains one sulfonic group and a preformed azobenzene group. In the second stage, PCMO complexes were heated at temperatures of 180-200 °C following the feasibility of a heat-induced reaction of the ammonium-sulfonate ion pair into a sulfonamide bond.

Heat-induced reactions involving two functional groups offer a less investigated way for the formation of covalent bonds in the electrostatically or H-bonding stabilized polymer complexes and polyelectrolyte multilayers [101-104]. Primary or secondary amine groups, on the one hand, and carboxylic groups, on the other hand, are usually involved in such reactions with the transformation of the ammonium-carboxylate ion pairs, or of the $-NH_2$....HOCO-hydrogen bonds, into amide groups by heating at appropriate temperatures.

Poly(allylamine hydrochloride) (PAH) and poly(2-aminoethyleneacrylamide hydrochloride) (PAEAH) were employed in the preparation of the PCMO complexes in the first stage, the structures of the starting polycations and of the corresponding PCMO complexes being collected in Figure 6. PAEAH has been prepared by the aminolysis-hydrolysis reaction of the nitrile groups contained in a copolymer of acrylonitrile with α-methylstyrene (5 mol %), with ethylenediamine [105].

Formation of the PCMO complexes takes place by the ionic exchange reaction between the ammonium chloride groups present in the side chain of both polycations and MO. This reaction is accompanied by the NaCl removal and the ionic binding of the organic counterion with the formation of a new structure, which must be free of Cl^- counterions [106].

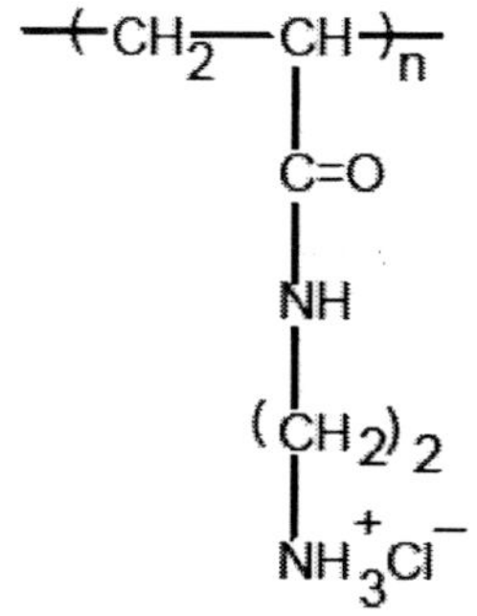

PAEAH

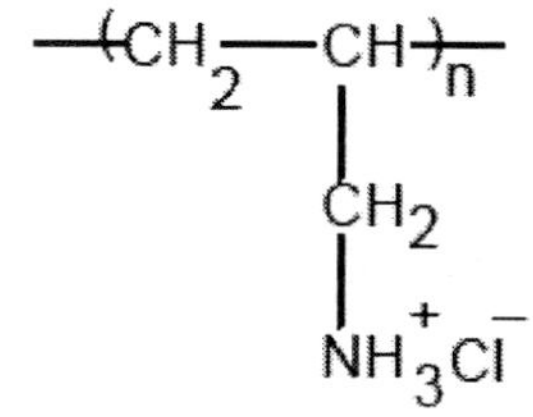

PAH

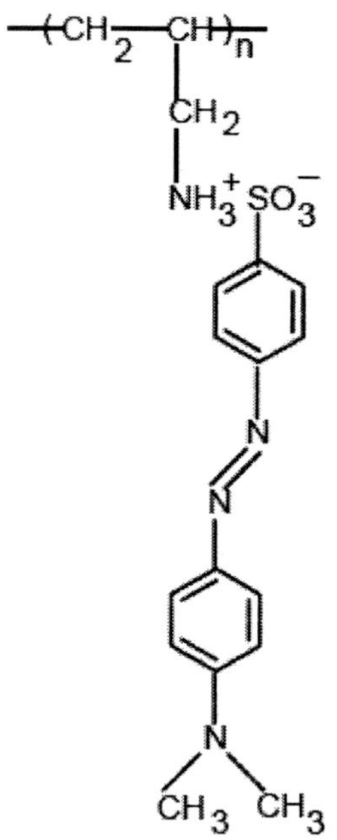

PCMO1

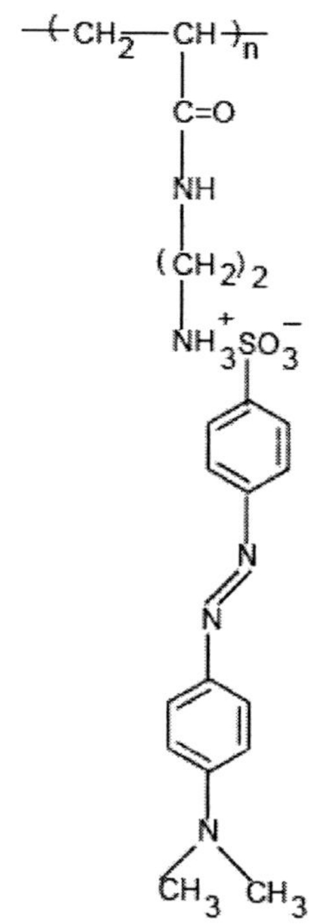

PCMO2

Figure 6. Structures of the polycations with primary amine groups as ammonium salts and of their complexes with MO.

PCMO complexes were characterized first by elemental analysis comparative with the starting polycations and MO, the results obtained being collected in Table 2. As can be seen, a lower value of ionic chlorine content was found both for the starting PAEAH and PAH comparative with theoretical values corresponding to the monomer units. Therefore, the maximum amount of the dye, which can be bound to the polycation chain was calculated on the basis of the real content of ammonium chloride units.

PCMO complexes were thermal treated in the range 180-200 °C to show if it is possible to transform the ionic binding in covalent linkage. The possible reaction takes place according to Scheme 1.

Table 2. Characterization by Elemental Analyses of the Starting Polycations and of Their Complexes with MO

	Cl_i, wt %		N, wt %		S, wt %	
Sample	Calcd.	Found	Calcd.	Found	Calcd.	Found
MO	0	0	12.8	12.3	9.8	10.4
PAH	38.0	32.0	15.0	16.5	-	-
PCMO1	0	0	15.5	12.4	8.8	9.3
PAEAH	22.63	16.1	17.87	19.5	-	-
PCMO2	0	0	16.7	12.3	7.6	7.4

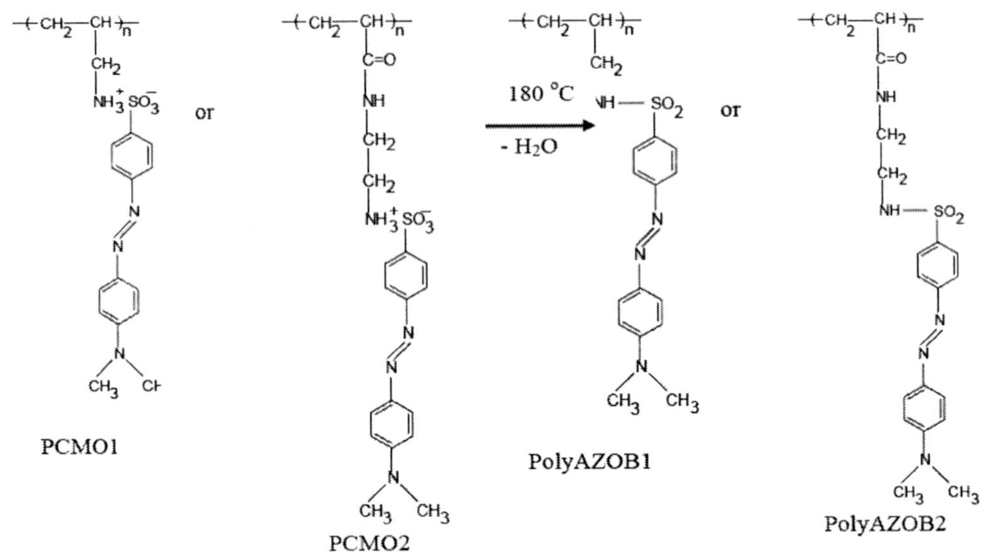

Scheme 1.

MO, polycations, and their PCMO complexes, on the one hand, and the thermal treated polycation/MO complexes, on the other hand, were also characterized by FTIR spectroscopy, ^{1}H-NMR spectrometry, dynamic thermogravimetry (TG) and derivative thermogravimetry (DTG).

4.1. FTIR Characterization

FTIR spectra of MO, PCMO complexes and the polymers with azobenzene moieties in the side chain, were recorded with a Digilab spectrophotometer Scimitar Series USA, resolution 4 cm^{-1}, in the range of 4000-400 cm^{-1} by KBr pellet technique and are presented in Figures 7a and 7b.

The spectrum of MO contains the following characteristic bands: 2900 cm^{-1} for CH_3, 2817 cm^{-1} for $-N(CH_3)_2$ (not shown in Figures 7a and 7b), 1604 and 1419 cm^{-1} for p-disubstituded benzene, 1365 cm^{-1} for aryl– $N(CH_3)_2$, 1184 and 1118 cm^{-1} for SO_3^- and 698 and 623 cm^{-1} for C-S bond. Besides these bands in the FTIR spectrum of PCMO2 (Figure 7b) some characteristic bands of pure PAEAH sample can be observed, as follows: 1660 cm^{-1}

(amide I), 1517 cm^{-1} (amide II), 1310 cm^{-1} (amide III), and 1446 cm^{-1} (NH_3^+). The spectrum of PolyAZOB2 contains new bands corresponding to the groups: $-SO_2-$ at 1390 and 823 cm^{-1}, and –NH– at 1316 cm^{-1}. The signals assigned to NH_3^+ groups diminished after thermal treatment, but still exist. This shows that after two hours of thermal treatment the electrostatic bonds were not completely transformed in covalent bonds.

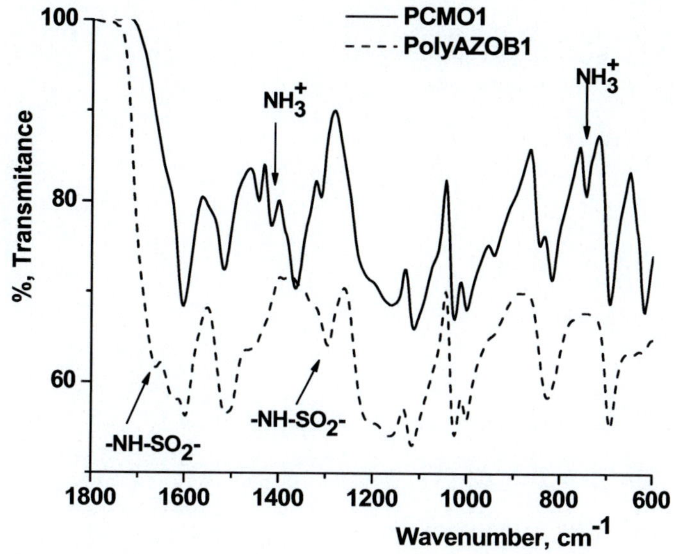

Figure 7a. FTIR spectra of the PAH/MO complex (PCMO1) and of the corresponding polymer with azobenzene groups in the side chain (PolyAZOB1).

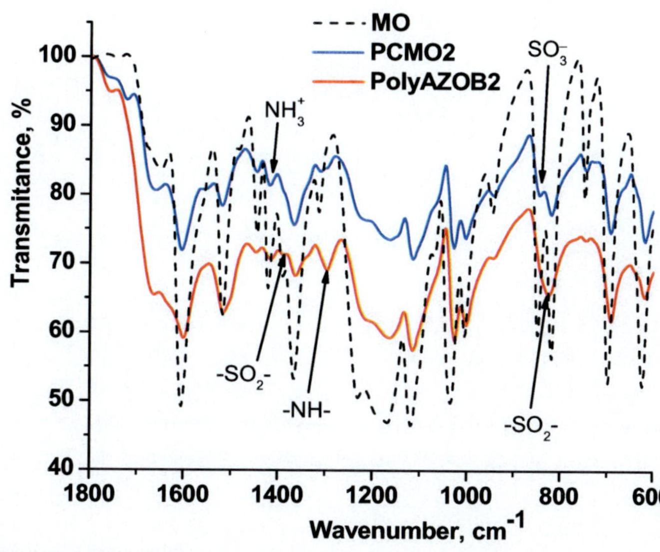

Figure 7b. FTIR spectra of the MO, PAEAH/MO complex (PCMO2) and of the corresponding polymer with azobenzene groups in the side chain (PolyAZOB2).

On the other side, the spectrum of PolyAZOB1 (Figure 7a) shows that the bands ascribed to NH_3^+ group completely disappeared, i.e., the electrostatic bonds were quantitatively transformed in covalent bonds after two hours of thermal treatment.

4.2. ^{1}H-NMR Characterization

The ^{1}H-NMR spectra of PAH, PAH/MO and PolyAZOB1 were performed at 60 MHz on a JNM-C-60 HL spectrometer at ambient temperature in D_2O.

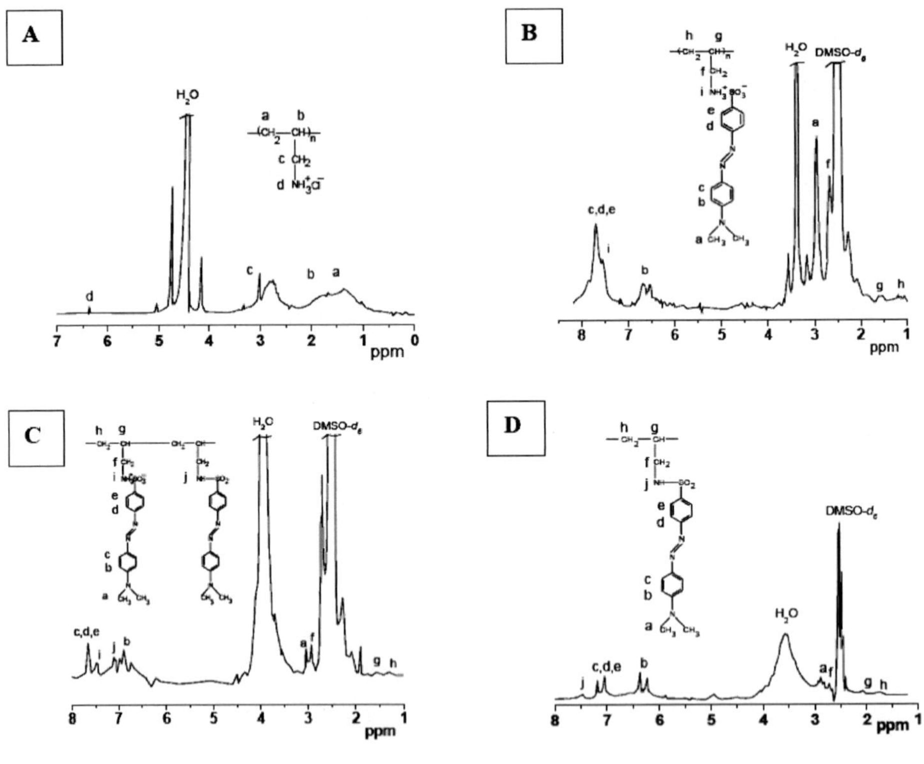

Figure 8. ^{1}H-NMR spectra of PAH (A), PAH/MO complex before (B) and after one hour (C) and two hours (D) of thermal treatment.

The ^{1}H-NMR spectra of MO, PAEAH and PAEAH/MO complex before and after the thermal treatment were performed at 400 MHz on a Bruker NMR spectrometer, model Avance DRX 400, by using TMS as an internal standard and DMSO-d_6 as a solvent for all samples, at ambient temperature. ^{1}H-NMR spectra collected in Figures 8 and 9 give clear evidences on the formation of PCMO complex and of the PolyAZOB structure. One can observe that ^{1}H-NMR spectrum of PCMO1 complex (Figure 8B) contains characteristic signals of the starting polycation (1.2 and 1.7 ppm assigned to the protons from the main chain and at 2.7 ppm for protons from the side chain) and of MO (6.7 and 7.7 ppm assigned to protons from p-disubsituted benzene and 2.9 ppm for protons from the methyl groups) [107].

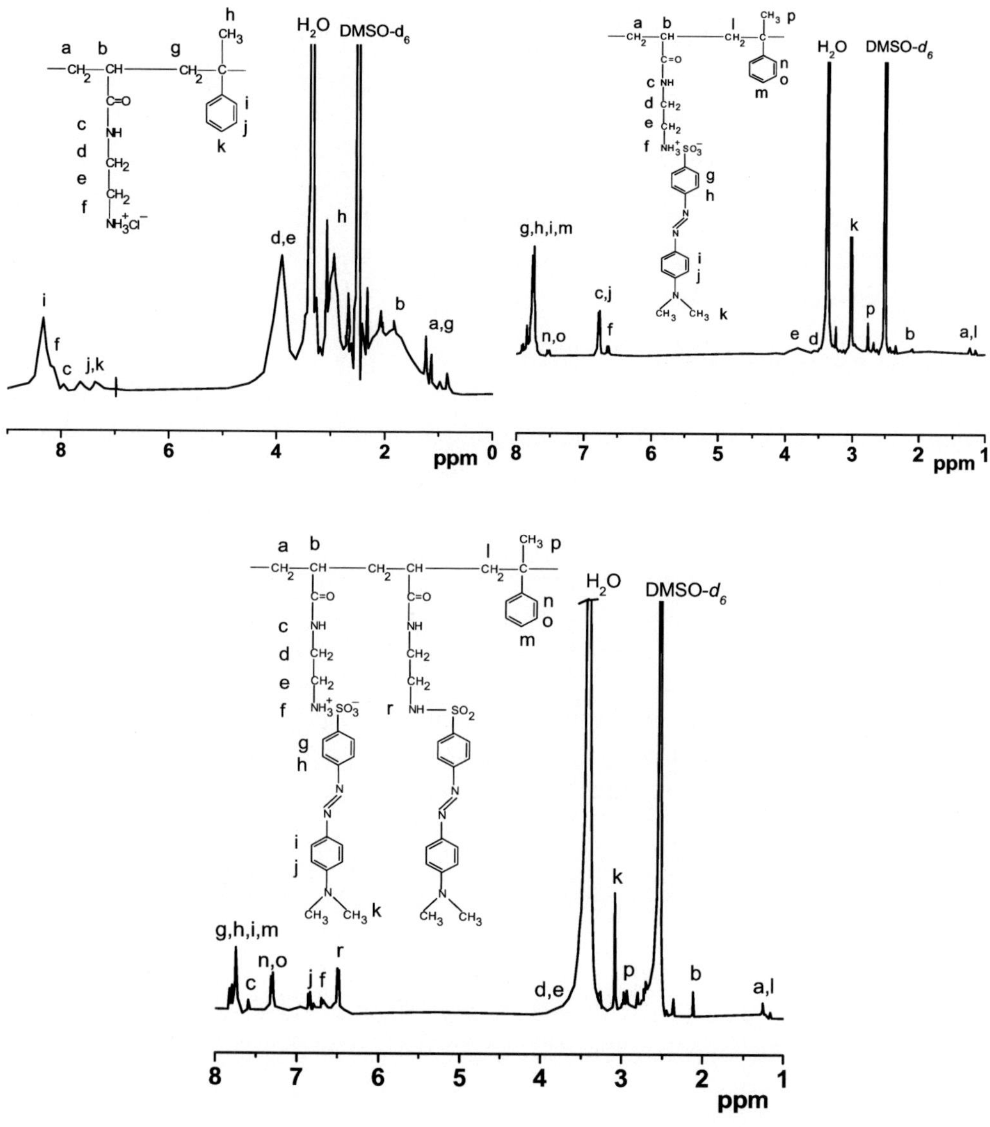

Figure 9. ^{1}H-NMR spectra of PAEAH, PAEAH/MO complex before and after two hours of thermal treatment.

The electrostatic interaction between PAH and MO is sustained by the presence of the signal at 7.5 ppm assigned to protons from NH_3^+ groups. After half an hour of thermal treatment at 180 °C, the spectrum presented a weak signal at 7.2 ppm (not shown here), assigned to the proton from $–NH–SO_2–$ groups, this suggesting that part of the electrostatic bonds have been transformed in covalent bonds. After one hour of thermal treatment the signal at 7.2 ppm increased and the signal at 7.5 ppm decreased, this showing that the content of covalent groups increased. Finally, after two hours at 180 °C the disappearance of the peak assigned to NH_3^+ groups was observed, the spectrum containing only the signal assigned to the $–NH-SO_2-$ covalent bond. Also, the signals assigned to protons from MO diminished this suggesting that some of the organic counterions were removed after two hours of thermal treatment.

Figure 9 presents the ^{1}H-NMR spectra of PAEAH, PCMO2 and PolyAZOB2. The signals characteristics for both starting materials can be identified in the spectrum of the polycation/dye complex. The signal ascribed to NH_3^+ groups, at 6.6 ppm is also present. After two hours of thermal treatment at 180 °C this signal decreased and we can observe a new signal, at 6.4 ppm, assigned to the proton from –NH–SO_2– group [106].

4.3. UV-vis Characterization

Electronic absorption spectra were recorded on a SPECORD M42 spectrophotometer in DMSO solutions using 10 mm quartz cells. Figures 10 and 11 show the influence of the polycation structure and of the thermal treatment duration on the UV-vis absorption spectra of polycation/MO complexes comparative with MO spectrum, at ambient temperature.

The presence of azobenzene chromophore in the PAH/MO complex is supported by the maximum at 430 nm, similar with MO. Half an hour thermal treatment at 180 °C led to the decrease of the maximum, but the shape of the curve remained the same. A very abrupt decrease of the characteristic maximum of MO was evident at longer thermal treatment duration. As Figure 10 shows, two hours of thermal treatment induced dramatic structural changes of the PAH/MO complex.

A different behavior was observed when PAEAH was used as polycation, i.e., the polycation with a longer spacer between the polymer backbone and the ammonium groups. As Figure 11 shows, a red shift of the characteristic maximum of MO from 430 to 435 nm was observed first in PAEAH/MO complex before any thermal treatment. This band gradually decreased in intensity by thermal treatment, a small blue shift being observed after one and two hours of thermal treatment at 180 °C.

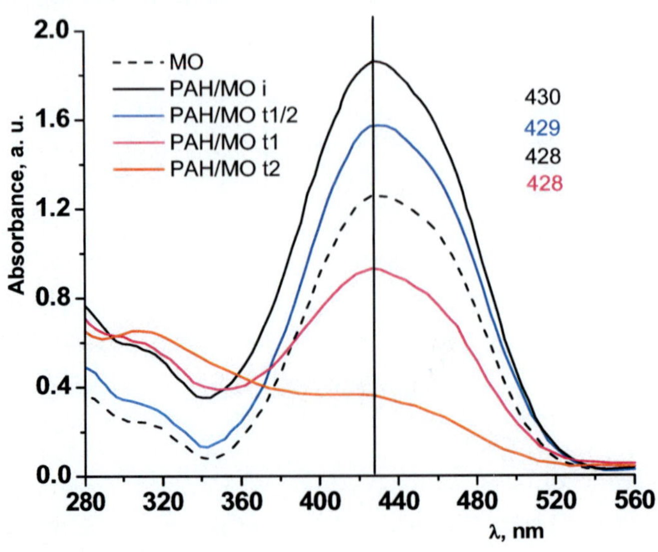

Figure 10. Absorption spectra of MO, PAH/MO complex and of the thermal treated complex, in DMSO 0.001%, as a function of the thermal treatment duration: ½ h (t1/2); 1 h (t1); 2 h (t2).

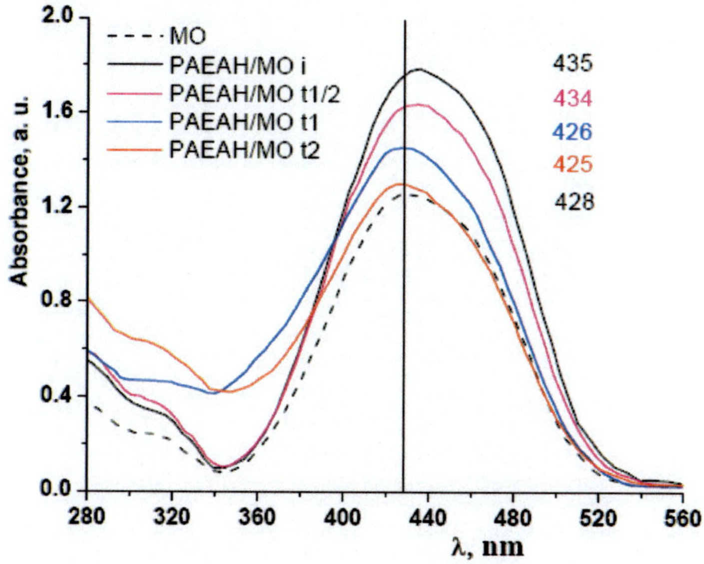

Figure 11. Absorption spectra of MO, PAEAH/MO complex and the thermal treated complex in DMSO, 0.001%, as a function of the thermal treatment duration: ½ h (t1/2); 1 h (t1); 2 h (t2).

The influence of the polycation structure on the thermal behavior of the polycation/MO complex is demonstrated by the shape of the absorption spectrum after two hours at 180 °C, which is almost similar with that of MO, unlike the corresponding spectrum of the PAH/MO complex (Figure 10). The increase of the spacer flexibility allows the azobenzene side chains a higher mobility, preventing thus the secondary degradation reactions which could accompany the main reaction of the covalent bonds formation.

4.4. Thermogravimetric Analysis

Thermogravimetric analysis (TGA) has been often used as a tool to characterize the thermal behavior of polymers as a function of their molecular structure [108,109] even for very complicated structures like the crosslinked copolymers [110,111], and as a sensitive method to follow water loss by physical [112] or chemical [106] processes. In our investigations, TGA was used to show the influence of the polycation structure on its thermal stability and to evidence the heat-induced reaction with the formation of sulfonamide bonds from sulfonate-ammonium ion pairs. TGA measurements were performed on 50 mg powder sample at a heating rate of 9 °C/min up to 900 °C, DTG sensitivity of 1/10, in air, using an Paulik-Paulik-Erdey Derivatograph, Budapest; α-Al_2O_3 was used as the reference material. The main characteristics of the thermal behavior were: the initial decomposition temperature, T_i (°C), the temperature corresponding to the maximum rate of decomposition T_m (°C), the final decomposition temperature T_f (°C), and the weight loss, W, (%). Both MO and all polymers showed a first degradation stage related to the loss of bound water at a temperature range of 60 - 150 °C, with T_m between 60 °C and 90 °C, the weight loss varying between 1 and 7%, which was not considered for the determination of degradation parameters. Some quantitative results concerning the degradation stages II, III and IV of the polycations and polycation/MO complexes before and after the thermal treatment, comparative with MO, were collected in Table 3.

Table 3. TGA Results of the Starting Polycations and of the Corresponding Polycation/MO Complexes Comparative with MO

Sample	T_i, °C	T_m, °C	T_f, °C	W, %	E_a, kJ/mol	n
Stage II						
MO	175	306	330	10	63.4	0.0
PCMO1	200	210	235	2	183.8	0.0
PAEAH	115	143	160	3	112.4	1.2
PCMO2	175	205	215	1	161.3	0.0
PolyAZOB2	180	233	240	1	102.3	1.2
Stage III						
MO	330	495	530	33.5	86.1	1.1
PAH	235	350	630	88	66.2	1.7
PCMO1	230	280	430	31	169.9	1.8
PolyAZOB1	210	270	380	30	71.8	2.0
PAEAH	160	353	460	51	47.2	1.3
PCMO2	215	260	450	29	158.3	1.6
PolyAZOB2	240	292	445	31	208.7	4.0
Stage IV						
MO	530	615	700	31	218.8	1.5
PCMO1	480	605	700	48	171.31	1.3
PolyAZOB1	380	610	705	59	87.6	0.8
PAEAH	460	550	725	36	141.9	1.9
PCMO2	450	630	700	54	135.1	1.0
PolyAZOB2	445	640	760	55	135.6	0.9

The effect of MO as a counterion on the thermal stability of the polycation/MO complexes comparative with the corresponding polycations was very clearly demonstrated by the values of the weight loss. The second decomposition stage was very much influenced by the polycation structure, being also characterized by relatively low weight loss, 1-10%. We suppose that, in the case of PC/MO complexes, this stage is correlated with the chemical reaction which takes place around 180 °C, with the result transformation of the sulfonate-ammonium ionic bonds into covalent linkages, according to Scheme 1. This stage was less evidenced for the two hours thermal treated polycation/MO complexes, and support the results obtained by previous methods.

As Table 3 shows, the main decomposition stages are III and IV. As can be observed, the third stage is characterized by the highest weight loss values for MO and both starting polycations, and the fourth stage is the most significant for the polycation/MO complexes. PAH seems to have the lowest thermal stability, the stage III being characterized by a weight loss of 88%, at a maximum degradation temperature of 350 °C for this polycation. All polycation/MO complexes exhibited higher thermal stability comparative with MO and the starting polycations and shows the contribution of the counterion nature on the thermal behavior of polycation. Thus, PCMO1 complex shows a weight loss of 2% , in the range of

200 - 235 °C, with T_m of about 210 °C. This small weight loss could be caused by a physical process controlled by diffusion, but could be also ascribed to the elimination of a water molecule from the PCMO1 complex, according to the Scheme 1. The last assumption is sustained by the disappearance of this stage in the thermogravimetric curves of PolyAZOB1. The next two decomposition stages, similar for PCMO1 and PolyAZOB1, correspond to their effective thermal degradation.

The thermogravimetric data of PCMO2 show a first decomposition stage, between 175 - 225 °C, with a weight loss of 1%, and T_m of about 215 °C, similar with the first decomposition stage of PCMO1. This stage was also observed in the PolyAZOB2 thermogravimetric curves, the weight loss being of 1%. This behavior suggest that for PCMO2, only part of the ammonium sulfonate ion pairs have been transformed in covalent bonds after two hours of thermal treatment at 180 °C. Both PCMO complexes were thermogravimetrically analyzed with a DTG sensitivity of 1/5, up to 250 °C to demonstrate that a chemical process took place in the range 175 - 225 °C [85]. A wight loss of 2% at T_m of about 205 °C was evident for PCMO1 and a weight loss of 1% at T_m of about 220 °C was evident for PCMO2, and these confirmed the results obtained at a DTG sensitivity of 1/10. Table 3 contains also the values of the activation energy, E_a, and the reaction order, n, calculated according to the Coats and Redfern method [113]. The highest values of E_a, in the second stage, were found for the PCMO1 and PCMO2 complexes, 183.8 kJ/mol and 161.3 kJ/mol, respectively, and support the chemical process discussed above.

The variation of activation energy, E_a, with conversion degree (α) was determined by Reich-Levi method [114], where $\alpha = W_t/W_\infty$, with W_t – weight loss at the "t" moment of the process and W_∞ - final weight loss of the decomposition stage. The E_a values for the studied samples were plotted versus α in Figure 12, for the temperature range of 160 - 240 °C, in Figure 13, for the temperature range of 200 - 400 °C, and in Figure 14, for the temperature range of 400 - 700 °C.

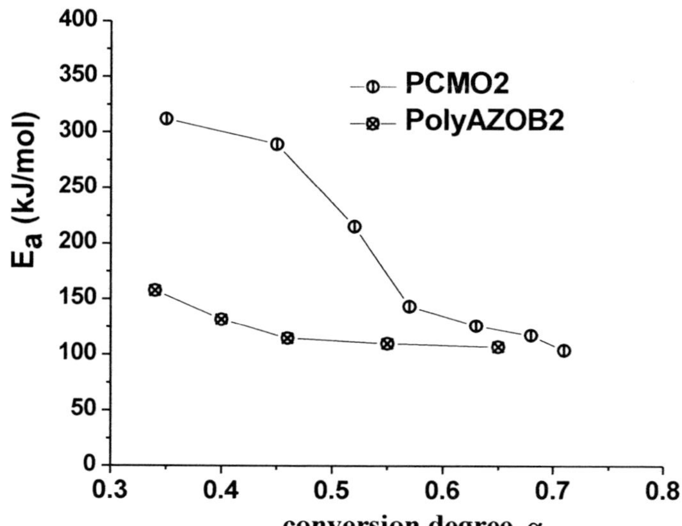

Figure 12. Plots of activation energy versus conversion degree in the temperature range of 160 - 240 °C.

Figure 12 illustrates the differences in the thermal behavior between PCMO2 and the PolyAZOB2. The much higher values of E_a found for PCMO2 in this stage of thermal decomposition support the chemical reaction, which take places in the complex in two hours of thermal treatment at 180 °C. In the same range of temperature, polyAZOB2 shows a low decrease of the Ea with the increase of α from 0.33 up to 0.46 and an almost constant value for higher values of α.

Figure 13 shows that the E_a - α curves for MO and PAEAH, in the temperature range of 200 – 400 °C, are characterized by approximately constant values of E_a with the increase of α, and the E_a values for PC/MO complexes decrease especially in the $0.4 < \alpha < 0.55$ range. The E_a values suggest that with the increase of α, the thermal degradation is favoured about for all samples investigated, the stronger bonds being observed in PCMO1 and PCMO2 complexes.

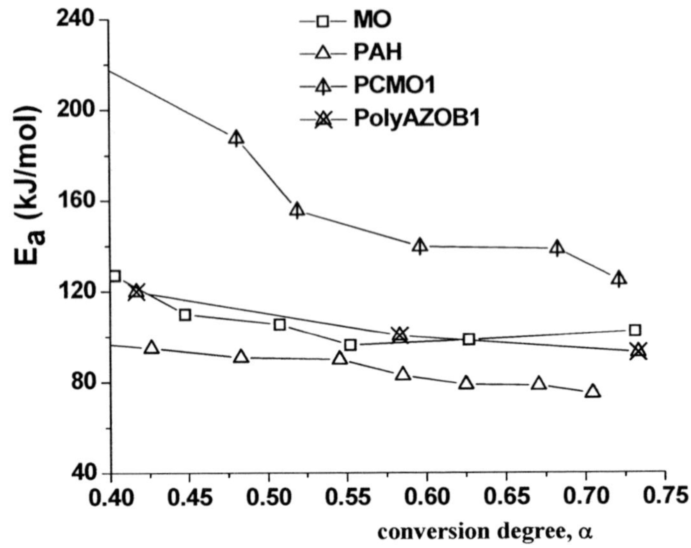

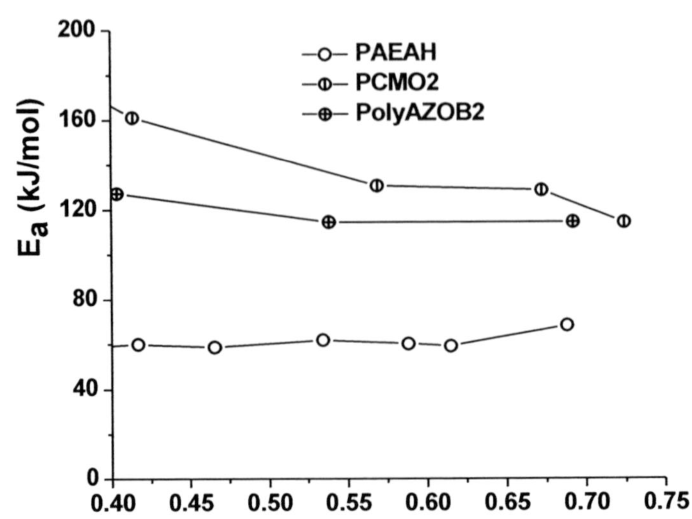

Figure 13. Plots of activation energy versus conversion egree in the temperature range of 200 - 400 °C.

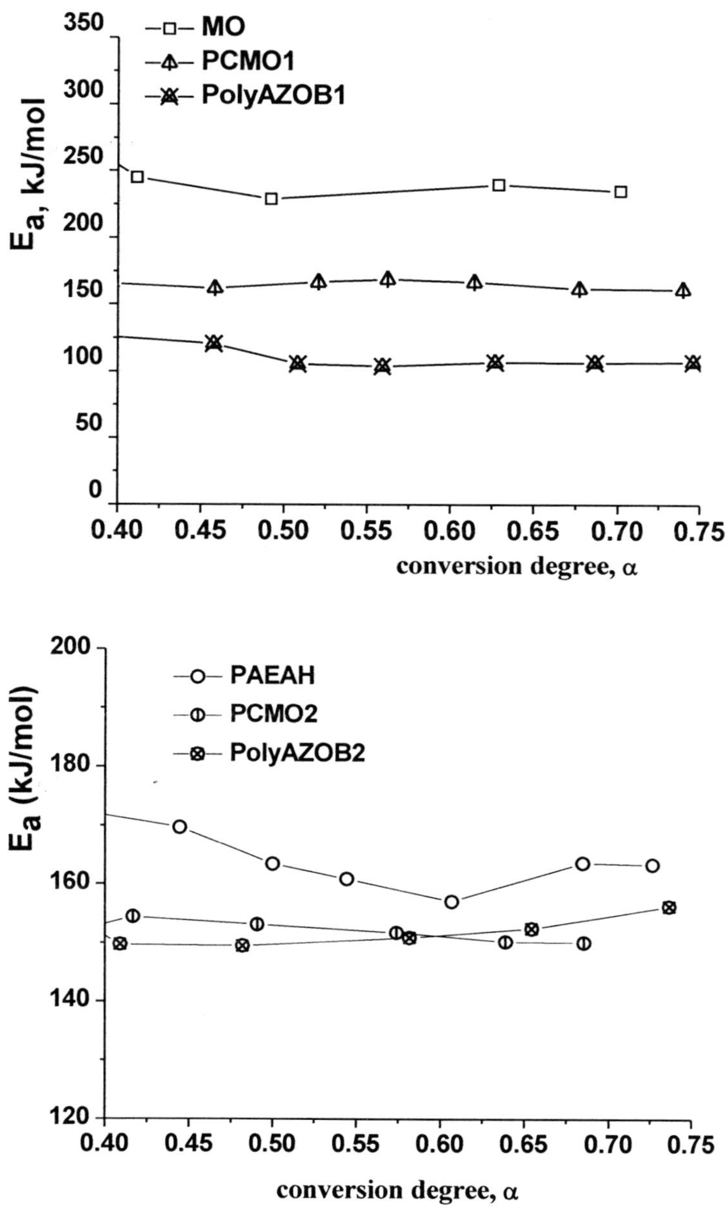

Figure 14. Plots of activation energy versus conversion degree in the temperature range of 400-700 °C.

The aspect of the E_a - α plots in the temperature range of 200-400 °C (Figure 13) let us to suppose that the depolymerization is the main process in this stage. In general, changes in the activation energy are associated with changes in the mechanism of degradation reactions. This behavior was less important for MO and polycations in this stage. At the same time, higher values of E_a were observed for both PC/MO complexes in comparison with the starting polycations suggesting differences in terms of the thermal stability. The E_a values for PAH/MO complex were higher than those of the PAEAH/MO complex. The curves show also that E_a decreases with the increase of α, in the same way for both PCMO complexes.

In the temperature range of 400 – 700 °C (Figure 14), almost constant values of E_a vs. α were found for PCMO complexes and a non-linear variation only for PAEAH. In this degradation range, the decrease of E_a could be also caused by the loss of the degradation compounds formed by the rupture of the functional groups attached to the polymer backbone.

5. Conclusion

This review summarizes the selected original articles which have brought new contributions in the field of molecular and supramolecular systems containing azobenzene chromophores, and the own contributions of the authors focused mainly on the polycation/azo dye complexes in aqueous solution as a function of the dye structure, the solid state properties of polycation/azo dye complexes, and the heat induced reactions of the complexes formed between MO and two polycations bearing primary amine groups.

The study of the interaction between some strong polycations bearing quaternary ammonium salt centers in the backbone, and a large variety of multicharged azo dyes, such as: CSMOO, PSS, DB1 and DR80 revealed a strong influence of both the polycation and the dye structure, on the one side, and of the ionic strength, on the other side, on the dye aggregation mode. The chromotropic properties of polycations and the metachromatic properties of the dyes have been deeply investigated. The solid state polycation/azo dye complexes were prepared and characterized by elemental analyses, DSC and WAXD.

Poly(allylamine hydrochloride) (PAH) and poly(2-aminoethyleneacrylamide hydrochloride) (PAEAH) were employed in the preparation of the polycation/MO complexes followed by a heat induced transformation of the ion pairs into sulfonamide covalent bonds. The feasibility of this reaction and the influence of the polycation structure on the properties of the polycation/MO complexes and of the new polymers bearing azobenzene groups as side chains have been demonstrated by FTIR, ^{1}H-NMR, UV-Vis spectroscopy and TGA.

Acknowledgement

The financial support of this research from the Project CEEX-MATNANTECH 40/2005 is gratefully acknowledged.

References

[1] Todorov, T.; Nikolova, L.; Tomova, N. *Appl. Opt.* 1984, *23*, 4309-4312.

[2] Natansohn, A.; Rochon, P. *Chem. Rev.* 2002, *102*, 4139-4175.

[3] El Halabieh, R. H.; Mermut, O.; Barrett, C. J. *Pure Appl. Chem.* 2004, *76*, 1445-1465.

[4] Rau, H. In: *Photochromism - Molecules and Systems*; Dürer, H.; Laurent, H. B.; Eds.; Elsevier, Amsterdam, 1990, pp 165-192.

[5] Menzel, H.; Rüther, M.; Stumpe, J.; Fischer, T. *Supramolecular Sci.* 1998, *5*, 49-59.

[6] Imai, Y.; Naka, K.; Chujo, Y. *Macromolecules* 1999, *32*, 1013-1017.

[7] Wang, C., Weiss, R. G. *Macromolecules* 2003, *36*, 3833-3840.

[8] Decher, G.; Hong, J. D. *Ber. Bunsenges. Phys. Chem.* 1991, *95*, 1430-1434.
[9] Toutianoush, A.; Saremi, F.; Tieke, B. *Mater. Sci. Eng. C* 1999, 8-9, 343-352 .
[10] Hong, J.-D. ; Park, E.-S. ; Park, A.-L. *Langmuir* 1999, *15*, 6515-6521.
[11] Kuiper, J. M.; Engberts, J. B. F. N. *Langmuir* 2004, *20*, 1152-1160.
[12] Balasubramanian, S.; Wang, X.; Wang, H. C.; Yang, K.; Kumar, J.; Tripathy, S. K. *Chem. Mater.* 1998, *10*, 1554-1560.
[13] Lenahan, K. M.; Wang, Y.-X.; Liu, Y.; Claus, R. O.; Heflin, J. R.; Marciu, D.; Figura, C. *Adv. Mater.* 1998, *10*, 853-855.
[14] Dante, S.; Advincula, R.; Frank, C. W.; Stroeve, P. *Langmuir* 1999, *15*, 193-201.
[15] Robello, D. R. J. *Polym. Sci. Part A Polym. Chem.* 1990, *28*, 1-13.
[16] Wang, X.; Balasubramanian, S.; Kumar, J.; Tripathy, S. K. *Chem. Mater.* 1998, *10*, 1546-1553.
[17] Dhanabalan, A.; Balogh, D. T.; Riul Jr., A.; Giacometti, J. A.; Oliviera Jr., O. N. *Thin Solid Films* 1998, *323*, 257-264.
[18] Natansohn, A. ; Rochon, P.; Meng, X.; Barrett, C.; Buffeteau, T.; Bonenfant S.; Pézolet, M. *Macromolecules* 1998, *31*, 1155-1161.
[19] Schab-Balcerzak, E.; Sapich, B.; Stumpe, J. *Polymer* 2005, *46*, 49-59.
[20] Kanbara, T.; Oshima, M.; Imayasu, T.; Hasegawa, K. *Macromolecules* 1998, *31*, 8725-8730.
[21] Yang, S.; Jacob, M. M.; Lian, L.; Cholli, A. L.; Kumar, J.; Tripathy, S. K. *Macromolecules* 2001, *34*, 9193-9196.
[22] Iftimie, G.; Natansohn, A.; Rochon, P. *Macromolecules* 2002, *35*, 365-369.
[23] Yoshida, T.; Kanaoka, S.; Aoshima, S. *J. Polym. Sci. Part A Polym. Chem.* 2005, *43*, 5138-5146.
[24] Freiberg, S.; Lagugne-Labarthet, F.; Rochon, P.; Natansohn, A. *Macromolecules* 2003, *36*, 2680-2688.
[25] Freiberg, S.; Lagugné-Labarthet, F.; Rochon, P.; Natansohn, A. *Can. J. Chem.* 2004, *82*, 1-10.
[26] Petrova, T. S.; Mancheva, I.; Nacheva, E.; Tomova, N.; Dragostinova, V.; Todorov, T.; Nikolova, L. *J. Mat. Sci. Mat. Electronics* 2003, *14*, 823-824.
[27] Martinez-Ponce, V.; Petrova, T. S.; Tomova, N.; Dragostinova, V.; Todorov, T.; Nikolova, L. *J. Opt. A Pure Appl. Opt.* 2004, *6*, 324-329.
[28] Advincula, R.; Aust, E.; Meyer, W.; Knoll, W. *Langmuir* 1996, *12*, 3536-3540.
[29] Dante, S.; Advincula, R.; Frank, C. W.; Stroeve, P. *Langmuir* 1999, *15*, 193-201.
[30] Park, M.-K.; Advincula, R. C. *Langmuir* 2002, *18*, 4532-4535.
[31] Suzuki, I.; Ishizaki, T.; Tomonori, H.; Anzai, Jun-ichi *Macromolecules* 2002, *35*, 577-580.
[32] Wang, H.; He, Y.; Tuo, X.; Wang, X. *Macromolecules* 2004, *37*, 135-146.
[33] Suzuki, I.; Ishizaki, T.; Inoue, H.; Anzai, Jun-ichi *Macromolecules* 2002, *35*, 6470-6474.
[34] Takase, H.; Natansohn, A.; Rochon, P. *Polymer* 2004, *44*, 7345-7351.
[35] Chun, C.; Ghim, J.; Kim, M.-J.; Kim, D. Y. *J. Polym. Sci. Part A Polym. Chem.* 2005, *43*, 3525-3532.
[36] Zhao, Y. *Pure Appl. Chem.* 2004, *76*, 1499-1508.
[37] Stiller, B.; Geue, T.; Morawetz, K.; Saphiannikova, M. *J. Microscopy* 2005, *219*, 109-114.

[38] Kim, M. J.; Chun, C.; Nakayama, T.; Kim, D. Y. *Jpn. J. Appl. Phys.* 2006, *45*, L169-L171.

[39] Guo, B.; Liu, J.; Jia, Y. J.; Wang, G. M.; Zhang, Q. J. *J. Opt. Adv. Mater.* 2005, *7*, 1017-1021.

[40] Tong, B. X.; Wang, G.; Yavrian, A.; Galstian, T.; Zhao, Y. *Adv. Mater.* 2005, *17*, 370-374.

[41] Tabiryan, N.; Serak, S.; Dai, X.-M. *Optics Express* 2005, *13*, 7442-7448.

[42] Patton, D.; Park, M.-K.; Wang, S.; Advincula, R. C. *Langmuir* 2002, *18*, 1688-1694.

[43] de Souza, N. C.; Zucolotto, V.; Silva, J. R.; Santos, F. R.; dos Santos Jr., D. S.; Balogh, D. T.; Oliviera, Jr., O. N.; Giacometti, J. A. *J. Colloid Interface Sci.* 2005, *285*, 544-550.

[44] Tuo, X.; Chen, D.; Cheng, H.; Wang, X. *Polym. Bull.* 2005, *54*, 427-433.

[45] Laschewsky, A.; Mayer, B.; Wischerhoff, E. Arys, X.; Jonas, A.; Kauranen, M.; Persoons, A. *Angew. Chem. Int. Ed.* 1997, *36*, 2788-2791.

[46] Fischer, P.; Laschewsky, A. *Macromolecules* 2000, *33*, 1100-1102.

[47] Cochin, D.; Passmann, M.; Wilbert, G.; Zentel, R.; Wischerhoff, E.; Laschewsky, A. *Macromolecules* 1997, *30*, 4775-4779.

[48] Fischer, P.; Laschewsky, A.; Wischerhoff, E.; Arys, X.; Jonas, A.; Legras, R. *Macromol. Symp.* 1999, *137*, 1-24.

[49] Lee, T. S.; Ahn, H.; Park, W. H.; Sohn, B.-H.; Kim, M.-J.; Kim, D.-Y. *J. Polym. Sci. Part A Polym. Chem.* 2003, *41*, 1196-1201.

[50] Holland, N. B.; Hugel, T.; Neuert, G.; Catanni-Scholz, A.; Renner, C.; Oesterhelt, D.; Moroder, L.; Seitz, M.; Gaub, H. E. *Macromolecules* 2003, *36*, 2015-2023.

[51] Tian, Y.; Watanabe, K.; Kong, X.; Abe, J.; Iyoda, T. *Macromolecules* 2002, *35*, 3739-3747.

[52] Cui, L.; Zhao, Y.; Yavrian, A.; Galstian, T. *Macromolecules* 2003, *36*, 8246-8252.

[53] Ravi, P.; Sin, S. L.; Gan, L. H.; Gan, Y. Y.; Tam, K. C.; Xia, X. L.; Hu, X. *Polymer* 2005, *46*, 137-146.

[54] Qian, J.; Hamada, K.; Mitsuishi, M. *Dyes and Pigments* 1994, *25*, 167-178.

[55] Behnke, M.; Tieke, B. *Langmuir* 2002, *18*, 3815-3821.

[56] Dragan, S.; Ghimici, L.; Cristea, M.; Airinei, A. *Acta Polym.* 1999, *50*, 260-266.

[57] Dragan, S.; Timpu, D. *J. Polym. Sci. Part A Polym. Chem.* 2003, *41*, 264-272.

[58] Dragan, S.; Cristea, M. In: *Recent Research Developments in Polymer Science*; Pandalai, S. G.; Ed.; Transworld Research Network: Trivandrum, Kerala, 2003, vol 7, pp 149-181.

[59] Dragan, S.; Ghimici, L.; Grigoriu, A.; Coman, D. *Bull. Inst. Pol. Iasi* 1998, *44*, 89-95.

[60] Maeda, S.; Mori, T.; Sasaki, C.; Kunimoto, K.-K.; Kuwae, A.; Hanai, K. *Polym. Bull.* 2005, *53*, 259-267.

[61] Cooper, Th. M.; Campbell, A. L.; Crane, R. L. *Langmuir* 1995, *11*, 2713-2718.

[62] Ariga, K.; Lvov, Y.; Kunitake, T. *J. Am. Chem. Soc.* 1997, *119*, 2224-2231.

[63] Yoo, D.; Wu, A.; Lee, J.; Rubner, M. F. *Synth. Met.* 1997, *85*, 1425-1426.

[64] Linford, M. R.; Auch, M.; Möhwald, H. *J. Am. Chem. Soc.* 1998, *120*, 178-182.

[65] Koetse, M.; Laschewsky, A.; Verbiest, T. *Mater. Sci. Eng. C* 1999, *10*, 107-113.

[66] Caruso, F.; Lichtenfeld, H.; Donath, E.; Möhwald, H. *Macromolecules* 1999**,** *32*, 2317-2328.

[67] Bertrand, P.; Jonas, A.; Laschewsky, A.; Legras, R. *Macromol. Rapid Commun.* 2000, *21*, 319-348.

[68] Dragan, S.; Schwarz, S.; Eichhorn, K.-J.; Lunkwitz, K. *Colloids Surf. A.* 2001, *195*, 243-251.

[69] Advincula, R. C.; Fells, E.; Park, M. *Chem. Mater.* 2001, *13*, 2870-2878.

[70] Dragan, S.; Schwarz, S. *Macromol. Symp.* 2002, *181*, 155-166.

[71] R. C. Advincula in "Handbook of Polyelectrolytes and Their Applications" (Tripathy, S.; Kumar, J.; Nalwa, H. S.; Eds.; American Scientific Publishers, Los Angeles, vol. 1, 2002, pp 66-98.

[72] Advincula, R.; Park, M.-K.; Baba, A.; Kaneko, F. *Langmuir* 2003, *19*, 654-665.

[73] Dragan, S.; Schwarz, S. *Prog. Colloid Polym. Sci.* 2003, *122*, 8-15.

[74] Nicol, E.; Habib-Jiwan, J.-L.; Jonas, A. M. *Langmuir* 2003, *19*, 6178-6186.

[75] Takahashi, M.; Kobayashi, K.; Takaoka, K. *Langmuir* 2000, *16*, 6613-6621.

[76] Sato, K.; Suzuki, I.; Anzai, J. *Langmuir* 2003, *19*, 7406-7412.

[77] Zucolotto, V.; He, J.-A.; Constantino, C. J. L.; Barbosa Neto, N. M.; Rodrigues Jr., J. J.; Mendonca, C. R.; Zilio, S. C.; Li, L.; Aroca, R. F.; Oliviera Jr., O. N.; Kumar, J. *Polymer* 2003, *44*, 6129-6133.

[78] Burke, S. E.; Barrett, C. J. *Macromolecules* 2004, *37*, 5357-5384.

[79] Quadrifoglio, F.; Crescenzi, V. *J. Colloid Interface Sci.* 1971, *35*, 447-459 .

[80] Takagishi, T.; Kozuka, H.; Kuroki, N. *J. Polym. Sci. Polym. Chem. Ed.* 1983, *23*, 447-455.

[81] Tanaka, H.; Sugimoto, T.; Hayashi, A.; Kuroki, N. *J. Polym. Sci. Part A Polym. Chem.* 1983, 21, 2311-2322.

[82] Handel, T. M.; Cohen, H. L.; Tan, J. S. *Macromolecules*, 1985, *18*, 1200-1206.

[83] Kim, W. S.; Seo, K. H. *J. Polym. Sci. Part A Polym. Chem.* 1989, *27*, 389-392.

[84] Dawydoff, W.; Linow, K. J.; Philipp, B. *Acta Polym.* 1991, *42*, 592-596.

[85] Vishalakshi, B. *J. Polym. Sci. Part A Polym. Chem.* 1995, *33*, 365-371.

[86] Wu, T. S.; Chen, K. M. *J. S. D. C.* 1993, *109*, 153-158.

[87] Oka, H.; Kimura, A. *J. S. D. C.* 1995, *111*, 311-315.

[88] Grigoriu, A.; Coman, D.; Dragan, S.; Ghimici, L. *Bul. Inst. Pol. Iasi* 1999, *45*, 57-60.

[89] Kugel, R. In: *Focus on Ionic Polymers*; Dragan, E. S.; Ed.; Reasearch Signpost: Trivandrum, Kerala, 2005, pp 93-116.

[90] Murakami, K. *Bull. Chem. Soc. Jpn.* 1993, *66*, 2808-2813.

[91] Wassmer, K.-H.; Schroeder, U.; Horn, D. *Makromol. Chem.* 1991, *192*, 553-565.

[92] Dragan, S.; Cristea, M.; Airinei, A.; Simionescu, B. C. *Macromol. Rapid Commun.* 1997, *18*, 541-545.

[93] Kugel, R. W. In: *Structure-Property Relations in Polymers*; Urban, M. W.; Craver, C. D.; Eds.; Advances in Chemistry 236, American Chemical Society, Washington, DC, 1993, pp 507-533.

[94] Buss, V.; Eggers, L. In: *Encyclopedia of Spectroscopy and Spectrometry*; Linden, J. C.; Tranter, G. E.; Holmes, J. L.; Eds.; Academic, NY, 2000, pp 388-396.

[95] Hochberg, G. C. *Colloid Polym. Sci.* 1994, *272*, 409-415.

[96] Peyratout, C.; Donath, E.; Daehne, L. *J. Photochem. Photobiol. A Chem.* 2001, *142*, 51-57.

[97] Gummow, B. D.; Roberts, G. A. F. *Macromol. Chem. Rapid Commun.* 1985, *6*, 381-386.

[98] Terayama, H. *J. Polym. Sci.* 1952, *8*, 243-253.

[99] Tashiro, T. *J. Appl. Polym. Sci.* 1990, *39*, 2279-2291.

[100] Lin, X.; Zhong, A.; Chen, D.; Zhou, Z.; He, B. *J. Appl. Polym. Sci.* 2003, *87*, 369-374.

[101] Baranovsky, V. Yu.; Kazarin, L. A.; Litmanovich, A. A.; Papisov, I. M. *Eur. Polym. J.* 1984, *20*, 191-194.

[102] Dai, J.; Jensen, A. W.; Mohanty, D. K.; Erndt, J.; Bruening, M. L. *Langmuir*, 2001, *17*, 931-937.

[103] Mengel, C.; Esker, A. R.; Meyer, W. H.; Wegner, G. *Langmuir* 2002, *18*, 6365-6372.

[104] Balachandra, A. M.; Dai, J.; Bruening, M. L. *Macromolecules* 2002, *35*, 3171-3178.

[105] Dragan, S.; Barboiu, V., Petrariu, I.; Dima, M. *J. Polym. Sci. Polym. Chem. Ed.* 1981, *19*, 2869-2880.

[106] Dragan, E. S.; Mihai, M.; Airinei, A. *J. Polym. Sci. Part A Polym. Chem.* 2006, *44*, 5898-5908.

[107] Pretsch, E.; Bühlmann, P.; Affolter, C. *Structure Determination of Organic Compounds*; Springer: Berlin, Germany, 2000, pp 161-244.

[108] Berwig, E.; Severgnini, V. L. S.; Soldi, M. S.; Bianco, G.; Pinheiro, E. A.; Pires, A. T. N.; Soldi, V. *Polym. Degrad. Stab.* 2003, *79*, 93-98.

[109] Bianco, G.; Soldi, M. S.; Pinheiro, E. A.; Pires, A. T. N.; Gehlen, M. H.; Soldi, V. *Polym. Degrad. Stab.* 2003, *80*, 567-574.

[110] Vlad, C. D.; Dinu, M. V.; Dragan, E. S. *Polym. Degrad. Stab.* 2003, *79*, 153-159.

[111] Uhl, F. M.; Levchik, G. F.; Levchik, S. V. *Polym. Degrad. Stab.* 2001, *71*, 317-325.

[112] Wang, Y.; Kawano, Y.; Aubuchon, S. R.; Palmer, R. A. *Macromolecules* 2003, *36*, 1138-1146

[113] Coats, A. W.; Redfern, J. T. *Nature* 1964, *201*, 68-69.

[114] Reich, L.; Levi, D. W. *Makromol. Chem.* 1963, *66*, 102-113.

In: New Trends in Ionic (Co) Polymers and Hybrids
Editor: Ecaterina Stela Dragan, pp. 31-64
ISBN 978-1-60021-611-4

Chapter 2

APPLICATION OF POLYELECTROLYTES IN PHASE SEPARATION PROCESSES

Luminita Ghimici, **Ionel Adrian Dinu** and **Ecaterina Stela Dragan***
"Petru Poni" Institute of Macromolecular Chemistry,
Aleea Grigore Ghica Voda 41A, 700487-Iasi, Romania

ABSTRACT

Polyelectrolytes are charged polymers which have found important applications in the major fields of science and engineering such as chemistry, physics, biology, chemical and materials engineering. The review presents some of the more important and recent experimental and theoretical data regarding the application of PELs in processes which imply the phase separation (*solid-liquid and liquid-liquid*) in aqueous systems. Selected papers in this area are briefly discussed with particular attention given to the influence of several parameters related to the characteristics of the suspended particle (concentration, size), the polyelectrolyte (type of functional groups, structure, molar mass, charge density, dose) and the environment (ionic strength, pH, temperature) on the polymers efficiency in the flocculation process.

Keywords: flocculation, suspension, emulsion, polyelectrolytes, flocculation mechanism, kinetic, dual flocculation, sludge dewatering.

LIST OF ABBREVIATION

PEL	Polyelectrolyte
PAM	Poly(acrylamide)
PEO	Poly (ethylene oxid)

* E-mail: lghimici@icmpp.ro; Telephone: +40.232.217454; Fax: +40.232.211299.

PVA	Poly(vinyl alcohol)
PAA	Poly(acrylic acid)
PDADMAC	Poly(diallyldimethylammonium chloride)
PVAm	Poly(vinylamine)
PEI	Poly(ethyleneimine)
PATC	Poly[2-(acryloyloxy)ethyl(trimethyl)ammonium chloride
POME	Palm oil mill effluent
P(MS-α-MeSty	Poly(maleic acid-co-α-methylstyrene)
PVBTMAC	Poly(vinylbenzyltrimethylammonium)
NaPA	Poly(sodium acrylate)
NaPAMPS	Poly(sodium 2-acrylamido-2-methylpropane sulfonate)
NaPSS	Poly(sodium styrenesulfonate)
CP6R	Crystal Ponceau 6R
P4R	Ponceau 4R

1. INDRODUCTION

Polyelectrolytes (PELs) comprise the class of macromolecules carrying ionic charges along the polymer chain. The combination of polymeric and electrolyte behavior gives them a huge importance in our daily life. Thus, in nature there are many macromolecules like polypeptides (thus, all proteins) and nucleic acids which are polyelectrolytes. In addition to the basic role played in the biological context, where their importance can be hardly estimated, PELs have found a number of important applications in the major fields of science and engineering such as chemistry, physics, biology, chemical and materials engineering. A fundamental knowledge of their properties in solution and an understanding of the basic phenomena involved in the application of these polymers such as *the solubilization, the adsorbtion and interaction* with different particles, substrates, small ions and oppositely charged macromolecules, *the change of rheological properties of aqueous systems* etc. are essential in developing high performance formulations for achieving optimum efficiency.

The solubilization effect of some ionic polymers having a certain hydrophile/hydrophobe molar ratio can be used for solubilization of low molecular compounds (hydrocarbons, dyes, etc.) or dispersion of insoluble particles to a stable system with water [1].

It is well known that a small amount of polyelectrolyte can have a dramatic effect on the stability of a colloidal dispersion; both *stabilizing* and *destabilizing* effects may occur. Colloidal suspensions used in several industrial applications such as paper making, ceramic processing, paints, detergents, cosmetics, food concentrates, and pharmaceutical products need hydrodynamic stabilization. The most effective mechanism of *stabilization* of these colloidal systems against particle irreversible aggregation results from inter-particle repulsive forces arising from *the adsorption* of suitable chemical species, like PELs, onto the particle surfaces [2-4]. The *destabilization* effect of PELs is also considered in phase separation processes in aqueous systems: solid-liquid separations (coagulation and flocculation of colloids, sludge dewatering, retention aids) and liquid-liquid separation (oil-water emulsions) [1]. Because of PELs tendency to *adsorb* onto solid surfaces they are employed in surface modification through the application of PELs-based coatings and adhesives [1,5-11]. Many applications of the ionic polymers in hydrometallurgy, recovery of trace metal ions, wastewaters treatment, etc. are based on the *interaction* with metallic ions of different valency [12,13]. One of the most important features of PELs is also their *complexation* capability when mixed with other macromolecules of opposite sign leading to stoichiometric or nonstoichiometric polyelectrolyte complexes and polyelectrolyte multilayers. These materials with very interesting properties have found applications in ionic polymeric sensors and artificial muscles, in preparation and modification of membranes for materials separation (gases, pervaporation separation of alcohol/water mixtures, microfiltration of protein solutions, and separation of ions in aqueous solution), electrical and optical devices, etc. [14-17].

The widely extended range of applications of polyelectrolytes modified with low amounts of a hydrophobic monomer (amphiphilic polyelectrolytes) as thickening agents, rheology modifiers and viscosity enhancers is a consequence of their ability to drastically *change the rheological properties* of aqueous solutions and suspensions. This property is exploited in fields like enhanced oil recovery, drag reduction polymers, latex paint systems, in the health and personal care industry for shampoos, conditioners, deodorants and body lotions [18-22].

Of great importance are the applications of ionic polymers in biomedical area such as pharmaceutical formulations, anticancer drugs and gene terapy [23-33]. An important group of polyelectrolytes, basically polycations, shows a high anti-bacterial activity, both in solution and as protective coatings against pathogenic bacteria, yeast and fungi [34-41].

It is not possible to exhaust all of the literature and topics which reflect the growing scientific and industrial interest on the applications of PELs in a short review. This chapter focuses only on a few of the more important and recent experimental and theoretical data regarding the application of PELs in processes which imply the phase separation (*solid-liquid and liquid-liquid*) in aqueous systems.

2. General Aspects of Phase Separation Processes

The increased environmental pollution caused by various industrial activities and natural events leads to a deterioration of the quality of drinking water and requires the removal of a variety of pollutants from surface waters, household (domestic) and industrial waste waters (textile, pulp and paper, food, metals processing, petroleum refining, etc). Among the natural pollutants in surface waters, especially rivers, suspended clay particles resulting from the erosion of soil cause turbidity. Contaminants from the industrial waste waters are of great diversification including clays, metal ions, cellulose-derived substances, thickener agents, dispersing agents, detergents, dyes, oil, grease, etc. If the suspension is allowed to stand quietly, the large particles settle out first. But, there is a portion of very fine material that settles only very slowly or not at all. If these colloidal particles aggregate into larger entities for which gravitational forces dominate, suspension stability is lost. For a long time an efficient way of achievement this, was the coagulation of suspensions by using of multicharged inorganic additives, termed coagulants agents, e.g. hydrolyzable salts of aluminum and iron (aluminum sulfate and ferric chloride) [42-44]. However, their use has emphasized some drawbacks related to the structure and properties of flocs (fragile, difficult to separate and dewater, low resistance to shear) or sludges (voluminous), etc. [44].

The use of organic polymers in combination with coagulants or, if possible, without them has become a standard procedure since 1950's to improve the solid-liquid separation and dewatering efficiency due to: (1) their versatile tailorability by controlling molar mass (*M*), molar mass distribution, chemical structure, nature of the functional groups, charge density; (2) smaller amounts are needed, 1-6 ppm (handling facilitated); (3) their relative insensitivity to pH; (4) lower sludge volume and better flocs properties (larger, stronger, stable, low-density - flotation facilitated); (5) shorter settling time [42,45-47].

Polymers used in water treatment may be natural (gum guar, starch, dextran, chitosan etc.) and synthetic, their selection depending on the application where they are considered for use. For water treatment, synthetic polymers are used almost exclusively. In the pulp and paper industry, on the other hand, the consumption of natural polymer is nearly twice than that of synthetic polymers [48]. In many situations, such as the clarification of water it has been empirically established that the most important feature of the flocculants is its charge density, whereas for the treatment of sewage-sludges both a high molecular weight and charge density are required [49].

Synthetic organic flocculants are water-soluble polymers which are prepared by vinyl addition, condensation polymerization, ring opening, and cyclopolymerization reactions [45].

Depending on the charge nature they can be classified as:

1) *nonionic*: poly(acrylamide) – PAM, poly(ethylene oxid) – PEO, poly(vinyl alcohol) - PVA, poly(N-vinylpyrrolidinone) [45,50-52];
2) *anionic:* salts of poly(acrylic acid) and its copolymers – PAA, sulfonate polymers and copolymers, maleic acid copolymers with nonionic monomers, derivatives of phenol-formaldehyde resin [45,53-56];
3) *cationic*: polyalkyleneimine and polyhydroxyalkyleneamine, poly(diallyldimethylammonium chloride) – PDADMAC and its copolymers; polymers and copolymers of quaternized tertiaryaminoalkyl esters and amides with acrylamide, poly(1-aminoethylene) like, polyvinylamine (PVAm) and related polymers [45,46,57-71]. The consumption of cationic PELs has risen dramatically because many contaminants from waste waters have negative surface charge [72,73].

Akers proposed the next scheme for the action of polymers as flocculants [74]:

- the dispersion of the flocculant solution in the liquid phase of the suspension;
- the diffusion of the flocculant to the solid-liquid interface;
- the adsorption of the flocculant to the solid; the conformation of polymer chain adsorbed at a solid-liquid interface is explained in terms of trains of segments that attach to the surface, loops of segments that extend into solution between trains, and tails of the dangling end of the chain [52];
- the collision of a particle bearing an adsorbed flocculant molecule with another particle;

- the adsorption of the flocculant molecule to the second particle and thus formation of an inter-particle polymer bridge;
- further bridge forming collisions, leading to the formation of a flocs.

In the past, the terms flocculation and coagulation have been used synonymously in connection with formation of aggregates, but later La Mer proposed a distinction between the two terms [75]: coagulation is the process where the colloidal suspensions are destabilized by reducing repulsive forces around the particles, whereas flocculation refers to the process whereby destabilized suspended solids are brought together to form larger aggregates which are easily separated by succeeding separation processes, such as sedimentation, flotation and filtration. According to NATO Advanced Study Institute flocculation is defined to include the process of particle destabilization and collection into larger aggregates [76].

Jar-test is an extremely useful technique for determining parameters such as flocculant optimum dosage, sample volume, stirring speed, floc time and the chemical conditions (ionic strength, pH) in the flocculation process. This apparatus could accommodate six beakers with the contents of each beaker being stirred simultaneously at the same speed with a six-spindle arrangement of steel paddles. The efficiency of flocculation process can be assessed in several ways: by measuring turbidity, suspended solids (SS), the chemical oxygen demand (COD) of the supernatant before and after the treatment with polymer, the settling rate of the flocculated material, the compactness of the sludge, or the ease of filtration, etc. [66].

The choice of a polyelectrolyte to be used as flocculant is potentially complicated and requires a comprehensive and extensive testing programme to find the most appropriate product for a given task. In the market place, selection of a flocculant depends on optimizing the cost-to-performance ratio, i.e., achieving desired performance at minimum cost. At the same time, because of the environmental concerns, there have been increasingly severe government regulations. Thus, for economic as well as environment reasons, it is a key point of further research in the separation processses to develop novel selective materials for a great variety of contaminants and to demonstrate their efficiency and selectivity in the frame of the treatment of real raw waters of different compositions.

3. Flocculation Mechanisms

The flocculation process is a complex one, the controversy concerning the different models for its mechanism being long standing. Ruehrwein and Ward first proposed a model in which a single polymer chain was bridging between two or more particles [77]. Since then, some more important studies about the flocculation process have been developed [78-91]. As a result, the fundamental mechanisms of flocculation have been classified as follows:

1. the bridging flocculation, where tails or loops of a few polymers with high affinity to the particle surface make bridges between two or more particles simultaneously;

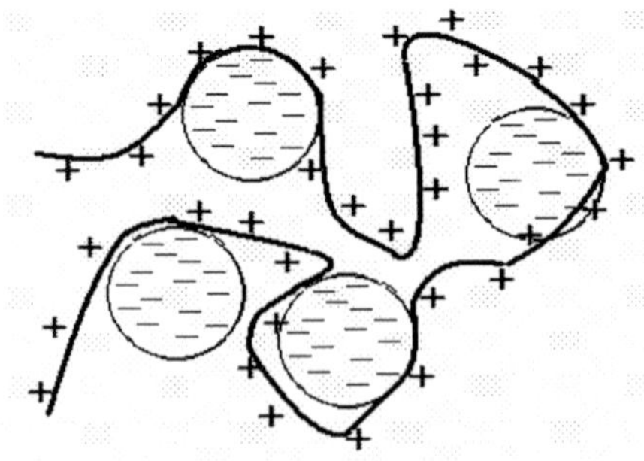

2. the patch (mosaic) flocculation when aggregation occurs as a result of the electrostatic attraction between oppositely charged regions on partially covered particles;

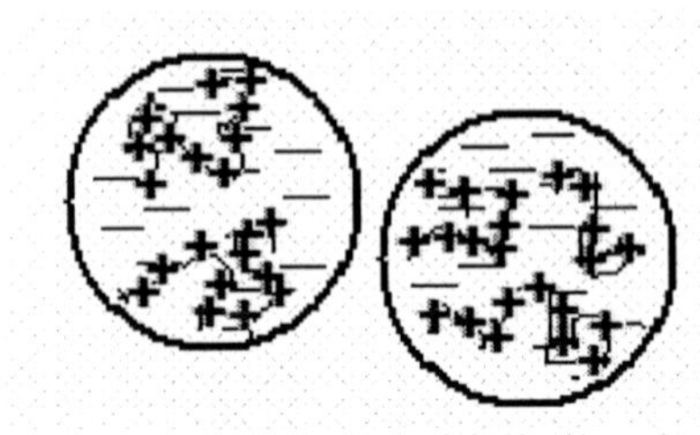

3. the flocculation by charge neutralization, where the particle surface charges are neutralized by the oppositely charged groups of the chain so that the particles attract each other by van der Waals forces;

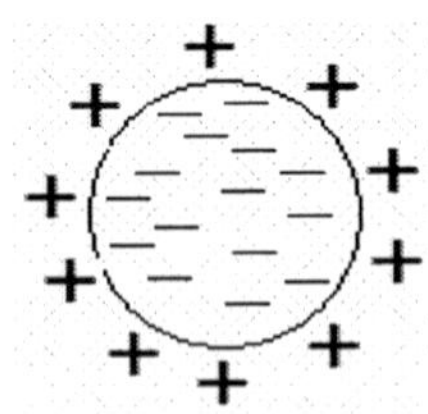

4. the sweep flocculation associated with flocculation by inorganic flocculants like hydrolyzing metal salts which in appropriate conditions (pH) precipitate and trap colloidal particles.

These flocculation mechanisms can act alone or in combination, depending on the properties of particles and polymers in solution. Flocculation process in the presence of nonionic and ionic flocculants runs in different ways.

Generally, for nonionic polymers and linear PELs with low charge density and high molar masses the main mechanism is the bridging one, while PELs with medium and high charge density and high molar masses cause aggregation according to the charge patch or neutralization models [46, 92, 93].

Although the various experimental approaches have given useful information on the flocculation mechanism, it is very difficult to predict how this process could take place in a practical system because of the many factors involved in a real situation.

4. KINETIC ASPECTS

The formation of flocs by a polymeric flocculant is often kinetically controlled [94-96]. The kinetic aspects include:

1. *mixing and diffusion* of polymer molecules with colloidal particles;
2. *adsorbtion* of polymer chains or segments onto the particle surface as a result of several types of interactions (hydrogen bonding, hydrophobic, electrostatic forces);
3. *re-conformation* of adsorbed chains or segments from their initial state to an equilibrium conformation;
4. *collision* between particles with adsorbed polymers to form flocs;
5. *breakage of aggregated* particles under the influence of applied shear.

These processes occur at different rates which depend on a number of factors. Furthermore, they do not occur sequentially, but to a large extent, concurrently, which makes the whole system difficult to analyze [97].

Mixing should be carried out as soon as possible after dosing the polymer into suspension because the polymer must be distributed throughout the system and adsorbed onto the colloidal particles. Poor mixing can lead to uneven distribution of polymer among the particles, some of which may receive an overdose, leading to restabilization of the suspension [97]. It has been suggested that, in a dilute suspension system, the mixing and diffusion of polymers with colloidal particles are often quick and therefore are not rate-determining steps in the formation of flocs [98].

Polymer adsorption onto colloidal particles is considered to be fast, strong, and virtually irreversible, because the polymer is not removed by washing with water [42,58]. Adsorption isotherms for cationic PELs onto suspended particles generally start with a region where the uptake of polymer is almost complete and then round off to a plateau [99]. The rounding off the isotherm is attributed to the polydispersity of the molar masses of the polymers involved. In some cases the plateau is not reached, probably because of an excess of polymer adsorbed onto particles [100].

Pelssers *et al.* assumed that adsorbed polymer chain would either be in an “active” or “inactive” state, the former corresponding to the extended, non-equilibrium condition in which bridging contact are possible [101]. Relaxation to a flatter, inactive configuration occurs, after a while. The rate of this relaxation process depends on the nature of the polymer and on the particle-polymer interaction. Relaxation time of a few seconds for PEO of several million molar mass [101] to many minute or even hours for quaternized polyvinyl pyridine of lower molar mass [102,103] have been reported.

The flocculation of particles can occur if the repulsive forces caused by double-layer interactions are reduced, allowing particles collision. The particles collision may be determined by simple Brownian motion, by fluid movement giving rise to velocity gradients or by particle motion due to an external force (e.g. gravity) [104]. Collision resulting from simple Brownian motion is called perkinetic flocculation; it depends only on the temperature and the concentration of particles. Orthokinetic flocculation results when aggregation is promoted by externally induced velocity gradient [42,104]. If there is no surface force repulsion between the particles, then every collision leads to aggregation and the process is called rapid flocculation. If a significant repulsion exists, then only a fraction of the collisions results in aggregation [104]; the process is called slow flocculation. Adsorption, re-conformation and collisions are predominant factors affecting flocculation and have recently drawn great attention from both theoretical and experimental point of view [95,105-108]. In some cases, flocs can be formed even if the collision occurs before the adsorbed polymer chain reaches an equilibrium state. This type of flocculation was firstly described as non-equilibrium flocculation [109,110]. A few experimental investigations of this aspect have been reported [110,111]. A simplified model, based on the frequency of particle – polymer collisions due to the combination of Brownian motion and shear/turbulence, has been applied by Hogg to provide insight into the nature of the process and the role of some important variables, such as polymer molar mass, particle size distribution, solid concentration and mixing conditions [112]. Some of his conclusions are as follows:

- the polymer adsorption on the surfaces of suspended particles is a non-equilibrium process; it occurs rapidly within a time scale of second or less;
- the collision processes tend to favor adsorption on larger particles (or flocs);
- the high molar mass polymer are especially ineffective for stable dispersions of very fine particles because of an enhanced preference for adsorption on larger particles and the smaller number of molecules available. Low molecular weight are more effective in such cases but still require very high dosage levels for thoroughly dispersed, stable suspensions.

The breakage of the aggregated particles can occur under high shear conditions. Claesson *et al.* suggest that the floc structure can be understood by considering the interparticle forces operating in the system [46]. Low charge density polymers give weakly attractive forces, thick adsorbed layers, dense and shear resistant flocs, whereas high charge density polymers produce stronger attractive forces and rigid flocs with an open structure. Breakage of such flocs is reversible, whereas bridging mechanism gives flocs which do not readily re-form after breakage at high shear [97]. The reversibility of the flocs depends not only on the flocculation mechanism but it also depends on the polymer dosage, as the covering grade affects the flocculation mechanism. The flocs obtained by the addition of a polyethyleneimine (PEI) dosage between 64 and 125 ppm are reversible, because the predominant flocculation mechanism is charge neutralization, whereas flocs induced by the addition of doses lower than 64 ppm are partially reversible [64].

Anyway, while the major removal mechanisms of the pollutants during coagulation and flocculation have been well studied, little information on the physical properties of flocs (floc size, compaction, and strength) is generally available [113]. One property that may have a significant impact at a water treatment work (WTW) is the potential for flocs to regrow after being broken this leading to an improved removal process. Unit processes at WTW are generally designed to minimize floc breakage; however, often in practice this is not the case, with regions of high shear being prevalent [114].

The coagulation of kaolin particles, alum and polyaluminum chloride has been shown to have only limited re-growth of flocs reaching only a third of their previous size after shear; PDADMAC showed near complete regrowth while a copolymer of acrylamide and cationic monomer showed total regrowth [115,116]. From these studies results that the particular coagulant used and, therefore, the coagulation mechanism involved has a considerable impact on the floc regrowth potential.

The application of fractal geometry is now a well-established means of describing the complicated structure of floc aggregates [117-119].

5. Parameters which Influence the Flocculation Process

The great diversification of the contaminants in wastewaters and the variety of technological treatment processes in which flocculants are used, require systems with different properties. The performance of a polymer in flocculation depends on several parameters related to the characteristics of:

- the particle: concentration, size;
- the polyelectrolyte: type of functional groups, structure, molar mass (*M*), charge density, dose;
- the environment: ionic strength (*I*), pH, temperature;
- the conditions of mixing: time, type of stirrer, size and shape of the reaction vessel [57,120].

The knowledge of these parameters is crucial for determining the usability of polymers as flocculants, therefore some of them will be discussed successively in the following. However, the choice of a PEL for a specific type of water is confirmed under service conditions.

5.1. Particles Properties

5.1.1. Concentration

Particle concentration may range from 50 mg/L in raw water systems to 10-20% in mineral suspensions in the mining industry. Generally, the higher solid concentrations (>1%) leads to the increase of collision rates that favor flocculation by bridging, whereas the solids in lower concentrations have lower collision rates leading to a flattened equilibrium configuration which favors the patch mechanism [48].

5.1.2. Size

The size of the dispersed particles commonly ranges from the sub-microns (e.g. clay particles down to 0.01 μm), through the micron (as bacteria and algae) up to 20 μm or even more in the mineral industry [42]. So far, the particle-size effect on the flocculation efficiency has not been well understood [48]. Iler has investigated the flocculation of carefully sized silica suspensions with quaternary ammonium methacrylate polymer and found that for particles of less than 40 nm diameter, the amount of polymer necessary to flocculate unit weight of silica is inversely proportional to the particle diameter, while, for particles larger than 40 nm, the amount of polymer is inversely proportional to the square of the diameter [121].

The study of the relationship between the kaolin particle dimension and the molar mass of PDADMAC showed that for a 0.09 μm sample the flocculation process was efficient using a polycation sample of $M_w = 3.5 \times 10^4$ g.mol^{-1} whereas flocculation of a 9.8 μm kaolin sample exhibited a steady increase in efficiency as M_w increased [122].

5.2. Polyelectrolyte Properties

5.2.1. Charge Type

Among the three groups of polyelectrolytes, only the cationic ones can destabilize the coloidal suspension by electrostatic interactions leading to their coagulation [81,123]; the nonionic and anionic polyelectrolytes act by means of the interactions between the functional groups of polymers and those on the particle surface [124]. The high-molar-mass cationic flocculants are often able to guarantee the required floc stability for the mechanical separation process whereas, the nonionic and anionic polymer cause accelerated deposition of the sediments in certain effluents [125].

Various cationic starch derivatives with high substitution degree, *DS*, were investigated regarding their flocculation properties [126,127]. The flocculation efficiency was noticed to depend on the amino group type which followed the order: quaternary > tertiary > secondary > primary.

5.2.2. Molar Mass (M_w)

The typical molar mass of the polymers used in the flocculation process ranges from 10000 g.mol^{-1} to 10 millions g.mol^{-1}. The effect of M_w on the flocculation is best discussed as a function of the flocculation mechanisms [81,92,128]. Generally, for systems in which bridging mechanism predominates, an increase in M_w improves flocculation, irrespective of the charge density [128]. In some cases, like that of the high molar mass unmodified PEI, the flocculation mechanism is not clear yet. Many authors agree that PEI acts by charge neutralization [129-131], but there are several studies about the PEI adsorption on different particles that show that the interaction between PEI and particles also depends on many variables such as the pH, the charge density, the ionic strength [92]. The flocculation of kaolin slurries with PEI of lower M_w (2×10^3 g.mol^{-1}) revealed that the charge neutralization was predominant one [132].

Although a high M_w is often mentioned as a required condition for bridging flocculation, molecular structure is also important, since at the same M_w, linear polymer are usually more effective flocculants than the branched ones due to larger persistent length providing more chances for macromolecules to adsorb on neighboring particles [133].

The M_w effect is less understood in systems where patch or neutralization mechanisms prevail. In these systems it was found that the optimum dose of flocculant is independent of M_w [81,92]. A match in spacing between PEL functional groups and oppositely charge surface sites [134,135] as well as chain stiffness [134,136] should also be considered as parameters affecting significantly the adsorption and flocculation behavior.

Das and Somasundaran have investigated the flocculation/dispersion dual behavior of alumina-PAA system as a function of M_w of the flocculant [53]. They have found that low M_w polymers, traditionally used as dispersants, can flocculate alumina at ultra low concentrations.

For M_w values ranging from 2000 to 5000 g.mol^{-1}, polymer behaved either as flocculant or dispersant dependent on the concentration range under study.

The influence of M_w on the floculation properties was also followed by Arita *et al.* in the case of poly[2-(acryloyloxy)ethyl(trimethyl)ammonium chloride] (PATC), M_w: 7.8 x 10^6, 2.7 x 10^6, 1.25 x 10^7 g.mol^{-1} [58]. They have observed that it takes a long time (10 days) for cationic polymers of high M_w to relax completely in water, even though the dissolving process is enhanced by mixing. It seems that polymers relaxed in solutions are fundamentally spherical in shape but their structure becomes fragile as the size increases, so highly swollen polymers (M_w = 1.25 x 10^7 g. mol^{-1}) are easily extended by external disturbances to form various structures (lump, flock, fibril) at their adsorption on the surfaces.

The flocculation of clay suspension in water with polycations (PAM type and PDADMAC) was also reported [137,138]. The PAM type polycations with high M_w acted by bridging (broad flocculation window), whereas the strong polycation PDADMAC (with relatively low M_w) showed typical patch behavior.

The study of Ariffin *et al.* concerning the influence of the cationic PAM molar mass (five samples) on their flocculating performance in the treatment of palm oil mill effluents (POME) showed that the polyelectrolyte adsorbtion increased as M_w increased [139]. The high molar mass sample (1.5 x 10^6 g. mol^{-1}) was the most effective polymer as it obtained a high removal efficiency with a dosage as low as 60 mg/L at pH 3 of POME. Very high molar mass sample (over 5 milion g mol^{-1}) produced poor flocs formation because the polymer did not dissolve readily but formed gel lumps known as 'fish eyes'.

5.2.3. Charge Density

It is a well known fact that the solution properties of polyelectrolytes are strongly influenced by the linear charge density of the chains which affects their conformations and hence, the flocculation ability. As we have already mentioned, it was found that polyelectrolytes with high charge density and high M_w led to the particle aggregation in accord with the patch model [92], whereas those with low charge density and high M_w favored the aggregation according to the bridging mechanism [93].

The interaction between humic acid with a series of synthetic cationic PELs having different charge density was studied by Kam and Gregory [140]. The degree of removal of humic acid was considerably better with polycations of higher charge density. The PELs molar mass (over a range from about 50000 to15 million g.mol^{-1}) appeared to have no effect on the results.

The flocculating efficiency of some cationic polyelectrolytes with different charge density and hydrophobic side chains, obtained by condensation polymerization of epichlorohydrin with dimethylamine, N,N-dimethyl-1,3-diaminopropane and primary amines, hexyloxypropylamine, (polymer type PCA_5H) and decyloxypropylamine (polymer type PCA_5D) [141] (Scheme 1) on a suspension of montmorillonite was also studied [59].

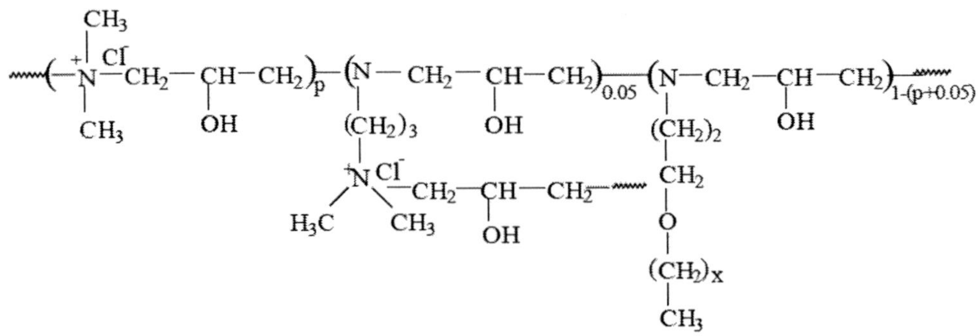

p = 0.94, x=5, polycation PCA_5H_1; p = 0.93, x=5, polycation PCA_5H_2;
p = 0.94, x=9, polycation PCA_5D_1

Scheme 1.

It was noticed the maximum purification degree of 85% for PCA_5D_1, 80% for PCA_5H_1 and 78% for PCA_5H_2 was reached after a settling time of 100 min. for all the polycations. Other evidence was that for the same hydrophobic amine (hexyloxypropylamine), the flocculating efficiency was greater for PCA_5H_1 than PCA_5H_2. PCA_5H_1 is a polycation with a greater charge density than PCA_5H_2 as the primary alkylic amine content is 1 mol%, only. Thus the electrostatic interactions appeared as dominant factor influencing the flocculation process. On the other hand, for the same content of hydrophobic amine, 1%, the turbidity values for PCA_5D_1, the polycation with longer nonpolar side chain - x = 9, were lower than those of PCA_5H_1, x = 5.

The flocculation behavior of some cationic starches (water soluble - SCS and water-dispersible –DCS) with *DS* from 0.21 to 0.57 [142] and *DS* from 0.25 to 1.54 [143] has been examined in the destabilization of kaolin dispersion. The best flocculation efficiency was exhibited by SCS, whose *DS* varied from 0.28 to 0.37 and DCS with *DS* of 0.32. With a further increase of *DS* the flocculation efficiency of DCS decreased by half or more and restabilization of kaolin dispersion occured with increasing the amount of DCS added. The water-dispersible cationic starches exhibited a higher flocculation efficiency compared with the soluble cationic starches. Bratskaya *et al.* found that flocculation has occurred long before the kaolin surface charge neutralization was reached and the efficiency of phase separation increased with increasing *DS* [143]. The flocculation mechanism of the kaolin dispersion with cationic starch derivatives was suggested to be a combination of bridging and charge patch processes. Addition of KCl aqueous solution (1, 10, 100 mM) proved to enhance the flocculation effectiveness, in comparison with water, through lowering the required polymer dose and broadening the flocculation window. Addition of electrolyte led to the surface charge screening and consequently, to the decrease of electrostatic repulsion between particles and hence, to attraction of flocculant in a broader concentration range. The same authors compared the flocculation efficiency of cationic potato starch derivatives, synthetic polycations (cationic polyacrylamides - Praestol) and chitosan, having close charge density (*DS* about 0.6) on a kaolin suspension [144]. They found that the dispersions destabilization was reached at the same minimum doses for all polycations, but in a broader concentration range for cationic polysaccharides due to the difference in conformations of flexible and semi-rigid polymers upon adsorption. It was also observed that the dependence of optimum flocculation dose on *DS* was close to a linear function for a series of Praestol flocculants but it deviated from linear dependence for cationic starch with *DS* > 0.7; the authors assumed that not all cationic sites of the starch derivatives of high *DS* were accessible for interactions due to the steric hindrances. Higher doses of cationic starch and synthetic polycation were required in diluted dispersions to reach complete phase separation, and maximum turbidity reduction by chitosan and commercial cationic starch do not exceed 60% (at pH=8). However, decrease of pH to 5.5 resulted in drastic increase of chitosan efficiency, thus proving domination of electrostatic contributions to its flocculation performance.

5.2.4. Molecular Architecture

Polymer structure also influences the adsorbtion and flocculation process. Using a Monte Carlo simulation, Balasz and Siemasko compared the adsorbtion of comb and linear polymers [145]. They concluded that the comb polymers, which contain "stickly" teeth, have greater polymer-surface affinity than the equivalent linear chains. This result was confirmed by Li *et al.* who found a better flocculation of a dilute titanium dioxide (TiO_2) suspension using graft copolymers of acrylamide and diallyldimethylammonium chloride compared with the homopolymers of PAM and PDADMAC [146].

Singh *et al.* summarized the various investigations carried out using flocculants based on polysaccharides and concluded that grafted polysaccharides were shear stable, more efficient flocculating and settling aids than linear polymers [47]. The reason for better flocculating power of the graft copolymers over the linear polymers was explained by the higher ability of the dangling grafted chains to bind colloidal particles than the linear ones. Of all the grafted polysaccharides, PAM grafted amylopectin having fewer but longer branches gave the best performance in a variety of synthetic and industrial effluents (kaolin, coal suspensions, etc.) [47,146-149].

Time-resolved size measurements using laser diffraction and rheology measurements showed the important role of the macromolecule structure (location of the charged groups) in the bridging flocculation of clay by some acidic polysaccharides [150]. The flocculation was more efficient when carboxylate groups were located in the side chain, well exposed toward clay particles, than those located on the backbones.

Caskey and Primus performed flocculation tests on a 3% kaolin slurries with linear and branched anionic PAM [151]. The linear polymers were more effective than the branched ones. The mean end-to end length of the adsorbed polymer molecules was more important than the total amount of polymer adsorbed.

The flocculation of nanosized anionic particles (silica) by some linear or branched cationic polymers based on PAM was affected by the molecular architecture of a polymer [152]. Only in the case of linear polymers the degree of aggregation of the particles was important.

5.2.5. Dose

The effect of any polymer on the flocculation process is also influenced by the polymer dose. The optimum amount of polymer required for flocculation will depend on which variable is the most important, because, for example, the optimum amount to produce maximum sedimentation is often different than the optimum amount required for clarification [153]. The flocculation efficiency decreases at lower or higher doses than the optimum dose. Thus, at lower polymer doses, the bridging does not take place while at higher doses redispersion of the particles happens due to the repulsive electrostatic forces which appear by adsorption of a greater number of the like charged chains onto the same particle.

La Mer showed in his model that the flocculation process is efficient when 50% of particle charge is covered with polymer ('half-surface coverage' concept) [154].

Related to the optimal dosage of flocculants in the solid separation of suspensions, Suzuki *et al.* proposed a new flocculation model that takes into account the bridging between a flocculant-adsorbed surface on one particle and a bare surface on another particle [155]. To verify the validity of the model, flocculation experiments were carried out using quartz particles and two kinds of polymers (methylated egg ovalbumin and a combination of methyl glycol chitosan and potassium polyvinyl sulfate) as flocculants. As a result, it was revealed that one of the model parameters, α, plays the most important role in flocculation phenomena, and it represents the fraction of particles surface covered by flocculant at an optimum concentration.

The flocculating efficiency of two cationic polyelectrolytes with quaternary ammonium salt groups in the main chain, with different branching degrees (PCA_5 called CS-2A and PCT_{20} called CS_2) on a bentonitic clay suspension, in comparison with PAM was followed [156]. The cationic polyelectrolytes were obtained by the condensation polymerization of epichlorohydrin (ECH) with dimethylamine and N,N-dimethyl-1,3-diaminopropane (DMDAP) (PCA_5) and triethylenetetramine (TETA) (PCT_{20}) respectively, as polyfunctional amines [157,158]. Their structures are shown in Scheme 2. An increase of the PCA_5 concentration from 0.1 to 0.6 mg/L led to the increase of the clarity degree from 75% to 100% after a settling time of 120 min, whereas with PCT_{20} a flocculation efficiency of 100% was not obtained in the same concentration range, even after a settling time of 24 h. The maximum purification degree of 92% was reached after a settling time of 120 min. An increase of the PCT_{20} concentration from 0.1 to 0.2 mg/L led to an increase of the purification from 78% to 92%; further increase of polymer concentration caused a decrease of the clarity from 92% to 83%, i.e. a stabilization effect upon the suspension occurred. The higher flocculation efficiency of PCA_5 than PCT_{20} was explained by the higher charge density and the smaller branching degree of the former polycation (the polyfunctional amine content (DMDAP) is 5 mol %, only). When PAM was used as flocculant a complete of the turbidity remove was achieved at a concentration of 0.4 mg/L, after a settling time of 120 min. From the flocculating efficiency point of view, the difference that appears between the studied polymers could be explained taking into account the action mechanism of polymers upon the particles: the main effect of cationic polymers, characterized by a high charge density, upon the suspension is the coagulation while, PAM interacts with the suspension particles by the H bonds established between the amide groups and the OH groups on the particle surface. In this case the bridges formation is easier because the M_w of PAM is much higher (about 1×10^6 $g.mol^{-1}$) than that of the cationic polymers (about $1\text{-}2 \times 10^5$ $g.mol^{-1}$).

The polymer dose influence on the flocculation process was also evidenced when polycations with different lengths of hydrophobic side chains (PCA_5H_1 and PCA_5D_1 see Scheme 1) were used to separate a montmorillonite suspension [159]. The increase of the polycation dose from 0.58 to 1.74 mg/g montmorillonite led to a faster floc formation and also to a faster sedimentation velocity. This result could be caused by the increase of the effective collision number between particles. The last value of the polycation dose (1.74 mg polycation/g montmorillonite) seemed to be close to the optimal dose for flocculation, in the case of PCA_5H_1 at least, at 25°C.

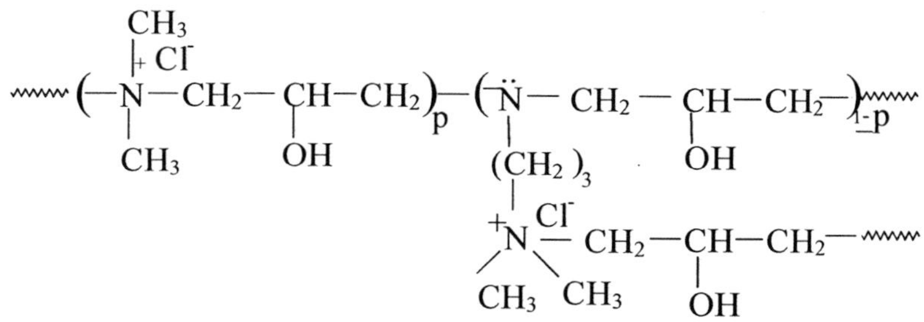

polycation PCA_5 (CS-2A), p = 0.95

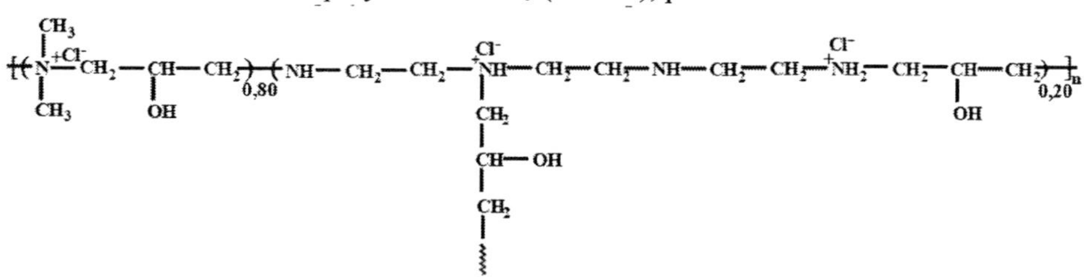

polycation PCT_{20} (CS-2)

Scheme 2.

Li *et al.* investigated the flocculation of a dilute titanium dioxide (TiO_2) suspension using homopolymers and graft copolymers of AM and DADMAC [146]. No significant flocculation was observed up to 10 mg polymer/g TiO_2 for PAM homopolymer. When pure PDADMAC with $M_w = 8.4 \times 10^4$ g.mol^{-1} was employed an optimal flocculating dose of 1.25 mg polymer/g TiO_2 was found. In the case of graft copolymers, three situations have been noticed. 1. At doses < 2.5 mg polymers/g TiO_2 flocculation was rapid. 2. At doses > 5 mg polymers/g TiO_2 there appeared to be a delay period before flocculation occurred. The length of delay period was a strong function of polymer dose: the more polyelectrolyte was added, the longer the delay period observed in the dose range of 5-30 mg polymers/g TiO_2. 3. When the doses were > 50 mg polymers/g TiO_2, no flocculation was observed.

Maximum flocculation of a cellulose fiber suspension by a cationic copolymer consisting of acrylamidopropyltrimethylammonium chloride (APTAC) and AM occured at a polymer dose of 0.4 mg.g^{-1} [160]. Above this addition level, the flocculation effect was reduced and dispersion took place when more than 1.2 mg. g^{-1} polyelectrolyte was added.

The research results on filler (calcium carbonate) flocculation and floc properties induced by modified PEI ($M_w = 1.3 \times 10^6$ g.mol^{-1}) showed that there were two PEI doses that produced a maximum flocculation due to different flocculation mechanisms [64]. Thus, when the polymer dose was lower than 50 ppm, bridging flocculation was the predominant mechanism and formed flocs were not reversible while, patching flocculation prevailed for polymer doses around 50 ppm. Neutralization flocculation was predominant for polymer doses of 64 ppm or higher. PEI doses of 100 ppm or higher produced a slow flocculation due to the PEI chains redistribution and of the electrostatic repulsion between the adsorbed polymer chains.

The effect of a cationic polyacrylamide (CPAM) dose on flocculation of a ground calcium carbonate (GCC) suspension was investigated by Blanco *et al.* [161]. The optimal dosage of CPAM was determined using a laser technology, M500L focused beam reflectance measurement (FBRM) probe, by adding the polymer gradually to the suspension and monitoring the mean chord size. The optimal CPAM dosage was 7 mg/L which corresponded to a low coverage grade of the particles (around 10% of the maximum CPAM adsorption). This and the fact that it was not necessary to neutralize the ζ potential to induce particle aggregation indicated that the flocculation mechanism was the bridging one. At the optimal dosage, the flocs were slightly instable because of the potential flattening of the polymer but they were reversible.

Effects of experimental parameters such as the dosage of chitosan in powder and flake form, contact time, mixing rate, settling time and pH were studied by Ahmad *et al.* in order to obtain the optimum conditions for the adsorption of residue oil from palm oil mill effluents (POME) [120]. FT-IR spectra and SEM micrographs of chitosan before and after adsorption were presented to prove that the residue oil had been adsorbed by chitosan. The greater the dosage of chitosan employed, the more effective was the removal of residue oil. It was found that the dosage of chitosan powder and flake needed to achieve ca. 99% removal of residue oil from POME was 0.5 g/L and 2 g/L, respectively. The greater dosage of chitosan in the flake form, relative to the powder form, to achieve the same oil removal efficiency was supposed to be due to the surface area. Powdered chitosan has an extra loose pore structure relative to the flake variety thereby facilitating the adsorption process. In addition, the amino groups in the flake exhibited an added steric hindrance compared to the situation in the particles and also a more restricted pore structure or a smaller pore volume.

5.3. Solution Properties

5.3.1. Ionic Strength (I)

Charges on a polyelectrolyte chain are increasingly screened when the ionic strength of the solution increases. The screening effect alters the conformation and the stiffness of polyelectrolytes in solution. This affects the kinetics of desorption and flocculation.

The increase of I can determine both the decrease and the increase of adsorbed amount of polymer on the opposite charged particles surface. The decrease in adsorbed amount of polymer with I increase, is explained by a screening of the electrostatic attraction between the species [162], while the increase is caused by the formation of loops and tails at the particle surface which favors bridging flocculation [163].

Nyström *et al.* emphasized the positive role of the NaCl presence in the flocculation of calcium carbonate dispersions induced by a cationic starch $DS = 0.2$, where the particle bridging is likely the predominant mechanism [61].

In the systems where the electrostatic patch mehanism predominates, the effect of ionic strength is less understood [48]. An increase in electrolyte concentration (NaCl) led to the increase of both the adsorption of PDADMAC, having an M_w of about 240000 g.mol^{-1}, on the barium chromate particles, and the sedimentation rate [62].

The adsorption and flocculation of a fiber suspension by a cationic polyacrylamide (C-PAM) increased with increasing I at low ionic strengths (1-10 mM NaCl) and then decreased at higher ionic strengths (10-100 mM NaCl) [160]. The increase in adsorption at low values of I was explained by an increasing accessibility to the fiber interior as the dimensions of the polyelectrolyte decreased. At higher I values, the adsorption of the polyelectrolyte decreased due to screening of the electrostatic attraction and competition for the surface sites between the polyelectrolyte segments and small ions.

The breakage of a model food emulsion consisted of a sunflower oil/water mixture, a surfactant (sodium dodecyl or tetradecyl sulfate) and NaCl by a natural polyelectrolyte, chitosan, has been studied by Pinotti *et. al.* as a function of different factors, like ionic strength, pH, size of the drops in the emulsion, relative concentration of oil and surfactant and type of emulsifier [164]. The experiments showed that the increase of NaCl concentration from 10^{-3} to 10^{-1} mol/L led to a diminished chitosan dose necessary to produce destabilization and flocculation of the emulsion (65, 48 and 33 mg/L for the emulsions containing 10^{-3}, 10^{-2} and 10^{-3} mol/L NaCl, respectively). Similar results were obtained when the same model system was treated with a cationic polyacrylamide [165].

5.3.2. pH

The solution pH influences the chains conformation and hence, the flocculation process [48,159,166,167]. For example, the polystyren latex particles have been flocculated by PVAm at pH values of 3, 9 and 10 [48]. At pH 3, where 95% of amino groups are protonated, the optimum concentration was of 0.178 mg/L, and increased at 0.605 mg/L at pH 9, and at 3.35 g/mL at pH 10. At pH 3, the flocculation followed the patch mechanism while at pH 10, where the polymer is almost uncharged, the bridging mechanism prevailed. At pH 9 the flocculation took place by means of both mechanisms.

The influence of pH on the flocculation efficiency of the montmorillonite by the polycation PCA_5H_1 showed this polycation is more effective in the weak basic (8.40) and basic (11.15) pH than in acidic pH (3.15) [159]. This behavior can be primarily explained by the specific and complex characteristics of the active sites of montmorillonite (Al-OH and Si-OH) which could be ionized in the basic range. The reduced influence of the pH variation on the flocculation efficiency in the basic range showed that the polycation is working not only as a salt but also as an ammonium quaternary base.

The studies of the waste water treatment with polyelectrolytes which contain arsonic acid groups $AsO(OH)_2$ [166] and zwitterionic polymers [167] showed that the highest solids removal efficiency took place at pH 2.

pH adjustments were undertaken to study their effect in removing residue oil using chitosan [120]. It was found that cca. 99% removal was achieved at pH values in the range 4.0–5.0. This led to the conclusion that the negative and positive charged forms of POME were stabilized at pH values between 4.0 and 5.0, providing the best conditions for the adsorption of residue oil by chitosan.

5.3.3. Temperature

The temperature influence on flocculation process is not clear. In some cases like the removing of some dyes from the waste water using two comercial polycations, the temperature increasing led to the decrease of the efficiency of this process [168], unlike the case of the flocculation of barium chromate with PDADMAC or its copolymer with acrylic acid (M_w = 240000 g.mol^{-1}), where the sedimentation rate increased with increasing temperature [62].

The results regarding the influence of the temperature (5, 15, 25 °C) on the separation kinetics of the montmorillonite particles as a function of the polycation structure (PCA_5H, PCA_5D – see Scheme 1) compared with the coagulation with $Al_2(SO_4)_3$, showed that the sedimentation velocity of the montmorillonite particles was more significantly influenced by temperature in the case of the coagulation process, increasing with the increase of temperature, as expected [159]. This behavior was explained by the decrease of medium viscosity and by the decrease of the electrokinetic potential of the mineral particles along with the temperature increase. The temperature influence on the flocculation process by the polycations was less evident (polycation dose of 0.58 mg polycation/g montmorillonite). A common characteristic for all the polycations was that, for less than about 60 min of settling, the lowest turbidity was found at 5 °C and the highest at 25 °C. The turbidity found at 15 °C was in between, both before and after 60 min of settling time. This different behavior could be explained by the reduced possibilities of the chain reconformations at low temperature; the polyelectrolyte is adsorbed from the beginning in a coiled shape which could promote the collision of more colloidal particles. The abrupt decrease of turbidity in the first minutes of settling found at 5 °C was followed by a levelling off, whereas at higher temperatures (15 and 25 °C) the flocculation process continued as a result of the higher flexibility and more possibilities for reconformations of the polycation chains. The efficiency of hydrophobic interactions was also evident, the polycation PCA_5D_1 with the longest alkyl chain, showing the highest efficiency on the removal of montmorillonite from the aqueous suspensions at every temperature; the polycations PCA_5H_1 and PCA_5H_2 had a comparable efficiency in this process under the same conditions.

The interaction between starch of varying *DS*, 0.2, 0.35, 0.35, 0.5 and calcium carbonate was investigated at three different temperatures : 25, 35, 45 °C [61]. Increasing the temperature from 25 to 45 °C caused changes in the adsorption and flocculation behavior of the dispersions, especially in systems where starch with *DS* of 0.2 was present. For the starches of highest *DS*, the influence of temperature was minor due to the stronger electrostatic repulsions. It was assumed the reason for the temperature dependence of flocculation could be the chain conformational changes due to their dehydration.

6. Dual Flocculation

In the water treatment, the systems composed of either two oppositely charged PELs [137,138,169-177] or a polymer and oppositely charged microparticle [160,178-180] has been shown to impart improvements of the flocculation process. The so-called dual system was first described in the literature by Moore *et al.* [181], and Britt *et al.* [182] who found that a combination of a low molecular weight cationic polymer and a high molecular weight anionic polymer, added in sequence, produced synergism in the flocculation of paper pulp. Up to now, many different systems have been applied in the paper [175-177] and textile industry [63], flocculation of cellulose and cellulose mixtures [171,172], amidine particles [183], clay [138], alumina [169,170,173], montmorillonite [184], kaolin suspensions [185], etc.

The investigation of dual flocculant systems is of growing importance since combinations of different polymers demonstrated some advantages compared with single flocculant polymer: a higher overall level of flocculation, less sensitivity to process variations and polyelectrolyte concentration, improved possibilities to control and optimize the flocculation, good sludge dewatering, superior retention, and shear – resistant flocs, etc. [137,169,172,173,175,176]. These effects are obtained by a combination of charge patch formation and bridging mechanisms. The disadvantage, on the other hand, is that the higher number of components makes it more difficult to predict the flocculation behavior [175]. Thus, despite the frequent use of dual flocculation a lack of fundamental understanding of the polyelectrolytes mode of action and all the parameters controlling their efficiency still exists. Besides all factors involved in the system with single flocculant, in the dual system comes into play other parameters, such as: molar ratio of the cationic and anionic charges, various modes of polymer addition (premixed or sequently added, as well as the order of polyelectrolytes addition), etc.

Flocculation of cellulosic suspensions by the dual system composed of PDADMAC and poly(maleic acid-co-α-methylstyrene) (P(MS-α-MeSty) increased sharply with increasing amount of polyanion and reached a maximum at a molar ratio (n^-/n^+) of about 0.55 [171]. It was also found that only a certain concentration of PDADMAC led to the improvement of flocculation; the authors concluded that the process is not a flocculation in a classical sense (e.g. patching and bridging) but is a complex flocculation with a highly charged polycation and a polyanion of relatively low molar mass. The addition of a preformed nonstoichiometric polyelectrolyte complex led to a good flocculation, too, with small quantities of a positively charged complex ($n^-/n^+ = 0.3$) or large quantities of a polyelectrolyte complex with a molar ratio of about 0.54. Flocculation in the former case has taken place by electrostatic attraction between oppositely charged parts of the particles, whereas in the latter case, cellulose and mineral material were encapsulated into the complex. In comparison with step-by-step flocculation, the preformed complex gave better results, especially for small quantities of the polycation.

Polycation-dye-polyanion (PC/D/PA) complexes obtained by the interaction between nonstoichiometric polycation/dye (PC/D) complexes with polyanions have been investigated by Dragan *et al.* [186]; these assemblies may increase the efficiency of dye removal from aqueous solution, especially in the case of dyes with high solubility in water. Poly(sodium acrylate) (NaPA), poly(sodium 2-acrylamido-2-methylpropane sulfonate) (NaPAMPS) and poly(sodium styrenesulfonate) (NaPSS) as polyanions and Crystal Ponceau 6R (CP6R) and Ponceau 4R (P4R) with two or three sulfonic groups as dyes were used for this study. Polycations differed in their content of the quaternary ammonium salt groups (N,N-dimethyl-2-hydroxypropylene ammonium chloride). The stoichiometry and the stability of the tricomponent complexes depended on the structure and molar mass of polyanion, the polycation structure, the dye structure and the P/D molar ratio.

For this investigation, a nonstoichiometric dicomponent complex PCA_5/CP6R, with a molar ratio P/D of 7.7 has been prepared. The interaction of this complex with different polyanions has been followed by UV-Vis spectroscopy, by the variation of the relative absorbance (A_t/A_d) as a function of the molar ratio [PA]/[PC]. As Figure 1 shows, increasing the polyanion amount caused a very small decrease of the A_t/A_d until the stoichiometry of the PC/D/PA complex, when an abrupt decrease of the A_t/A_d was observed, accompanied by the phase separation. The influence of polyanion structure was clearly observed by the position of the complex stoichiometry. Thus, for the two strong polyanions, NaPAMPS (M_v = 1400000 g.mol^{-1}) and NaPSS (M_v = 140000 g.mol^{-1}), the molar ratio [PA]/[PC] corresponding to the complex stoichiometry, marked by the minimum of the A_t/A_d ratio is close to 0.7, unlike the weak polyanion, NaPA (M_v = 217600 g.mol^{-1}), when the minimum of the A_t/A_d ratio was close to 0.9. Considering the P/D ratio used in the preparation of the PCA_5/CP6R nonstoichiometric complex, whose value was 7.7, the composition of PC/D complex would be 1 mol positive sites of polycation for 0.13 mol of CP6R.

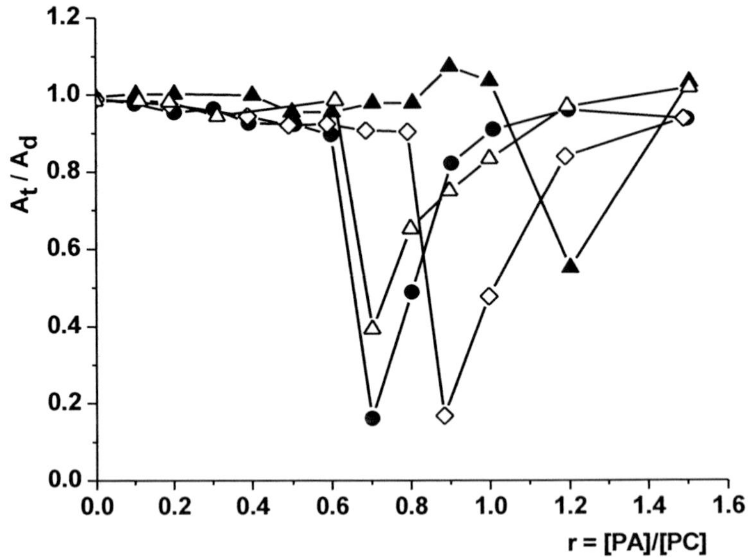

Figure 1. Relative absorbance (A_t/A_d) as a function of molar ratio [PA]/[PC] in the formation of PC/CP6R/PA complexes: (,) PC = PCA_5 and PA = NaPAMPS; (8) PC = PCA_5 and PA = NaPSS; (M) PC = PCA_5 and PA = NaPA; (7) PC = PCA_{20} and PA = NaPSS.

Taking into account that both sulfonic groups of this dye participate in the complex formation, the following theoretical structure of the complex can be assumed: 1 mol positive charges of polycation/0.26 mol SO_3^- groups. This ratio suggests a theoretical composition of the tricomponent complex of about 1 mol positive charges of polycation/0.26 mol SO_3^-/0.74 mol anionic groups of polyanion. This shows that the amount of polyanion needed to achieve the stoichiometry of the tricomponent complex was very close to the theoretical one only for the strong polyanions being much higher for the weak polyanion. In the last case, a higher amount of polyanion has been included in the tricomponent complex, probably because of the difficulties in the charge compensation of the opposite charges having different ionic strength. Formation of the PCA_5/CP6R/PA complexes with different polyanions suggests as the main driving force in the complex formation the electrostatic interactions between the polyanion and the free positive charges of the nonstoichiometric polycation/dye complex.

Influence of the polycation can be also observed in Figure 1 when PCA_{20} was used as polycation. Influence of the branched structure of the second polycation has been noticed first by the high increase of the A_t/A_d ratio just before the stoichiometry and by the very high content of polyanion (NaPA) needed to achieve the minimum of the relative absorbance, (about 1.2). A high amount of the dye was excluded from the complex before the end point, when the branched polycation was used.

The polyanion in excess added after the complex stoichiometry caused the exclusion of the dye from the complex environment, this behavior being a consequence of the PC/PA complex formation, and has been observed also for other systems. The rhythm of the dye exclusion depended on the polyanion structure and it is illustrated by the relative concentration of free dye as a function of PA/PC in Figure 2. The presence of the free dye in solution only after the end point shows the cooperative binding of the dye, which was not disturbed until the stoichiometric point was reached.

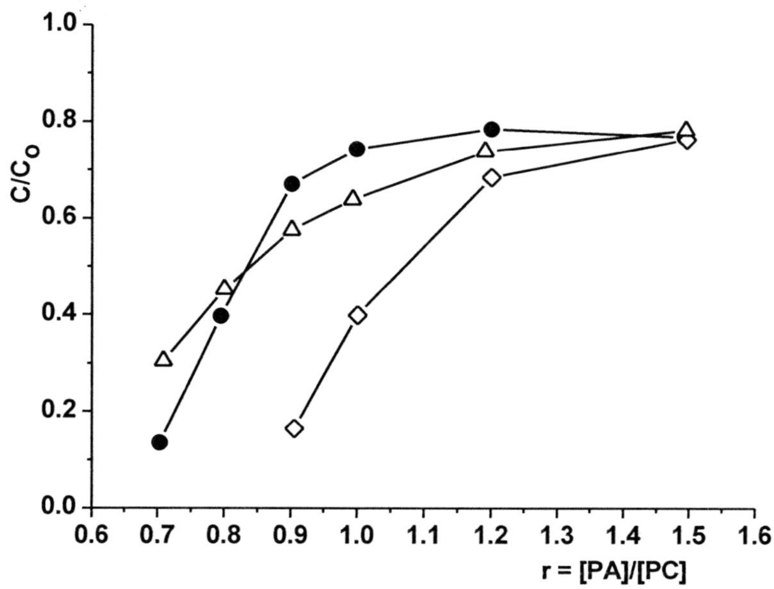

Figure 2. Relative concentration of free dye (C/C_0) as a function of molar ratio [PA]/[PC] in the formation of PCA_5/CP6R/PA complexes: (,) NaPAMPS; (8) NaPSS; (M) NaPA.

Zemaitaitiene *et al.* have examined the influence of the polyanion (sodium carboxymethylcellulose- CMC, sodium carboxymethylstarch - CMS)/polycation (PDADMAC, poly(vinylbenzyltrimethylammonium) -PVBTMAC) molar ratio (n^-/n^+) on the removal of various anionic and cationic dyes from textile waste water and observed that the dual system enhanced the flocculation of dyes irrespective of the (n^-/n^+) molar ratio and the dyes type [63]. The main difference in their flocculation behavior was the value of the optimal (n^-/n^+) molar ratio. Thus, in general, the anionic dye was removed by addition of a greater amount of cationic polymer (n^-/n^+ decreases) comparative with the basic dye. The optimal (n^-/n^+) molar ratio was also controlled by the medium pH value. In alkaline medium the range of optimal (n^-/n^+) molar ratios became wider, but the amounts of the cationic polymer needed for flocculation was greater than in neutral solutions. In acidic medium, the systems with polysaccharides as anionic component had a positive influence on the removal of dyes.

Flocculation of alumina was investigated with polystyrene sulfonate (PSS) and cationic polyacrylamide (CPAM) as polymers alone and in combination with each other [169]. When it was used alone, PSS showed some limited flocculation, while CPAM did not produce any flocculation of the alumina powder. Enhanced flocculation was obtained with double flocculants due to the polymer-polymer interactions (interparticle bridging). It was found that when the two polymers were added in premixed form, the flocculation was not as good as that obtained when polymers were added sequentially regardless the mode of polymer addition. The same results were obtained by Nyström and Rosenholm using cationic starch and anionic sodium polyacrylate (NaPA) for flocculation of semi-dilute calcium carbonate dispersions [176]. The PELs were also added pre-mixed or sequentially to the calcium carbonate dispersion. The flocculation with mixtures of polyelectrolytes was dependent on the charge density of the starch and not substantially affected by the sequence of addition of the PELs.

The order of addition of the two polyelectrolytes - highly charged polycation (PDADMAC)/ poly(acrylamide-co-sodium acrylate) with different charge densities and very high molar masses (14 million g mol^{-1}) tested as flocculants for clay suspensions and natural waste waters from gravel pits influenced the flocculation mechanism [138]. The addition of the polycation followed by the polyanion, i.e., a combination of patching and bridging mechanisms was the most effective. The flocculation behavior also depended on the concentration of polycation and the molar ratio of anionic and cationic charges. In the case of clay suspension the optimum ratio [−]/[+] for getting a fast sedimentation with clear supernatant was about 0.2.

Fan *et al.* studied the flocculation of alumina fines with combinations of PAA and cationic copolymers of AM and dimethylaminoethyl acrylate (Percol) of various *Mw* and cationic charge densities [173]. In order to determine the role of the polymer conformation and its changes due to the presence of another polymer, the fluorescence label method was employed to monitor PAA conformation at the alumina/water interface under different polymer addition modes, such as the sequential addition of the two polymers comparative with co-addition. The authors found the sequential addition of the two polymers produced better flocculation than the co-addition; they assumed the polymer complexation took place after the adsorption of the first polymer (PAA), and thus, a maximum amount of the second polymer (Percol) was available for complexation coupled with bridging. Increase in the concentration of one of the two polymers was noticed to make the other polymer, no matter it played the role of anchoring or bridging, more stretched.

Lu and Pelton have dealt with the investigation of flocculation mechanism of an unusual dual system based on mixtures of PEO and synthetic polypeptides (PEY1), called cofactor, based on tyrosine and glutamic acid [177]. Flocculation experiments were carried out with precipitated calcium carbonate (PCC). The cofactor/PEO flocculation system is unusual because the adsorption of the polymer onto colloidal particles is no driven by electrostatic attraction. This is an advantage because, unlike the situation for cationic flocculants, the performance of cofactor/PEO is rather insensitive to the presence of dissolved anionic polyelectrolytes present in some types of papermaking suspensions. Cofactors are low molar mass water-borne phenolic polymers bearing negative charges and can be based on phenolic resins, tannin, lignin, vinyl phenol copolymers and tyrosine-rich polypeptide. They have proposed that PEO/cofactor complexes form in the aqueous phase and adsorb onto the surface of the colloidal particles. Flocculation would occur if PEO/cofactor complex on one particle bound to adsorbed complex on a second particle, i.e, if the complexes are sticky. As PEO/PEY1 to be sticky, they must simultaneously have unbound PEY1 and PEO segments. There have been also other publications discussing the flocculation properties of PEO/cofactor complexes [187,188].

Another kind of flocculation which implies two oppositely charged polyelectrolytes is heteroflocculation. Seppänen *et al.* have studied the heteroflocculation of kaolin particles by mixing two kaolin suspensions: one stabilized by a cationic polymer (linear co-polymers of AM and (3-metacrylamidopropyl)-trimethylammonium chloride (MAPTAC), and the other stabilized by a highly charged low molar mass anionic polymer (NaPA) [185]. This kind of flocculation resulted in flocs with an open structure and floc strength strongly dependent on the cationic polymer properties. Floc strength increased with decreasing charge density and increasing molar mass of cationic polymer; for a polymer with a charge density of 40 mol %, the floc strength also increased with the ionic strength of a solution. Higher floc strength obtained with polymers of low charge density was assumed to be due to a more flexible bond since these polymers had a more extended adsorption state than polymers of high density. They also took into consideration that the added amount of the polymer with low charge density was higher than that of a polymer with a high charge density. Maximum floc size was obtained at a mixing ratio where the flocs became slightly negatively charged. The adsorption of a two – component (cationic polyacrylamide – CPAM - and silica nanoparticle) system onto a model (silica) surface has been compared with the adsorbtion and flocculation of a fiber suspension [183]. An increase in ionic strength led to an increase in the adsorbtion of both polyelectrolyte and nanoparticles onto a silica surface. A maximum in flocculation of the fiber suspension was found at 0.4 mg g^{-1} of polyelectrolyte. Above this limit, the flocculation ability of the system decreased and dispersion finally occurred. The addition of silica nanoparticles to this system enhanced the degree of flocculation.

In order to improve the flocculation of wastewaters from gravel pits which contain finely dispersed inorganic particles as well as natural organic matter (NOM), like humic acid, Petzold *et al.* investigated the influence of humic acid on the separation process of clay suspension, used as model, by some cationic polymers (CPAM of very high M_w or PDADMAC of lower M_w-35000 g.mol); the dual systems composed of the above mentioned polycations and very high molar mass (several million) [poly-(acylamide-co-sodium acrylate)] with about 30% of negative charges as anionic component was also investigated [180]. The time dependence of the sedimentation behavior as well as the turbidity of the supernatant and the light absorption was monitored as a measure for humic acid removal. Polymers with high charge density, which form only small flocs, have been more effective in removing humic acids owing to complex formation and precipitation. It was shown that the complexes formed by the highly charged polycation PDADMAC and the weak polyanion humic acid were also able to collect the fine clay particles. So, excellent clarification and also effective removal of humic acid have been obtained, but only at the optimum flocculation concentration (small flocculation window). The best results were obtained with the dual system, a broad flocculation window as well as larger flocs, compared with monoflocculation, being noticed.

Cellulose and kaolinite clay particles whose surfaces have been modified by nonstoichiometric polyelectrolyte complexes, obtained from PDADMAC and sodium salt of poly(maleic acid-co-α-methylstyrene) (PMAMS), were investigated as sorbents for removal of anionic, nonionic and cationic surfactants from aqueous solutions [189]. A significant increase was noticed only in the case of oppositely charged anionic surfactant removal.

The flocculation behavior of some preformed non-stoichiometric polyelectrolyte complexes (PECs) obtained during the interaction between PDADMAC and native potato starch on aqueous solutions of dye (C.I. Acid blue 78) was investigated by Klimavičiūtė [190]. Such PECs contain free quaternary ammonium groups and are called cationized starches (CS). The effectiveness of dye precipitation and the amount of CS needed for this purpose depended on the composition of CS. When the amount of PDADMAC increased in CS, the level of residual dye decreased; the best precipitation of dye was achieved with CS containing 0.015-0.02 mol PDADMAC to anhydroglucose residue (mol/AGR) when only 3-

5% residual dye remained in solution, and the dye precipitated in a wide range of concentrations. The flocculation results with the preformed CS were compared with those obtained when CS formed in dye solution by a separate dosing of PDADMAC and starch. In all cases preformed CS flocculated the dye more effectively than that formed in the dye solution.

Flocculation efficiency of some nonstoichiometric interpolyelectrolyte complex dispersions (NIPECs), synthesized by the interaction between poly(sodium 2-acrylamido-2-methylpropanesulfonate) (NaPAMPS) and three strong polycations bearing quaternary ammonium salt centers in the backbone (PDADMAC, PCA_5, PCA_5D_1) on a stable monodisperse silica suspension have been tested comparative with the starting polycations [191]. PDADMAC and PCA_5 alone showed a very narrow range of flocculation and the residual turbidity corresponding to the optimum flocculation concentration remained about 0.4 g/mg for PDADMAC. As Figure 3 shows, the window of the optimum flocculation concentration was broader in the case of PCA_5D_1 compared with PCA_5 because of the higher hydrophobicity induced by the presence of 1mol % of decyloxypropylamine side chains. NIPECs dispersions revealed much larger window of flocculation.

The amount of PELs corresponding to the optimum flocculation concentration was higher because at a molar ratio of 0.7 the excess of positive charges is not so high. Likewise the residual turbidity was much lower than in the case of the single polycation since the hydrophobic parts from the complex itself had also a contribution. Such NIPECs can be used as specialized flocculants for wastes containing nanoparticles with low charge such as dispersed dyes from the textile factories or slurries from microelectronics which are not well separated by the comercial polycations up to date.

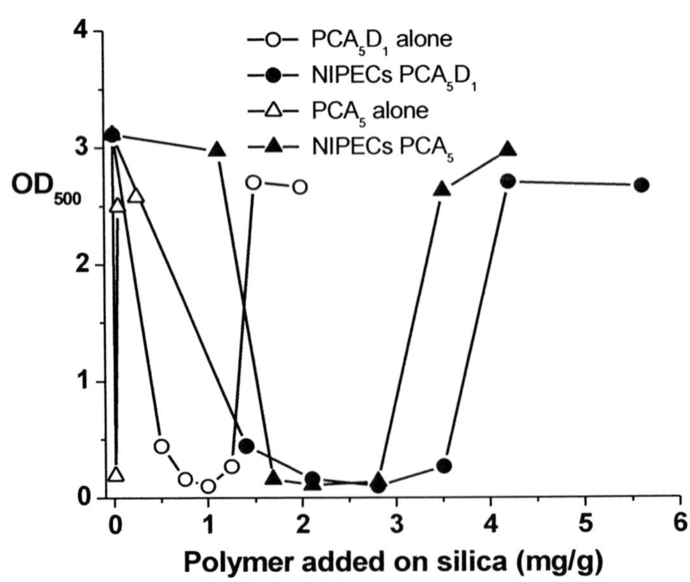

Figure 3. Flocculation test of silica nanoparticles with PCA_5 and PCA_5D_1 comparative with NIPECs prepared with NaPAMPS added on the polycations.

7. SLUDGE DEWATERING

The water purification processes produce a large amount of dilute sludge that has to be dewatered in order to reduce its volume which causes disposal problems and to facilitate convenient handling. The dewatering of the waste water sludge is a major concern of the water industry and therefore has received considerable research attention [93,125,192-196]. This operation not only reduces the sludge volume but also increases the caloric value of the product [196]. Dewatering is mainly performed by mechanical techniques based on centrifugation or filtration. The mechanical equipment used to dewater sludge includes sludge thickening tanks, centrifuges, vacuum filters and filter presses. However, a certain amount of water cannot be removed by mechanical means.

Vesilind suggested that there are at least four identifiable "types" of water in sludge [193]:

1. the free or bulk water: water not associated with and not influenced by suspended solids;
2. the interstitial water: water trapped in the crevices and interstitial spaces of flocs and organisms. Some interstitial water is held within the floc structure and can become free water if the floc is destroyed;
3. the vicinal water: multiple layers of water molecules held tightly to the particle surface by hydrogen bonding which creates the appearance of highly structured water molecules. Vicinal water is not free when the physical confinement is eliminated;
4. the water of hydration: water chemically bound to the particles and removable only by the expenditure of thermal energy.

Mechanical removal is possible only for the free and interstitial water while for the vicinal is difficult unless the sludge is conditioned. For the dewatering operation is very useful to know how much water will be available for removal. If it is possible to accurately measure water distribution, it might be possible to estimate the highest possible cake solids concentration that can be achieved for a given sludge. One approach might be to develop a relationship between the solids concentration and the sum of the bulk, interstitial and vicinal waters [193].

The separation of the solids from liquids in sludge by mechanical means is difficult and requires an improvement of the structure and strength of the sludge cake by addition of inorganic or organic flocculants. Generally, differences in sludge dewatering performance are attributed to a combination of sludge characteristics (extracellular polymeric substances - EPS, fine fraction, sludge viscosity, hydrophobicity, etc.) [197-199], as well as to the polyelectrolyte properties (molar mass, charge density, chemical structure, polymer dose) [93,196,200]. In addition, mixing conditions may also considerably determine the polyelectrolyte conditioning efficiency and dewatering performance [201].

Avey *et al.* have monitored the dewatering efficiency of some cationic polyelectrolytes obtained by homopolymerization or copolymerization of allyl-acrylate quaternary ammonium salts with AM and DADMAC. The results were compared to those obtained with a commercial polyelectrolyte Zetag 7557 (acryloxyethyltrimethylammoniumchlorideacrylamide copolymer) [93]. Although charge densities and molecular weights of the polymers were different, sludge dewatering curves for all tested polymers had similar shapes. The efficiency increased with the increase of the polyelectrolyte concentration. A plateau was observed after the polymer concentration of 15 mg/L; at this concentration the settling time ranged between 10 and 13 sec.

The influence of the cationic polyacrylamide with an average molar mass of 10^7 g.mol^{-1} and a charge density of 20% (T-3052) on the settling behavior of a clay slurry was studied by Chu *et al.* employing a computerized axial tomography scanner (CATSCAN) [195]. Because the flocculated sludge settles rapidly, most of the settling process was controlled by sediment compaction. The original slurry exhibited a slow settling velocity (the settling was not completed at 6000 s) and a turbid supernatant during settling while the flocculated slurries presented a much better settleability. The flocculated slurries usually settled faster with the increasing polyelectrolyte dose: the settling was completed within 4000 s at dose of 0.5 g/kg dried solid (DS), and 2000 s at 2-4 g/kg DS. The authors suggested charge neutralization played an essential role in the present flocculated system. Flocculation produced not only a fast-settling sediment but a clear supernatant.

Krentz *et al.* performed dewatering experiments on model suspensions (kaolin) and industrial suspensions (washings from gravel treatment plants and harbour sediment) using polyacrylamido-*co*-(trimethylammoniumpropyl chloride) (PTCA) samples with different degree of cationicity, τ [125]. For harbour sediment, a dewatering index, I_D, of 80 was achieved with the highly charged copolymer, PTCA3 ($\tau = 40\%$) at 30% lower flocculant dose than with weakly cationic copolymer PTCA1 ($\tau = 3$ %); for gravel washing, PTCA1 proved to be more effective: for comparable degrees of dewatering ($I_D = 80$) approximately 40% less flocculant was required than for PTCA3. In shear experiments on gravel washings and model suspensions (1:1 mixture of kaolin and titanium dioxide) with particles of differing size (d_{50} = 0.5 and 5.7 μm), low charge density flocculant exhibited an increased floc stability at lower concentration than is necessary to achieve maximum I_D values. For harbour sediment and model suspensions with unimodel particle size distributions, this stability did not occur until the doses used were higher than the concentrations needed to achieve maximum I_D values.

The influence of polyelectrolyte molar mass and charge density (copolymers of PAM and quaternized dimehylaminoethyl acrylate) on the dewatering performance of sludge samples with different solids concentrations and surface charge characteristics have been recently investigated [196]. Higher molar mass polyelectrolytes showed to yield clear flocculation already at low doses, due to the bridging effect. However, under these conditions, elastic flocs were created due to high internal electrostatic repulsion, as shown by the filter cake rebound after pressure release. Lower molar mass polyelectrolytes required higher doses to become effective flocculants. In sufficient dose, they caused considerable water release from the sludge, enhancing the filtration phase. In this case, the formed flocs tended to form less elastic filter cakes, which were explained by the higher degree of charge neutralization. Furthermore, the charge neutralization mechanism is supposed to lead to a lower porosity and a better structure of the flocs formed.

Some studies have been made using small-angle light scattering to get more information about the structure of kaolin sludge flocs before and after addition of the cationic polymer, T-3051 (PAM with a molar mass of 10^7 g.mol^{-1} and a charge density of 20% [202]. A two level model was proposed to interpret the structural changes of flocs after polyelectrolyte - induced flocculation. This model comprised of:

1) a primary floc (of fractal dimension D_S) consisting of primary particles and
2) a secondary floc (of fractal dimension D_F) consisting of the microflocs. After flocculation, the compactness of both the primary and the secondary flocs changed markedly.

The compactness of the interior structure of the primary flocs exhibited no correlation with the occurence of charge neutralization or with sludge dewaterability. More information about the fractal structure (i.e. decreasing effective density with increasing aggregate size) of the aggregates formed in water and wastewater treatment were summarized by Waite [203].

8. Conclusion

The application of polyelectrolytes in phase separation processes has received a great attention reflected in the scientists' efforts to develop novel materials with adjustable properties for a great variety of contaminants from waters and to demonstrate their flocculation efficiency. The fundamental and experimental studies greatly contribute to the deeper understanding of the flocculation process. However, *flocculation optimisation practices in industries are still based on trials*, therefore, further research is still necessary for a better clarification of the polyelectrolytes structure-properties-flocculation efficiency interdependencies.

Acknowledgements

This study was supported by the Romanian National Research of Excellence Program-CEEX, Grant 62/2005 - Advanced Treatment Technologies for Recycling Industrial Wastewater/RIWATECH.

References

[1] Dautzenberg, H.; Jaeger, W.; Kötz, J.; Philipp, B.; Seidel, C.; Stscherbina, D. *Polyelectrolytes: Formation, Characterization and Application*, Carl Hanser: München, 1994, pp 272- 327.

[2] Boris, D. C.; Colby, R. H. *Macromolecules* 1998, *31*, 5746-5755.

[3] Romero-Cano, M. S.; Martin-Rodriguez, A.; Chauveteau, G.; de las Nieves, F. J. *J. Colloid Interface Sci.* 1998, *198,* 273-281.

[4] Petterson, A.; Marino, G.; Pursiheimo, A.; Rosenholm, J. B. *J. Colloid Interface Sci.* 2000, *228,* 73- 81.

[5] Szucs, A.; Haraszti, T.; Dekany, I.; Fendler, J. H. *J. Phys. Chem. B* 2001, *105*, 10579-10587.

[6] Webber, G. B.; Wanless, E. J.; Butun, V.; Armes, S. P.; Biggs, S. *Nano Letters* 2002, *2*, 1307-1313.

[7] Kiriy, A.; Minko, S.; Gorodyska, G.; Stamm, M. *Nano Letters* 2002, *2*, 881-885.

[8] Minko, S.; Kiriy, A.; Gorodyska, G.; Stamm, M. *J. Am. Chem. Soc.* 2002, *124*, 3218-3219.

[9] Jaber, J. A.; Chase, P. B.; Schlenoff, J. B. *Nano Letters* 2003, *3*, 1505-1509.

[10] Falconnet, D.; Pasqui, D.; Park, S.; Eckert, R.; Schift, H.; Gobrecht, J.; Barbucci, R.; Textor, M. *Nano Letters* 2004, *4*, 1909-1914.

[11] Chanana, M.; Gliozzi, A.; Diaspo, A.; Chodnevskaja, I.; Huewel, S.; Moskalenko, V.; Ulrichs, K.; Galla, H.-J.; Krol, S. *Nano Letters* 2005, *5*, 2605-2612.

[12] Rivas, L. B.; Pereira, D. E.; Moreno-Villoslada, I. *Prog. Polym. Sci.* 2003, *28*, 173-208.

[13] Pekel, N.; Güven, O. *Polym. Bull.* 2004, *51*, 307-314.

[14] Kim, K. J.; Shahinpoor, M. In: *Handbook of Polyelectrolytes and Their Applications*, Tripathy, S. K.; Kumar, J.; Nalwa, H. S.; Eds.; American Scientific Publisher, Los Angeles, 2002, Vol. 3, pp 1-23.

[15] Thünemann, A. F.; Müller, M.; Dautzenberg, H.; Joanny, J-F., Löwen, H. *Adv. Polym. Sci.* 2004, *166*, 113-171.

[16] v. Klitzing, R.; Tieke, B. *Adv. Polym. Sci.* 2004, *165,* 177-210.

[17] Hammond, T. P. *Adv. Polym. Sci.* 2004, *16*, 1271-1293.

[18] McCormick, C. L.; Bock, J.; Schulz, D. N. In: *Encyclopedia of Polymer Science and Engineering*, 2nd ed.; Kroschwitz, J. I.; Ed.; John Wiley and Sons: New York, NY, 1989, Vol. 17, pp 730-784.

[19] Winnik, M. A.; Yekta, A. *Curr. Opin. Colloid Interface Sci.* 1997, *2*, 424-436.

[20] *Principles of Polymer Science and Technology in Cosmetics and Personal Care*, Goddard, E. D.; Gruber, J. V.; Eds.; Marcel Dekker: N. Y. 1999.

[21] Dickinson, E. *Colloids Surf. B* 1999, *15*, 161-176.

[22] *Associative Polymers in Aqueous Solution*; Glass, J. E., Ed.; ACS Symposium Series 765; American Chemical Society: Washington, DC, 2000.

[23] Bromberg, L. In: *Handbook of Polyelectrolytes and Their Applications*; Tripathy, S. K.; Kumar, J.; Nalwa, H. S.; Eds.; American Scientific Publisher, Los Angeles, 2002, Vol. 3, pp 23-46.

[24] Azzam, T.; Eliyahu, H.; Shapira, L.; Linial, M.; Barenholz, Y.; Domb, A. J. *J. Med. Chem.* 2002, *45*, 1817-1824.

[25] Le Bon, B.; Van Craynest, N.; Boussif, O.; Vierling, P. *Bioconjugate Chem.* 2002, *13*, 1292-1301.

[26] Zelikin, A. N.; Putnam, D.; Shastri, P.; Langer, R.; Izumrudov, V. A. *Bioconjugate Chem.* 2002, *13*, 548-553.

[27] Brown, M. D.; Schätzlein, A.; Brownlie, A.; Jack, V.; Wang, W.; Tetley, L.; Gray, A. I.; Uchegbu, I. F. *Bioconjugate Chem.* 2002, *11*, 880-891.

[28] Van Steenis, J. H.; Van Maarseveen, E. M.; Van Steenis, J. H.; Verrijk, R.; Crommelin, D. J. A.; Storm, G.; Hennink, W. E. *J. Controlled Release* 2003, *87*, 167-176.

[29] Crommelin, D. J. A.; Storn, G.; Jiskoot, W.; Stenekes, R.; Mastrobattista, E.; Hennink, W. E. *J. Controlled Release* 2003, *87*, 81-88.

[30] Verbaan, F. J.; Oussoren, C.; Snel, C. J.; Crommelin, D. J. A.; Hennink, W. E; Storm, G. *J. Gene Med.* 2004, *6*, 64-75.

[31] Funhoff, A. M.; Van Nostrum, C. F.; Lok, M. C.; Fretz, M. M.; Crommelin, D. J. A.; Hennink, W. E. *Bioconjugate Chem*. 2004, *15*, 1212-1220.
[32] Bos, G. W.; Kanellos, T.; Crommelin, D. J. A.; Hennink, W. E; Howard, C. R. *Vaccine* 2004, *23*, 460-469.
[33] Tan, J. F.; Ravi, P.; Too, H. P.; Hatton, A.; Tam, K. C. *Biomacromolecules* 2005, *6*, 498-506.
[34] Alemdar, A.; Butun, V. *J. Appl. Polym. Sci*. 2005, *95,* 300-306.
[35] Tan, S.; Li, G.; Shen, J.; Liu, Y.; Zong, M. *J. Appl. Polym. Sci*. 2000, *77*, 1869-1876.
[36] Li, G.; Shen, J.; Zhu, Y. *J. Appl. Polym. Sci*. 2000, *78*, 668-675.
[37] Korycka-Machala, M.; Ziolkowski, A.; Rumijowska-Galewicz, A.; Lisowska, K.; Sedlaczek, L. *Microbiology* 2001, *147*, 2769-2781.
[38] Kenawy, E.-R.; Abdel-Hay, F.I.; El-Shanshoury, A. E.-R. E.-S.; El-Newehy, M. H. J. *Polym. Sci. Part A Polym. Chem*. 2002, *40*, 2384-2393.
[39] Lin, J.; Qiu, S.; Lewis, K.; Klibanov, A. M. *Biotechnol. Prog*. 2002, *18*, 1082-1086.
[40] Cakmak, I.; Ulukanli, Z.; Tuzcu, M.; Karabuga, S.; Genctav, K. *Eur. Polym. J.* 2004, *40*, 2373-2379.
[41] Grunlan, J. C.; Choi, J. K.; Lin, A. *Biomacromolecules* 2005, *6*, 1149-1153.
[42] Kitchener, J. A. *Br. Polym. J.* 1972, *4*, 217-229.
[43] Gregory, J. In: *The Scientific Basis of Flocculation*; Ives, K. J.; Ed.; Sijtjoff and Noordhoff International Publishers, Alphen aan den Rijh, The Netherlands, 1978, pp 89-100.
[44] Beardsley, J. A. *J. Amer. Water Works Ass*. 1973, *65*, 85-92.
[45] Huang, Y-S.; Lipp, D. W. In: *Polymeric Materials Encyclopedia*; Salamone, J. C.; Ed.; CRC Press, USA, 1996, Vol 4, pp 2427-2440.
[46] Claesson, P. M.; Dahlgren, M. A. G.; Eriksson, L. *Colloids Surf. A* 1994, *93*, 293-303.
[47] Singh, R. P.; Tripathy, T.; Karmakar, G. P.; Rath, S. K.; Karmakar, N. C.; Pandey, S. R.; Kannan, K.; Jain, S. K.; Lan, N. T. *Current Science* 2000, *78,* 798-803.
[48] Rose, J. R.; St. John, M. R. In: *Encyclopedia of Polymer Science and Engineering*, 2nd ed.; Kroschwitz, J. I.; Ed.; John Wiley and Sons: New York, NY, 1987, vol.7, pp 211-233.
[49] Reuter, J. M. *Chem. Ing. Tech*. 1976, *48*, 389-394.
[50] Adachi, Y.; Wada, T. *J. Colloid. Interface Sci*. 2000, *229*, 148-156.
[51] Li, D.; Zhu, S.; Pelton, R. H.; Spafford, M. *Colloid Polym. Sci*.1999, *277*, 108-114.
[52] Kamibayashi, M.; Ogura, H.; Otsubo, Y. *J. Colloid Interface Sci*. 2005, *290*, 592-597.
[53] Das, K. K.; Somasundaran, P. *Colloids Surf. A* 2003, *223*, 17-25.
[54] de Vasconcelos, C. L.; Dantas, T. N. C.; Pereitra, M. R.; Fonseca, J. L. C. *Colloid Polym. Sci*. 2004, *282*, 596-601.
[55] Bajdur, W. M.; Sulkowski, W. W. *J. Appl. Polym. Sci*. 2003, *89*, 3000-3005.
[56] Scott, J. P.; Fawell, P. D.; Ralph, D. E.; Farrow, J. B. *J. Appl. Polym. Sci*. 1996, *62,* 2097-2106.
[57] Adachi, Y.; Matsumoto T.; Cohen Stuart M. A. *Colloids Surf. A* 2002, *207*, 253-261.
[58] Arita, T.; Kanda, Y.; Hamabe, H.; Ueno, T.; Watanabe, Y.; Higashitan, K. *Langmuir* 2003, *19*, 6723-6729.
[59] Ghimici, L.; Dranca, I. ; Dragan, S. ; Lupascu, T. ; Maftuleac, A. *Eur. Polym. J.* 2001, *37*, 227-231.

[60] Doherty, W. O. S.; Fellows, C. M.; Gorjian, S.; Senogles, E.,; Cheung, W. H. *J. Appl. Polym. Sci*. 2003, *90*, 316-325.
[61] Nystrom, R.; Backfolk, K.; Rosenholm, J. B.; Nurmi, K. *Colloids Surf. A* 2003, *219*, 55-66.
[62] Soponvuttikul, C.; Scamehorn, J. F.; Saiwan, C. *Langmuir* 2003, *19*, 4402-4410.
[63] Zemaitaitiene, R. J.; Zliobaite, E.; Klimaviciute, R.; Zemaitaitis, A. *Colloids Surf. A* 2003, *214*, 37-47.
[64] Fuente, E.; Blanco, A.; Negro, C.; Pelach, A. M.; Mutj, P.; Tijero, J. *Ind. Eng. Chem. Res*. 2005, *44*, 5616-5621.
[65] Mahvi, A. H.; Razavi, M. *J. Appl. Polym. Sci.* 2005, *2*, 397-399.
[66] Smith-Palmer, T. In: *Polymeric Materials Encyclopedia*, Salamone J. C.; Ed.; CRC Press, Vol. 4, 1996, pp 2440-2440.
[67] Gomitz, E.; Hahn, M.; Jaeger, W.; Dautzenberg, H. *Prog. Colloid Polym. Sci*. 1997, *107*, 27- 135.
[68] Ma, M; Zhu, S. *Colloid Polym. Sci*. 1999, *277*, 115-122.
[69] Whipple, W. L.; Maltesh C. *Langmuir* 2000, *16*, 3124-3132.
[70] Sakkayawong, N.; Thiravetyan, P.; Nakbanpote, W. *J. Colloid Interface Sci.* 2005, *286*, 36-42.
[71] Alemdar, A.; Bütün, V. *J. Appl. Polym. Sci.* 2005, *95*, 300-306.
[72] Clarke, M. C. *Br. Polym. J.* 1990, *22*, 47-.51.
[73] Letterman, R. D.; Pero, R. W. *J. Amer. Water Works Ass*. 1990, *82*, 87-97.
[74] Akers, R. J. *Filtration and Separation* 1972, *9*, 423-425.
[75] La Mer, V. K. *J. Colloid Sci*. 1964, *19*, 291-293.
[76] Ives, K. J. In: *The Scientific Basis of Flocculation*; Ives, K. J.; Ed; Sijtjoff and Noordhoff International Publishers, Alphen aan den Rijh, The Netherlands, 1978, pp 1-2.
[77] Ruehrwein, R. A.; Ward, D. W. *Soil Sci*. 1952, *73*, 485-492.
[78] Michaels, A. S. *Ind. Eng. Chem*. 1954, *46,* 1485-1490.
[79] La Mer, V. K. *Discuss. Faraday* 1966, *42*, 248- 254.
[80] O'Melia, C. R. In: *Physico-chemical Processes for Water Quality Control*; Weber, W. J., Jr.; Ed.; Wiley Interscience, New York, NY, 1972, pp 61-109.
[81] Gregory, J. *J. Colloid Interface Sci*. 1973, *42*, 448-456.
[82] Akesson, T.; Woodward, C.; Jonsson, B. *J. Chem. Phys*. 1989, *91*, 2461-2469.
[83] Bolto, B. A. *Prog. Polym. Sci*. 1995, *20*, 987-1041.
[84] Caruso, F.; Niikura, K.; Furlong, D. N.; Okahata, Y. *Langmuir* 1997, *13*, 3422-3426.
[85] Laurent, D.; Schlenoff, J. B. *Langmuir* 1997, *13*, 1552-1557.
[86] Lösche, M.; Schmitt, J.; Decher, G.; Bouwman, W. G.; Kjaer, K. *Macromolecules* 1998, *31*, 8893-8906.
[87] Caruso, F.; Lichtenfeld, H.; Donath, E.; Mohwald, H. *Macromolecules* 1999, *32*, 2317-2328.
[88] Watanabe, Y.; Kubo, K.; Sato, S. *Langmuir* 1999, *15*, 4157-4164.
[89] Biggs, S.; Habgood, M.; Jameson, G. J.; Yan, Y. *Chem. Eng. J.* 2000, *80*, 13- 22.
[90] Nyström, R. J.; Rosenholm, J. B.; Nurmi, K. *Langmuir* 2003, *19*, 3981-3986.
[91] Yan, Y. D.; Glover, S. M.; Jameson, G. J.; Biggs, S. *Int. J. Miner. Process.* 2004, *73*, 161-175
[92] Lindquist, G. M.; Stratton, T. A. J. *Colloid Interface Sci*. 1976, *55*, 45-59.

[93] Avei, D.; Mol, N.; Dagasan, L. *Polym. Bull.* 2002, *48*, 353-359.
[94] Rose, D. R.; St. John, M. R. In: *Encyclopedia Polymer Science and Engineering*, 2nd Edition, Mark, H. F.; Ed.; John Wiley and Sons; New York, NY, 1990, Vol 7, pp 211-233.
[95] Van de Ven, T. G. M. *Adv. Colloid Interface Sci.* 1994, *48*, 121-140.
[96] Adachi, Y. *Adv. Colloid Interfaces* 1995, *56*, 1-31.
[97] Gregory, J. In: *Industrial Water Soluble Polymers*; Finch, C. A.; Ed.; Pentafir Associates; Aybesburg, UK, 1996, pp 62-75.
[98] Yu, X.; Somasundaran, P. *J. Colloid Interface Sci.* 1996, *178*, 770– 774.
[99] Van der Linden, C.; Van Leemput, R. *J. Colloid Interface Sci.* 1978, *67*, 63-69.
[100] Gill, R. I. S.; Herrington, T. M. *Colloids Surfaces* 1986, *22*, 51-58.
[101] Pelssers E. G. M.; Cohen Stuart, M. A.; Fleer, G. *J. Chem. Soc. Faraday Trans*. 1990, *86*, 1355 – 1361.
[102] Pefferkorn, E.; Elaissari, H. *J. Colloid Interface Sci.* 1990, *138*, 187- 194.
[103] Dahlgren, M. A. G.; Waltermo, A.; Blomberg, E.; Claesson, P. M.; Sjöström, L.; Akesson, T.; Jönsson, B. *J. Phys. Chem*. 1993, *97,* 11769-11775.
[104] Ives, K. J. In: *The Scientific Basis of Flocculation*; Ives, K. J.; Ed.; Sijthoff and Noordhoff International Publishers, Alphen aan den Rijn, The Netherlands, 1978, pp 37-61.
[105] Ray, D. T.; Hogg R. *J. Colloid Interface Sci.* 1987, *116***,** 256-268.
[106] Halab, L.; Ricard, A. *J. Colloid Interface Sci*. 1996, *179***,** 269-275.
[107] Barany, S.; Cohen Stuart, M. A. *Colloid Surf. A* 1996, *106*, 213-221.
[108] Hu, Q. H.; Qiao, S. Z.; Haghseresht, M. A.; Wilson, M. A.; Lu, G. Q. *Ind. Eng. Chem. Res*. 2006, *45*, 733-738.
[109] Simon, S.; Picton, L.; Le Cerf, D.; Muller, G. *Polymer* 2005, *46*, 3700-3707.
[110] Chaplain, V.; Janex, M. L.; Lafuma, F.; Graillat, C.; Audebert, R. *Colloid Polym. Sci*. 1995, *273*, 984-990.
[111] Pelssers E. G. M.; Cohen Stuart, M. A.; Fleer, G. *J. Colloid Surf.* 1989, *38*, 15-25.
[112] Hogg, R. *Colloid Surf. A:* 1999, *146*, 253-263.
[113] Jarvis, P.; Jefferson, B.; Parsons, S. A. *Environ. Sci. Technol*. 2005, *39*, 2307-2314.
[114] McCurdy, K.; Carlson, K.; Gregory, D. *Water Res*. 2004, *38*, 486-494.
[115] Yukselen M. A,; Gregory, J. *Water Supply* 2002, *2*, 95–101.
[116] Yukselen, M. A.; Gregory, J. *Int. J. Miner. Process* 2004, *73*, 251-259.
[117] Tang, S. *Colloids Surf.* 1999, *157*, 185-192.
[118] Waite, T. D. Cleaver, J. K.; Beattie, J. K. *J. Colloid Interface Sci*. 2001, *241*, 333-339.
[119] Wu, R. M.; Lee, D. J.; Waite, T. D.; Guan, J. *J. Colloid Interface Sci*. 2002, *252*, 383-392.
[120] Ahmad, A. L.; Sumathi, S.; Hameed, B. H. *Adsorbtion Sci. Tech*. 2004, *22*, 75-88.
[121] Iler, R. K. *J. Colloid Interface Sci*. 1971, *37*, 364- 373.
[122] Yorke, M. A. In: *Water-Soluble Polymers*; Bikales, N. M.; Ed.; Plenum Press, New York, 1973, pp 93-104.
[123] Cane, J. C.; La Mer; V. J.; Linford, H. B. *J. Am. Chem. Soc*. 1964, *86*, 3450-3453.
[124] Cress, O. L. *Polymer Topics* 1972, 333-339
[125] Krentz, D.-O.; Oelmeyer, G.; Laudan, J.; Kulicke, W.-M. *Colloid Polym. Sci*. 2003, *281*, 423-430.
[126] Khalil, M. I.; Aly, A. A. *Starch/Starke* 2001, *53*, 84-89.

[127] Khalil, M. I.; Ali, A. A. *Starch/Starke* 2001, 53, 323-329.
[128] Walles, W. E. *J. Colloid Interf. Sci.* 1968, *27*, 797-803.
[129] Alince, B. *Colloids Surf.* 1998, *39*, 279-288.
[130] Suty, S.; Luzakova, V. *Colloids Surf. A* 1998, *139*, 271-278
[131] Porubska, J.; Alince, B.; van de Ven, T. G. M. *Colloids Surf. A* 2002, *210*, 97-104.
[132] Gill, R. I. S.; Herrington, T. M. *Colloids Surf. A.* 1987, *28*, 41-48.
[133] Hocking, M. B.; Klimchuk, K. A.; Lowen, S. *J. Macromol. Sci. Rev. Macromol. Chem. Phys.*1999, *C39,* 177- 203.
[134] Ashmore, M.; Hearn, J. *Langmuir* 2000, *16*, 4906-4911.
[135] Ashmore, M.; Hearn, J.; Karpowicz, F. *Langmuir* 2001, *17*, 1069-1073.
[136] Netz, R. R.; Joanny, J. F. *Macromolecules* 1999, *32*, 9026-9040.
[137] Petzold, G.; Lunkwitz, K.; Schwarz, S. *Chem. Ing. Tech.* 2002, *74,* 438- 442.
[138] Petzold, G.; Mende, M.; Lunkwitz, K.; Schwarz, S.; Buchhammer, H. M. *Colloids Surf. A* 2003, *218*, 47- 57.
[139] Ariffin, A.; Shatat, R. S. A.; Norulaini, A. R. N.; Mohd Omar, A. K. *J. Appl. Polym. Sci.* 2004, *4*, 393-397.
[140] Kam, S. K.; Gegory, J. *Water Res.* 2001, *35*, 3557-3566.
[141] Dragan, S.; Ghimici, L. *Bull. Inst. Pol. Iasi* 1998, *XLIV,* 109–114.
[142] Sableviciene, D.; Klimaviciute, R.; Bendoraitiene, J.; Zemaitaitis, A. *Colloids Surf. A* 2005, *259*, 23-30.
[143] Bratskaya, S.; Schwarz, S.; Liebert, T.; Heinze, T. *Colloids Surf. A* 2005, *254*, 75-80.
[144] Bratskaya, S.; Schwarz, S.; Laube, J.; Liebert, T.; Heinze, T.; Krentz, O.; Lohmann, C.; Kulicke, W-M. *Macromol. Mater. Eng.* 2005, *290*, 778-785.
[145] Balasz, A. C.; Siemasko, C. P. *J. Chem. Phys.* 1991, *95*, 3798-3803.
[146] Li, D.; Zhu, S.; Pelton, R. H.; Spafford, M. *Colloid Polym. Sci.* 1999, *277*, 108-114.
[147] Rath, S. K.; Singh, R. P. *J. Appl. Polym. Sci.* 1997, *66*, 1721-1729.
[148] Rath, S. K.; Singh, R. P. *Colloids Surf. A* 1998, *139,* 129-135.
[149] Nayak, B. R..; Singh, R. P. *J. Appl. Polym. Sci.* 2001, *81*, 1776-1785.
[150] Labille, J.; Thomas, F.; Milas, M.; Vanhaverbeke, C. *J. Colloid Interface Sci.* 2005, *284,*149-156.
[151] Caskey, J. A.; Primus, R. J. *Environ. Prog.* 1986, *5*, 98-103.
[152] Larsson, A.; Walldal, S.; Wall, S. *Colloids Surf. A* 1999, *159*, 65-76.
[153] Mabire, F.; Audebert,R.; Quivoron, C. *J. Colloid Interface Sci.* 1984, *97,* 120-136.
[154] Healy, T. W.; La Mer V. K. *J. Colloid Interface Sci.* 1964, *19*, 322-332.
[155] Suzuki, A.; Maruyama, H.; Seki, H.; Kashiki, I.; Inoue, N. *Ind. Eng. Chem. Res.* 2006, *45*, 1123-1127.
[156] Lupascu, T. ; Dranca, I. ; Sandu, M. ; Dragan, S.; Ghimici, M. *Angew. Makromol. Chem.*, 1994, *220*, 11-19.
[157] Dragan, S.; Ghimici, L.; Carpov, A.; Chirica, E.; Maftei, M. *Mat. Plast.* 1990, *27*, 72-78.
[158] Dragan, S.; Ghimici, L. *Angew. Makromol. Chem.* 1991, *192*, 199-211.
[159] Dragan, S.; Maftuleac, A.; Dranca, I.; Ghimici, L.; Lupascu, T. *J. Appl. Polym. Sci.* 2002, *84*, 871- 876.
[160] Solberg, D.; Wagberg, L. *Colloids Surf. A* 2003, *219*, 161-172.
[161] Blanco, A.; Negro, C.; Fuente, E.; Tijero, J. *Ind. Eng. Chem. Res.* 2005, *44*, 9105-9112.

[162] Van de Steeg, H. G. M.; De Keizer, A.; Cohen Stuart, M. A.; Bijsterbosch, B. H. Nord. *Pulp. Res. J.* 1993, *1*, 34-40.
[163] Schwarz, S.; Lunkwitz, K.; Kebler, B.; Spiegler, U.; Killmann, W.; Jaeger, W. *Colloids Surf. A* 2000, *163*, 17-27.
[164] Pinotti, A.; Bevilacqua,A.; Zaritzky, N. J. *Food Eng.* 1997, *32*, 69-81.
[165] Pinotti, A.; Bevilacqua,A.; Zaritzky, N. *J. S. D.* 2001, *4*, 57-63.
[166] Zayas, T.; Percino, M.J.; Cardoso, J.; Chapela, V. M. *Polymer* 2000, *41*, 5505-5512.
[167] Anleu, E.; Cardoso, J.; Manero, O. *J. Appl. Polym. Sci.* 2000, *86*, 2693-2700.
[168] Koprivanac, N.; Jovanovic-Kolar, J.; Bosanac G.; Meixner J. *Microchemical Journal* 1992, *46*, 379-384.
[169] Yu, X.; Somasundaran, P. *Colloids Surf. A* 1993, *81*, 17-23.
[170] Yu, X.; Somasundaran, P. *J. Colloid Interface Sci.* 1996, *177*, 283-287.
[171] Petzold, G.; Lunkwitz, K. *Colloids Surf. A* 1995, *98*, 225-233.
[172] Petzold, G.; Buchhammer, H.-M.; Lunkwitz, K. *Colloids Surf. A* 1996, *119*, 87-92.
[173] Fan, A.; Turro, N. J.; Somasundaran, P. *Colloids Surf. A* 2000, *162*, 141-148.
[174] Bouyer, F.; Robben, A. ; Yu, W. L. ; Borkovec, M. *Langmuir* 2001, *17*, 5225-5231.
[175] Nyström, R.;Rosenholm, J. B.; Backfolk, K.; Nurmi, K. *J. Colloid Interface Sci.* 2003, *262*, 48-54.
[176] Nyström, R.; Rosenholm, J. B. *Colloids Surf. A* 2005, *252*, 135-141.
[177] Lu, C.; Pelton, R. *Langmuir* 2005, *21*, 3765-3772.
[178] Bolto, B.; Abbt-Braun, G.; Dixo, D. *Water Sci. Technol.* 1999, *40*, 71-80.
[179] Edzwald, J. K.; Tobiason, J. E. *Water Sci. Technol.* 1999, *40*, 63-70.
[180] Petzold, G.; Geissler, U.; Smolka, N.; Schwarz, S. *Colloid Polym. Sci.* 2004, *282*, 670-676.
[181] Moore, E. E. *Tappi* 1976, 120-122.
[182] Britt, K. W.; Dillon, A. G.; Evans, L. A. *Tappi* 1977, *60*, 102-104.
[183] Bouyer, F.; Robben, A.; Yu, W. L.; Borkovec, M. *Langmuir* 2001, *17*, 5225-5231.
[184] Magdassi, S.; Rodel, B. Z. *Colloids Surf. A* 1996, *119*, 51-56.
[185] Seppänen, R.; Strom, G.; Elftonson, J. *Colloids Surface A* 2000, *164*, 131-141.
[186] Dragan, S.; Dragan, D.; Cristea, M.; Airinei, A.; Ghimici, L. *J. Polym. Sci. Part A Polym. Chem.* 1999, *37*, 409-418.
[187] Lindstrom, T.; Glad-Nordmark, G. *J. Colloid. Interface Sci.* 1984, *97*, 62-67.
[188] Pelton, R. H.; Xiao, H.; Brook, M.; Hamielec, A. *Langmuir* 1996, *12*, 5756-5762.
[189] Lukaszczyk, J.; Lekawska, E.; Lunkwitz, K.; Petzold, G. *J. Appl. Polym. Sci.* 2004, *92*, 1510-1515.
[190] Klimavičiūté, R. *Fibres and Textiles* 2004, *12*, 74-78.
[191] Schwarz, S.; Dragan, S. *Macromol. Symp.* 2004, *210*, 185-192.
[192] Eriksson, L.; Alm, B. *Water Sci. Technol.* 1993, *25*, 27-33.
[193] Vesilind, P. A. *Industrial Waste Water* 1996, p.43- 46.
[194] Olmeyer, G.; Krentz, O.; Kulicke, W. M. *Chem. Eng. Technol.* 2002, *25*, 47- 50.
[195] Chu, C. P.; Ju, S. P.; Lee, D. J.; Tiller, F.M.; Mohanty, K. K.; Chang, Y. C. *Ind. Eng. Chem. Res.* 2002, *41*, 1227-1233.
[196] Saveyn, H.; Meersseman, S.; Thas, O.; Van der Meeren, P. *Colloids Surf. A* 2005, *262*, 40-51.
[197] Mikkelsen, L. H.; Keiding, K. *Water Res.* 2002, *36*, 2451-2462.
[198] Hashimoto, M.; Hiraoka, M. *Water Sci. Technol.* 1990, *22*, 143-152.

[199] Liu, Y.; Fang, H. H. P. *Crit. Rev Environ. Sci. Technol.* 2003, *33*, 237-273.
[200] Parker, S.; Kulicke, W. M.; Bohm, N.; Kotz, J.; Jaeger, W. *Angew. Makromol. Chem.* 1997, *250*, 15-30.
[201] Langer, S. J.; Klute, R.; Hah, H. H. *Water Sci. Technol.* 1994, *30*, 129-138.
[202] Wu, R. M.; Lee, D. J.; Waite, T. D.; Guan, J. *J. Colloid Interface Sci.* 2002, *252*, 383-392.
[203] Waite, T. D. *Colloids Surf. A* 1999, *151*, 27-41.

In: New Trends in Ionic (Co) Polymers and Hybrids
Editor: Ecaterina Stela Dragan, pp. 65-89
ISBN 978-1-60021-611-4

Chapter 3

SYNTHESIS AND APPLICATIONS OF SOME ORGANIC CHELATING SORBENTS

Maria Valentina Dinu and Ecaterina Stela Dragan*
"Petru Poni" Institute of Macromolecular Chemistry,
Aleea Grigore Ghica Voda 41A, 700487 Iasi, Romania

ABSTRACT

Organic chelating sorbents have been intensively investigated because they proved to be effective materials for the removal/recovery of polluting metal ions from industrial effluents, water supplies and mine waters. This review surveys the main contributions concerning organic chelating sorbents containing amidoxime and iminodiacetate functional groups. Also, the classification of organic chelating sorbents according to the structure of the functional group with specifications on the type of interaction between chelating groups and polymer matrix, as well as the nature and physical status of polymer matrix has been described in the paper. Two less explored approaches for the synthesis of crosslinked chelating polymers as beads containing iminodiacetic groups have been shown. The preparation of anion exchangers with primary amine groups by the aminolysis-hydrolysis reaction with 1,2-diaminoethane of the nitrile groups, contained in some porous copolymers of acrylonitrile-divinylbenzene, and by the Hofmann degradation of the amide groups contained in acrylamide-divinylbenzene copolymers, followed by the carboxymethylation of the primary amine groups to prepare chelating ion exchangers, has been presented. The influence of porogen nature, monomers dilution and crosslinking degree on the properties of the ion exchangers was examined. Structural characterization, functionalization-morphology correlations, the ion exchange properties, and potential applications for the ion exchangers thus prepared have been discussed.

Keywords: organic chelating sorbents, amidoxime functional group, iminodiacetate functional group, acrylonitrile/divinylbenzene porous copolymers, acrylamide-divinylbenzene copolymer, porogen, toluene, 2-ethyl-1-hexanol, porosity, ion exchange capacity.

---

* E-mail:: sdragan@icmpp.ro. Tel.: +40.232.217454; Fax: +40.232.211299.

LIST OF ABBREVIATIONS

AA	Acrylamide
AN	Acrylonitrile
AO-H	Amidoxime chelate fabrics
AAO	Anion exchangers with amidoxime groups
CIE	Chelating ion exchanger
CMC	Carboxymethylcellulose sodium salt
DETA	Diethylene triamine
DTTA	Diethylene triamine tetra acetic acid
DVB	Divinylbenzene
DMF	Dimethylformamide
EDA	Ethylenediamine
EDTA	Ethylenediamine tetraacetic acid
2EH	2-ethyl-1-hexanol
FPAO	Fibrous poly(methylacrylamidoxime)
GMA	Glycidyl methacrylate
IPNs	Interpenetrating polymer networks
MA	Methyl acrylate
MAA	Methacrylic acid
MFPA	Macroporous fibrous polymeric adsorbent
MMA	Methyl methacrylate
NaMCA	Sodium monochloroacetate
PAN	Poly(acrylonitrile)
PEG	Polyethlene glycol
Poly(DPAAm)	Poly(N,N'-dipropionitrile acrylamide)
PVP	Poly(N-vinyl-2-pyrolidone)
Ponceau 4R	3-hydroxy-4-(sulfonaphtylazo)-5,7-naphtalene disulfonic acid trisodium salt
St	Styrene
p-TSA	p-Toluene sulfonic acid
4-VPy	4-Vinyl pyridine

1. INTRODUCTION

Environmental contamination with heavy metals as pollutants gained lately more significance, because they are nonbiodegradable, i.e., have a high persistence, and can cause nervous system damage and even cancer, by their accumulation at certain levels. Heavy metals are released in waste waters from hydrometallurgy (metal finishing industry), battery manufacturing, alloy industries, mining activities, printing, and the utilization of fossil fuels, to mention only a few sources of such pollutants. A difficult problem encountered in the removal of the metal ions is that the target species are usually in low concentration and in complex mixtures. Numerous variables must be considered in practice, such as: salinity, pH, temperature and the presence of organic substances [1-6]. One central task is removal and recovery of non-ferrous metals, a selective process being required for this purpose.

Among various methods of heavy metal removal, such as: liquid-liquid extraction, precipitation, electrolytic concentration, membrane filtration, neutralization, ion exchange and adsorption, which have been proposed, the last is generally preferred because of its high efficiency, ease of handling, and availability of different sorbents [7]. Therefore, the most promising technique for removal of metal cations is their adsorption on organic ion exchangers containing chelating functional groups. Furthermore, the extraction of metal ions using organic ion exchangers containing chelating functional groups is a green analytical method since it does not involve the use of toxic chlorinated organic solvents, which are very frequently used in conventional liquid-liquid extraction tehnique or other methods [8].

Normally, the chelating groups attached to polymer networks contain two or more electron donor atoms, such as N, S, O, P, and As. Some of the criteria in the selection of chelating polymers are: specific and fast complexation of the metal ions as well as the reutilization of the chelating sorbents. The sorption mechanism would be through chelation instead of simple ion exchange, like in the common ion exchangers, i.e., the electrostatic exchange is accomplished by a rapid chemical reaction, leading to the establishement of strong covalent bonds. As a consequence their interaction with transition metals is so strong that elution is difficult and must be carried out with strong complexing agents [3]. The kinetics of the chemical reactions on the chelating resins are different from conventional ion exchangers because of the high degree of irreversibility of the reaction occurring between the chelating ligand and the metal ions, the process being controlled either by diffusion or a chemical reaction. The matrix properties can also influence the selectivity of the chelating resins.

An organic chelating sorbent consists in two main parts: chelating functional group(s) and the polymer matrix or support. The structure of polymer matrix and of the chelating groups, on the one side, and the interaction mode between the support and functional group, on the other side, determine the specific applications of the chelating sorbents. Organic chelating sorbents, by their high selectivity in binding metal ions and metal complex ions have a major implication in concentration control (preconcentration of the trace elements from solutions so dilute that no analytical determinations are possible), and in inorganic analytical chemistry. The selectivity of a certain chelating sorbent is attributed to the differences in the stability constants of the complexes formed with the chelating groups of the resin [1-3]. The selectivity can be modified by changing the environment characteristics (pH, temperature) or by the addition of different complexing agents. Numerous studies and reviews concerning the synthesis and characterization of the selective chelating sorbents, and the wide applicability of these resins in the removal of toxic metal and complex ions from industrial effluents, as well as in selective metal ion recovery processes, have been recently published [7-22]. This demonstrates the constant interest in finding new chelating materials with increased performances.

The classification of these materials may be done according to some criteria, as follows:

- the structure of the chelating functional groups;
- nature of the support;
- type of interaction between the support and the chelating groups;

– physical status of the polymer matrix.

According to the nature of the support, the organic chelating sorbents can be: synthetic and natural. Synthetic organic chelating sorbents are the most suitable materials because of their well defined physical forms (granules, beads and fibers) and their various chemical and morphological structures [9].

According to the type of interaction between the support and the chelating groups or chelating compounds the organic chelating sorbents can be classified in:

- chelating functional groups are covalently bound to the polymer matrix;
- chelating functional groups are included in the support by ionic interactions;
- chelating compounds are embedded (impregnated) in the polymer matrix by physical interactions.

The physical status of the crosslinked polymers as polymer matrix, usually can be as beads, fibres and foamed plastics.

The main types of chelating functional groups covalently bound to the polymer matrix were summarized in Table 1.

Table 1

Functional group	Structure	Applications	References
Amidoxime	$-C(NH_2)=NHOH$	Selective uranium extraction	[10,11,16, 23-38]
Hydroxamic Acid	-CO-NHOH	Selective retention of Fe(III) ions	[39-45]
Hydroxyamine	$-CH_2-N(CH_3)-CH_2-(CHOH)_4-CH_2OH$	Selective retention of boronic acid	[6,9]
Iminodiacetate	$-N-(CH_2-COONa)_2$	Selective removal of heavy metals	[46-52]
Aminophosphonate	$-CH_2-NH-CH_2-P(O)(OH)_2$	Decalcification of brine solutions	[6,9,53,54]
Fosfonic Acid	$-P(O)(OH)_2$	Preconcentration of uranium ions	[55]
Polyethyleneamine	$-NH-(C_2H_4-NH-)_x-H$	Removal of transition metals	[56-71]
Dithiocarbamate	$- CH_2-N(NH)_2CS_2H$	Selective removal of transition metals	[72-87]
Isothiouronium	$-S-C(NH_2)=NH$	Selective removal of Hg ion and noble metals	[88-92]
8 – chinoline	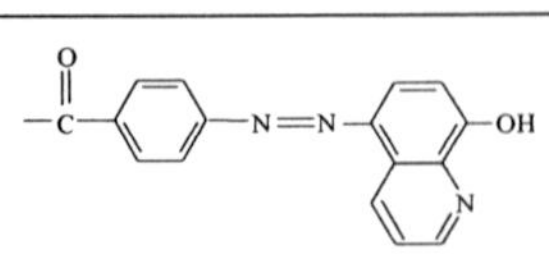	Selective removal of transition metals	[6,93]
Cyclic polyether		Separation of alkali and alkali-earth metals	[6,94]

Analytical chemistry and metal recovery employing chelating ion exchangers require properties like: high capacity, high selectivity, and fast kinetics. There is a growing interest in the development of selective ion exchangers for the metal cation recovery. Most of the commercial resins show a high capacity but a poor selectivity towards metal ions [9]. Selectivity of a chelating group for a certain metal ion could be obtained by applying multidentate ligands [20].

In order to achieve stable materials with chromatographic properties, organic–inorganic composite ion-exchange materials have been generated. The combination of the physical strength of an inorganic support with the high ion exchange capacity and kinetics of the organic ion exchangers could lead to an interesting class of hybrid ion exchangers with special applications. The organic–inorganic hybrid materials should be considered as next generation composite materials that will encompass a wide variety of applications. Efforts have been made to synthesize such hybrid ion-exchangers with a good ion-exchange capacity, high stability, reproducibility and selectivity for toxic heavy metal ions [95,96].

This chapter underlines some recent developments and trends emphasizing the synthesis and applications of two types of organic chelating sorbents considering also the original contribution of the authors in the field. The selected chelating sorbents mainly contain amidoxime and iminodiacetate functional groups. References come from the last 10 years, except the first contributions in the field and those of fundamental nature.

2. ORGANIC CHELATING SORBENTS CONTAINING AMIDOXIME FUNCTIONAL GROUPS

A careful analysis of the literature data shows that various kinds of sorbents containing the amidoxime functional group have been synthesized by the methods of suspension polymerization and irradiation-induced graft polymerization of poly(acrylonitrile) followed by functionalization with hydroxylamine in ethanol solution [10,16,23-38].

According to Egawa et al. [26], the reaction mechanism of the nitrile (CN) group with hydroxylamine is that presented in Scheme 1.

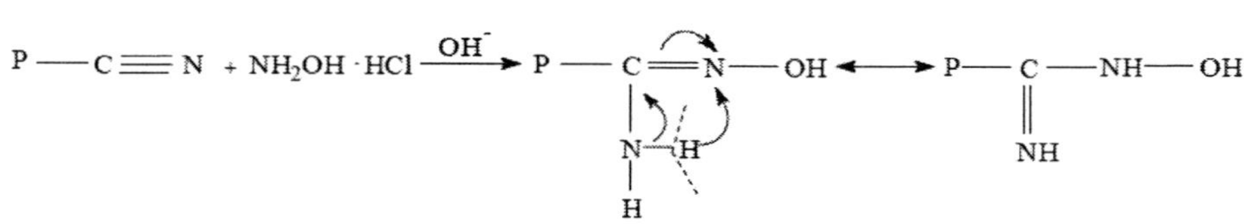

where P is the polymer backbone

Scheme 1.

The amidoximes exist predominantly in the *syn*-hydroxyamino form that is stabilized by an intramolecular hydrogen bond. The compounds behave in water as bases. The amidoximes may coordinate as a nonionic group, but they generally undergo a metal-assisted hydroxyl proton dissociation upon coordination. Two of these amidoxime groups capture a single metal ion. Figure 1 shows a schematic representation of the organic chelating sorbents containing amidoxime functional groups and a model for metal capture.

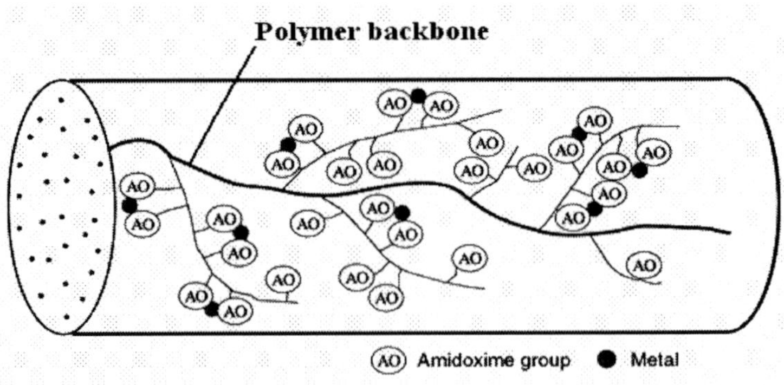

Figure 1. Schematic representation of amidoxime sorbent.

In general, the polymers used as matrix for ion-exchange resins are synthesized through free-radical suspension polymerization reactions to give spherical beads which can be readily functionalized. Porosity is introduced into the beads with porogens. The porogen can be a good solvent for the polymer, a non-solvent for the polymer, or a linear polymer. The pore size ranges from less than 50 Å (micropores) to more than 500 Å (macropores) [97,98].

Experiments involved in the adsorption of uranium from seawater with spherical resin or bead and fibrous or nonwoven fabric chelating sorbents containing amidoxime group indicate that it has a very high selectivity, strong adsorption ability and large loaded capacity for many metal ions, especially for U(VI) advantaged over others [10,25-29,37,38]. Furthermore, the amidoxime groups have a more hydrophilic character than that of the nitrile function and play a vital role in the environmental monitoring of toxic metals.

2.1. Recovery and Enrichment of Uranium from Seawater and Aqueous Media

With the development of nuclear power plant, new uranium resources, other than uranium ores, have been explored in order to meet the energy demand for the near future. Seawater containing many kinds of valuable elements is a buffered solution in the pH range of 7.5 to 8.3. The concentration of uranium in seawater is about 2.8 - 3.3 mg/m^3 and it exists mainly as the tricarbonatouranate (VI) ion, $UO_2(CO_3)_3^{4-}$ [99]. Although this concentration is very low, the estimated reserve of uranium dissolved in seawater is quite considerable. It may be almost a 1000-fold larger than the terrestrial resources. So, seawater is considered as an inexhaustible resource for uranium, which provides an additional head-end option to nuclear fuel cycle. Therefore, considering the anticipated depletion of uranium resource reserves in the future, the recovery of uranium from seawater is one of the effective pathways. Its recovery has gained great interest in many countries over the past four decades and with increased intensity since the past two decades [10,11,25-38].

Many types of inorganic [100,101] and organic chelating sorbents [10,11,16,25-38] have been studied for the recovery of uranium from seawater. The process using chelatings sorbents is throught considered to be the most effective method for recovering uranium because of the high selectivity for uranium, the ease handling, and the safety to the environment. Among them, polymeric sorbents containing an amidoxime group have been attracted one's wide attention because of the high loading of uranium and the rapid adsorption rate, which results from the high affinity of amidoxime group toward metal ions [10,11,16,23-38,102-104]. Moreover, this kind of chelating adsorbent possesses excellent mechanical strength for resisting to the friction of seawater and a low probability of entrainment from the adsorption unit [104]. Numerous amidoximated sorbents were designed and studied for the recovery of uranium from seawater and aqueous media [105-109]. Egawa et al. [10,25-29] in their pioneering work, synthesized macroreticular chelating resins containing amidoxime groups from acrylonitrile (AN)-divinylbenzene (DVB)-alkyl acrylate, alkyl methacrylate, or 4-vinylpyridine (4-VPy) copolymers as beads. It was found that the chelating resin containing amidoxime groups prepared from AN-DVB-methyl acrylate (MA) indicated the highest adsorption ability of uranium from seawater. Hydroxamic acid and carboxylic groups, in addition to amidoxime groups, were formed during the reaction of the copolymer beads with a methanol solution of hydroxylamine. The adsorption ability for uranium was greatly influenced by the pore structure (macropores) as well as by the swelling porosity (micropores). The resin prepared with 10 mol % of DVB and 10 mol % of MA had the highest adsorption ability for uranium. On the other hand, improved adsorption ability for uranium was not observed in the case of the macroporous resins prepared by the copolymerization of 4-VPy as the basic component.

Rivas et al. [32] showed that the sorption of uranyl ion strongly depends on the pH value of the solution. Poly(acrylamidoxime) formed stable complexes with uranyl ions as the pH increased; the highest value of 99% (2.5 meq/g dry resin) was found at pH 5. The sorption equilibrium was achieved relatively fast within 40 min.

A cyanomethylated styrene-divinylbenzene (St-DVB) copolymer has been prepared by the reaction of a chloromethylated St-DVB copolymer with sodium cyanide [11]. A PSt-based macroreticular resin containing a pair of neighboring nitrile groups in each monomeric styrene unit was obtained from the reaction of the cyanomethylated copolymer with 3-chloropropionitrile under phase transfer conditions. The resin was reacted with hydroxylamine to afford a chelating resin containing a pair of neighboring amidoxime groups in each monomeric styrene unit. The chelating resin showed high adsorption capacity for uranyl ions (0.72-1.23 μg UO_2^{2+}/mg resin). It was also found that the resin was an effective adsorbent for other heavy metal ions such as Hg^{2+} (1.53-2.01 μg Hg^{2+}/mg resin) and Pb^{2+} (0.73-1.78 μg Pb^{2+}/mg resin).

Amidoxime chelate fabrics were shown by Lin [110] and Sugo et al. [111,112] to be very useful for recovering uranium from seawater and enrichment or separation of noble metal ions from aqueous systems. The amidoxime chelate fabrics (AO-H) were prepared by radiation-induced graft polymerization of acrylonitrile onto a polyethylene hollow fiber, followed by chemical conversion of the produced cyano group to an amidoxime group. Distribution of the amidoxime group was uniform throughout the hollow-fibers. The fixed-bed adsorption column, charged with the bundle of AO-H fibers, was found to adsorb uranium from natural seawater at a sufficient high rate: 0.66 mg uranium per g of adsorbent in 25 days.

Zhang et al. [37] investigated the adsorption properties of uranium onto a new type of crosslinked poly(methylacrylamidoxime) chelating sorbent containing amidoxime chelating functional group (noted as FPAO) synthesized by grafting acrylonitrile (AN) and methacrylic acid (MAA) onto polypropylene. FPAO showed the strong adsorption and the quite high adsorption capacity towards uranium at pH value of 6.0. However, it might be probably difficult to apply it in the recovery of uranium from seawater because of the inappropriate pH value, which was lower than that of seawater, i.e., 7.5–8.3. So, development of a novel kind of adsorbent containing amidoxime chelating functional group, which matches with the pH value of seawater when adsorbing, was necessary. A more novel macroporous fibrous polymeric adsorbent (MFPA) containing amidoxime chelating functional group has been synthesized recently by the method of irradiation-induced graft polymerization of poly(acrylonitrile) (PAN) followed by functionalization with hydroxylamine in ethanol solution [38]. In comparison with the previous sorbent containing amidoxime functional group, the hydrophilicity and the complex ability of MFPA for uranium were improved significantly, especially the pH value employed in the adsorption process was very close to that of seawater. İt was found that U(VI) showed a strong adsorption onto the amidoxime functional group at pH = 8.0. Such an appropriate pH value in the adsorption of uranium towards MFPA was of great benefit to the application of MFPA in the actual recovery process. Also, Zhang et al. [38] described the possible adsorption procedure of U(VI) towards MFPA adsorbent, in terms of the effect of pH value on the adsorption process, as follows:

- When pH ≤ 1.0, UO_2^{2+} as a main cation species existed in an acidic solution and MFPA molecule was a proton acceptor (Scheme 2).

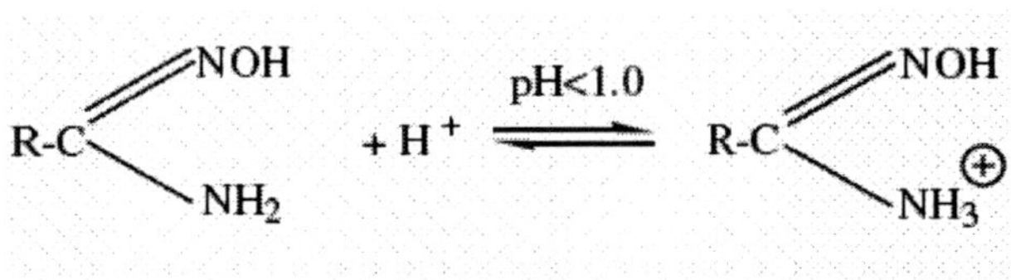

Scheme 2.

Such a protonation reaction resulted in a quick decrease in the reactivity of MFPA and ultimate termination of the complex reaction between MFPA adsorbent and UO_2^{2+}. Namely, UO_2^{2+} did not complex with MFPA at this pH range. So, in pH ≤ 1.0, it was found that U(VI) showed no adsorption towards MFPA.

- At pH ≥ 2, with an increase in basicity of the aqueous solution, R-C(NOH)NH_3^+ as a cation gradually converted to a neutral R–C(NOH)NH_2 molecule by reacting with OH^-, while UO_2^{2+} gradually hydrolyzed to $UO_2(OH)^+$ by connecting with OH^- (Scheme 3).

$$R{-}C(NOH)NH_3^+ \leftrightarrow R{-}C(NOH)NH_2 + H^+$$

$$UO_2^{2+} + OH^- \leftrightarrow UO_2(OH)^+$$

Scheme 3.

Since $UO_2(OH)^+$ was a complexible species with the neutral R–C(NOH)NH_2, so, U(VI) showed the adsorption towards MFPA adsorbent with an increase in pH value. However, the increase degree in the adsorption of U(VI) from pH 2.0 to 7.0 was not significant.

- When the pH value was in the range of 7.0 to 8.0, the experiment of the determination of the complex composition showed that $UO_2(OH)^+$ was strongly complexed with R–C(NOH)NH_2 and formed a complex of 1:2 type, UO_2 x 2(R–C(NO)NH_2), according to Scheme 4.

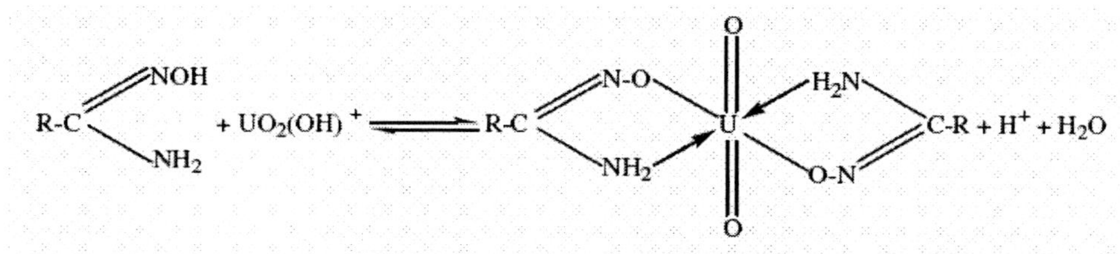

Scheme 4.

Therefore, in this pH region especially around pH 8.0, U(VI) was noticeably adsorbed by the MFPA adsorbent.

- In the pH range of 8.0 to 9.0, $UO_2(OH)^+$ reacted with OH^- further and gradually formed $UO_2(OH)_2$ or $UO_2(OH)_3^-$ described in Scheme 5. A significant decrease in the complexation of U(VI) with MFPA was observed by the formation of a colloidal precipitate of ($UO_2(OH)_2$ or $UO_2(OH)_3^-$). Also, the experiment showed that the adsorption of U(VI) towards MFPA adsorbent noticeably decreased in this pH range.

$$UO_2(OH)^+ + OH^- \leftrightarrow UO_2(OH)_2$$

$$UO_2(OH)_2 + OH^- \leftrightarrow UO_2(OH)_3^-$$

Scheme 5.

- At pH 9.0, a colorless colloidal precipitate $UO_2(OH)_2$ or $UO_2(OH)_3^-$ was observed. Because of the strong complexation of U(VI) with oxygen and nitrogen atoms contained in amidoxime functional group, U(VI) was strongly complexed in aqueous solution and adsorbed by MFPA, especially in weakly basic solution such as at pH 8.0. As a result, a lot of hydrogen ion produced as depicted in Scheme 4 was of great benefit to the complex formation of UO_2 x 2(R–C(NO)NH_2). Hence, the interaction between U(VI) and amidoxime functional group containing fibrous materials was assigned to the complex bonding formed between U(VI) with oxygen and nitrogen atoms contained in MFPA. Since the adsorption pH value, 8.0, was very close to that of seawater, i.e., in the range of 7.5 to 8.3, so, the hydrophilic MFPA adsorbent was promising to apply in recovery of U(VI) from seawater [38].

Sorbents containing both amidoxime and other hydrophilic groups have a much higher uranium uptake than an sorbent containing only amidoxime groups, because the amidoxime group is less hydrophilic than amide, carboxyl, and hydroxyl groups [107]. The water uptake, and so the uranium uptake, increased in proportion to the kind and the amount of hydrophilic groups because the diffusion of aqueous solution in more hydrophilic polymers is faster than in less hydrophilic polymers [108]. A new polymeric adsorbent has been synthesized starting from the interpenetrating polymer networks (IPNs) based on polyethylene glycol (PEG) and acrylonitrile (AN), and its adsorptive ability for uranium from aqueous media has been investigated [113]. The optimal experimental conditions to obtain an amidoximated IPN with the maximum uranium uptake having a certain geometrical shape and physical structure were determined as 50 °C of amidoximation temperature and 20 h of amidoximation time. Gamma spectrometry was found to be the most reliable method for the determination of the adsorption of uranium from natural media. The maximum adsorption capacity of the IPNs obtained at these optimal conditions was 540 mg U/g IPN/day. The results obtained by Güven et al. [113] showed again that the amidoxime polymers are highly effective for recovery of uranium from aqueous media.

Another polyamidoxime resin containing two amidoxime group per repeating unit was recently developed and studied concerning the uranium uptake [114,115]. The unique advantage of these polymers is that they contain double amidoxime groups per repeating unit, an additional diethylene spacer unit between neighboring amidoxime groups in each monomeric unit, being able to recover more efficiently uranium and other transition metal ions from seawater and aqueous media, at very low concentration levels (ppb). Adsorption capacity of newly developed adsorbent was investigated for uranyl, lead, copper, and cobalt ions from relatively concentrated (ppm) aqueous solutions under noncompetitive conditions. The adsorption capacity of amidoximated poly(N,N'-diproprionitrile acrylamide) poly(DPAAm) powder and amidoximated nonwoven fabric was investigated for UO_2^{2+}, Pb^{2+}, Cu^{2+}, and Co^{2+} metal ions at a high concentration level. The order of affinity based on the amount of metal ion uptake (mg metal ion/g dry polymer) was as follows: $UO_2^{2+} > Pb^{2+} > Co^{2+} > Cu^{2+}$, for amidoximated poly(DPAAm) powder, and $Pb^{2+} > UO_2^{2+} > Co^{2+} > Cu^{2+}$ for amidoximated nonwoven fabric at concentrations of 1000 ppm of metal ion. From the Langmuir equation, the adsorption capacities were found to be 400 mg UO_2^{2+}/g dry polymer and 250 mg UO_2^{2+}/g dry polymer, for amidoximated poly(DPAAm) and nonwoven fabric, respectively. Therefore, it was concluded that the adsorption capacities of these new poly(amidoxime) chelating polymers for uranyl ions are very high. These new adsorbents are more accessible for the adsorption of metal ions in aqueous systems than the conventional adsorbents having only one amidoxime group per repeating unit.

Pekel et al. recently reported on the interaction of UO_2^{2+} ions with amidoximated IPN as a function of the adsorbate and adsorbent amounts and temperature [116]. UO_2^{2+} ions adsorption on amidoximated poly(acrylonitrile)-poly(N-vinyl-2-pyrolidone) (PAN/PVP) IPN gels decreased with increasing temperature (considering only specific interactions). The Langmuir adsorption model was used to calculate the adsorption free energy (ΔG_{ads}) and adsorption entropy (ΔS_{ads}) values. UO_2^{2+} ion adsorption onto amidoximated IPN took place spontaneously via electrostatic interactions. The change in rate of adsorption as a function of temperature was followed, and the rate constants decreased with increasing temperature for the adsorption of uranyl ions. The energy of activation (E_a) of adsorption was +101.5 kJ/mol for the binding of uranyl ions with amidoxime groups on IPN hydrogels. A larger entropy change was observed in the complexation of amidoxime groups with uranyl ions, indicating that amidoxime groups selectively adsorb uranyl ions by an entropic effect besides the electrostatic interactions.

2.2. Other Applications of Organic Chelating Sorbents Containing Amidoxime Groups

Most of the investigations were focused on the uranium extraction from seawater or aqueous media, but there are some articles published about the uptake of transition metals by poly(amidoxime) resin. Verraest et al. [117] demonstrated that Inuline modified with an amidoxime group formed stable complexes with Cu(II). Colella et al. [23,24] also demonstrated that poly(acrylamidoxime) can be successfully used for the preconcentration of trace metals from aqueous solutions. Ferreira et al. [118] studied the complexation equilibrium of copper, lead, and zinc by using Duolite ES-346, which is a St-DVB copolymer containing an amidoxime chelating group per unit. An increase in the pH favored the metal ion complexation. Also, Lutfor et al. [31] prepared a poly(amidoxime) chelating resin from PAN grafted on sago starch and found that the sorption capacity of As^{3+} ion by the resin was pH dependent. The sorption capacity of As^{3+} was maximum 1.56 mmol/g at pH 6. Moreover, this resin has the advantage of a faster rate of equilibrium ($t_{1/2} < 9$ min), probably due to the new backbone polymer.

Chitosan [poly[β-(1→4)-2-amino-2-deoxy-D-glucose]], derived from chitin [poly[β-(1→4)-2-acetamido-2-deoxy-D-glucose]] by deacetylation, is undoubtedly one of the most promising and attracting resources in quantity. Among some interesting properties of chitosan, chelating ability arisen from its characteristic structure is especially noteworthy. This high adsorption capacity was ascribable primarily to its remarkable hydrophilicity in cooperation with the relatively high amino group content. Amidoximated chitosan-*g*-PAN copolymer was synthesized by a reaction between hydroxylamine and cyano group in chitosan-*g*-PAN copolymer prepared by grafting PAN onto crosslinked chitosan with epychlorohydrine [119]. The adsorption and desorption capacities for heavy metal ions were measured under various conditions. The adsorption capacity of amidoximated chitosan-*g*-PAN copolymer increased with increasing pH values, from 0.35 to 0.42 for Cu^{2+}, and from 0.17 to 0.35 mmol/g for Pb^{2+}, but decreased from 0.28 to 0.24 for Zn^{2+} and from 0.18 to 0.16 mmol/g for Cd^{2+} with increasing PAN grafting % from 0 to 170%. Stability constants of amidoximated chitosan-*g*-PAN copolymer (170 grafting %) were 17.80 for Cu^{2+}, 16.54 for Pb^{2+}, 14.25 for Zn^{2+}, and 13.46 for Cd^{2+}. The desorption capacity for all metal ions increased with increasing pH values. These results indicate that amidoximated chitosan-g-PAN copolymer could be a good adsorbent in the recycling of metal ions from seawater or waste water [119].

A new poly(acrylamidoxime) chelating resin was prepared via amidoximation of PAN grafted on the carboxymethylcellulose sodium salt (CMC) [36]. This resin exhibited high sorption capacity towards metal ions (42.3 mmol Zn^{2+}/g resin, 34.1 mmol Cu^{2+}/g resin). As Table 2 shows, this resin exhibited adsorption capacities much higher than those of either other biopolymer-based or fully synthetic resin. This superiority should be related to the CMC substrate. However, a comparison between metal ion sorption capacity (MSC, mmol/g) data of CMC-g-poly(amidoxime) and other polysaccharide-based poly(amidoxime) resins suggested that a ''synergistic effect of the glucopyranoside/carboxylate and amidoxime functional groups'' led to a very high binding capacity of the CMC-based resin.

Table 2. Metal ion sorption capacity (MSC, mmol/g) of the CMC-based chelating resin in comparison with that of some polysaccharide-based/synthetic resins [36]

Base	Resin	Metal	MSC, mmol/g	References
Cellulose	CMC-g-poly(amidoxime)	Zn(II)	42.3	[36]
		Cu(II)	34.1	[36]
	Cellulose-g-poly(amidoxime)	Cu(II)	0.7	[120]
	Cellulose-based IPN hydrogel[a]	Cu(II)	7.8	[121]
Starch	Starch-g-poly(amidoxime)	Cu(II)	3.0	[31]
	Starch-g-poly(hydroxamic acid)	Cu(II)	3.46	[42]
Chitosan	Crosslinked chitosan	Cu(II)	2.6	[122]
Full syntetic	Poly(amidoxime-co-hydroxamic acid)	Zn(II)	7.00	[22]
	Sulfomethylated crosslinked poly(AAm)	Cu(II)	4.07	[123]

[a]Interpenetrating polymer network composing hydroxypropylcellulose and poly(methacrylamide).

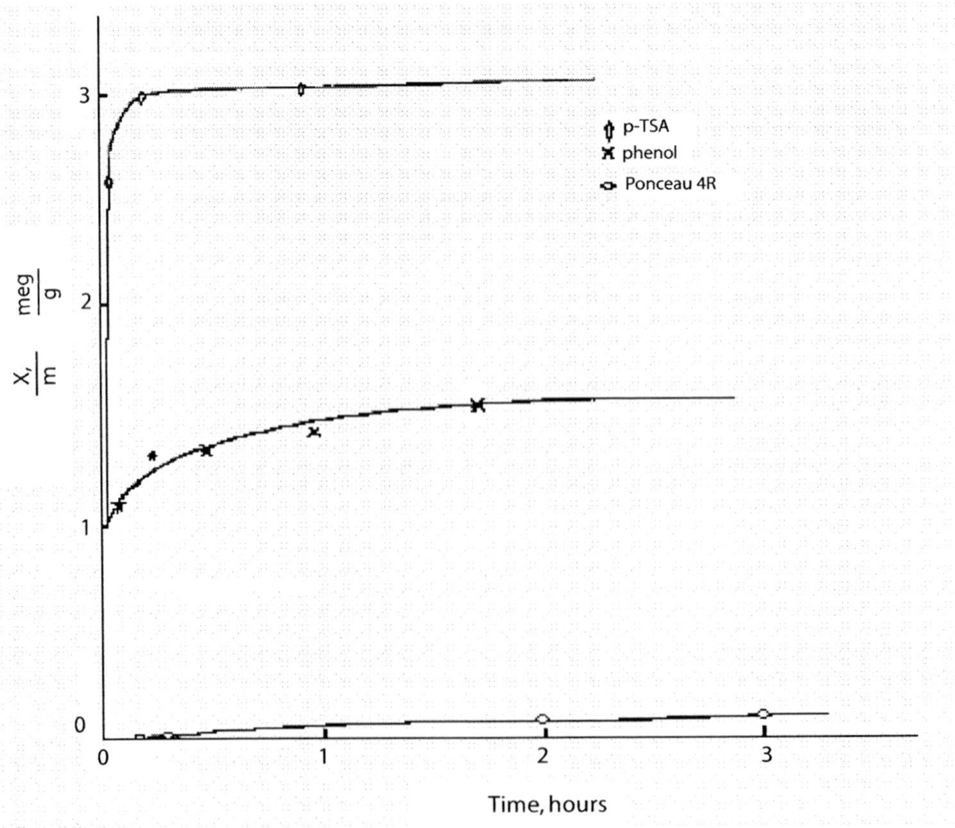

Figure 2. Sorption of p-TSA, phenol and Ponceau 4R on the resin AAO-10.

The synthesis of a new chelating ion exchanger with hydroxamic and amidoxime groups was performed by the aminolysis reaction of an ethylacrylate/AN/DVB copolymer with hydroxylamine in the presence of sodium ethoxylate. The retention properties of the new resin for Pb(II), Cd(II), Cu(II), Ni(II), Zn(II), Mn(II), and Fe(III) metal cations from aqueous solutions has been studied. The retention capacity depended on the contact time, the pH of the solution, and the types of counterions of the metal cations. The results showed that the resin preferentially retained the Pb(II) ions from the binary aqueous mixtures [22].

Dragan et al. [124] reported the sorption of some aromatic compounds on the ion exchangers with amidoxime groups (AAO) synthesized by the reaction of a porous AN-DVB copolymer with a solution of hydroxylamine in methanol, at 80 °C, for 2 h. The aromatic compounds used in the study were: phenol, *p*-toluenesulfonic acid (p-TSA), and 3-hydroxy-4(sulfonaphtylazo)-5,7-naphthalenedisulfonic acid trisodium salt (Ponceau 4R). In Figure 2 is shown the sorption of these compounds on the ion exchanger AAO-10 (total exchange capacity of 3.71 meq/g). The sorption of Ponceau 4R, which has the highest molar mass was lower, because of the small pore radius (329.8 Å) and the structure of the functional groups.

3. Organic Chelating Sorbents Containing Iminodiacetate Functional Groups

3.1. Literature Survey

Numerous papers [125–127] and reviews [128,129] are dealing with metal sorption and heavy metal uptake using polymer sorbents containing chelating groups. Ca (II) and Mg (II) ions are most abundant ions in water. Their presence in water causes the widespread use of softeners as detergent additives in laundry. As a result the process imparts additional wastes to the water by softening agents. Especially in steam generators, pretreatment of water by using chelating polymers can be beneficial and cost effective to give Ca (II) and Mg (II) free-water. The chemistry of calcium reveals that ethylenediamine tetra acetic acid (EDTA) is one of the most important chelating agents. EDTA is a chemically stable compound and able to chelate with many transition metal ions including Ca(II) and Mg (II). EDTA is such a powerful chelating agent that even $CaCO_3$ (water-insoluble) becomes soluble on standing in a solution. Based on these considerations, Bicak et al. have prepared crosslinked polystyrene sulfonamides with EDTA-like chelating groups. It has been shown that these polymers are very efficient in removal of Ca (II) and Mg (II) ions [46] and heavy metal ions [47]. By a similar analogy polymers with carboxymethylated oligoethylene imine pendant groups have been prepared using chloromethylated polystyrene resin for efficient sorption of metal ions [130]. In earlier reports polystyrene with imino diacetic acid functions have been shown to be useful in removal of many metal ions to some extent [131]. In all those studies, the basic requirements such as high chelating ability, inert backbone, and regenerability were established. The only drawback of polystyrene based resins are mechanical disintegration in long-term uses. To overcome this drawback Bicak et al. [132] have used glycidyl methacrylate (GMA)–methylmethacrylate (MMA)–divinyl benzene (DVB) terpolymer beads as reactive support. By incorporation of diethylene triamine (DETA) via ring opening of the epoxide groups and subsequent carboxymethylation with chloro acetic acid, the chelating polymer was obtained. Its chelating ability towards Ca (II) and Mg (II) ions and regenerability conditions were investigated. In conclusion, GMA based polymers with diethylenetriamine tetra acetic acid (DTTA) functions can be successfully used in removal of various metal ions including Ca(II) and Mg(II) ions. For instance based on 1.8 mmol g sorption capacity one can deduce that 1 kg of the dry resin can clean up to 600 l of water with 300 mg/ l $CaCO_3$ hardness, in each regeneration step.

3.2. Ion Exchangers Containing Iminodiacetic Groups Based on Porous AN-DVB Copolymers

In our investigations, AN-DVB copolymers were prepared by suspension polymerization method in the presence of three porogens: gasoline with t_f = 110 – 130 °C, toluene and 2-ethyl-1-hexanol (2EH). Anion exchangers with primary amine groups were obtained by the aminolysis-hydrolysis reaction of the nitrile groups with ethylenediamine (EDA), 8 h at 118 – 120 °C, at a molar ratio CN : EDA : H_2O of 1 : 10 : 2. To prepare a chelating ion exchanger (CIE) with iminodiacetate groups, the ion exchangers with primary amine groups were reacted with sodium monochloroacetate (NaMCA), in aqueous medium, at 70 °C and a pH of 7.5 – 8.0, with NaI as catalyst [49-52]. The anion and cation exchange capacity were determined as previously shown in [133] and [134], respectively.

The FTIR spectra were recorded on a DIGILAB spectrometer Scimitar Series, USA, rezolution of 4 cm^{-1}, using KBr pellet technique. The solvents uptake coefficients were determined by centrifuging at 500 g of the samples after an immersion for 48 h in the appropriate solvent. The apparent density (ρ_{ap}) was determined using a mercury pycnometer at 1.333×10^{-2} Pa. The real density (ρ_r) was measured in n-heptane. The porosity was calculated using the following equation:

$$P = [1 - (\rho_{ap}/\rho_r)] \times 100 \quad (1)$$

The retention properties for metal ions of the CIEs were determined by using a batch equilibrium procedure. Aqueous metal ion solutions (100 mL) with different concentrations were treated with about 2.0 g of ion exchangers at room temperature. The metal ion solution was $CuSO_4.5H_2O$ with different concentrations. The pH of the metal-ion solution was adjusted before equilibration over a range of 3-6 with a citrate-phosphate buffer solution. The resin was filtered off, and the concentration of the metal ion remaining in the filtrate was measured by UV-vis spectroscopy at 775 nm (that was the maximum absorption wavelength of $CuSO_4$). UV-vis spectroscopy was performed on a UV-vis Specord M42 Carl Zeiss Jena, Germany.

The concentration of metal ion bound was calculated with equation 2 [135]:

$$\text{Metal ion bound} = [(S_0 - S)V] / (W \times 1000) \quad (2)$$

where: S_0 and S are the concentrations of the metal ion in aqueous solution (mg/L) before and after the interaction with chelating resin, respectively, V is the volume of the aqueous phase (mL) and W is the amount of the dry resin (g).

Structural characteristics and solvent uptakes for some AN-DVB copolymers prepared in the presence of gasoline with a distillation range of 110-130 °C as porogen are presented in Table 3. Sample name includes: AN from acrylonitrile, the nominal crosslinking degree, which is considered identical with DVB weight %, and C or A, which allows to identify the starting copolymer and the anion exchanger, respectively. As Table 3 shows, the copolymer AN-6C, with the highest content in AN, has the highest values of solvent uptake, for both solvents (cyclohexane and toluene). These results were explained by the higher swelling in dimethylformamide (DMF) of the copolymers with a lower content in DVB. The good solvent (DMF) was removed from the copolymer with methanol, which is a nonsolvent for both the PAN sequences and the aromatic components. Thus, the swelling porosity was preserved. After about 12% of DVB, the swelling porosity had a lower contribution (Figure 4 later). The higher values of Q_T compared with Q_{cx}, at higher crosslinking degree, are explained by the increase of toluene affinity for the copolymers with a higher content in aromatic components.

Table 3. Some characteristics of the starting AN-DVB copolymers

Sample	DVB, %	AN, %	Q_{cx}, g/g	Q_T, g/g
AN-6C	6	90.33	0.2733	0.2689
AN-12C	12	85.25	0.2126	0.2093
AN-15C	15	75.9	0.1841	0.2154
AN-20C	20	67.6	0.1811	0.2191

Anion exchangers with primary amine groups were prepared by the reaction of AN-DVB copolymers with EDA, in the presence of water. The first information about the transformation of the nitrile groups into 2-aminoethylacrylamide groups were supplied by FTIR spectra. Thus, the characteristic band of nitrile groups from 2243 cm^{-1}, diminished but it was still present after 8 h of reaction with EDA of the AN-12C. The new bands at 1639, 1568 and 1291, cm^{-1} have been assigned as follows: the stretching vibrations of the C=O bond (amide I band), the deformation vibrations of the N-H bond (amide II band), the O-C-N bond (amide III band), respectively.

Anion exchange capacities, both volumetric (C_{sv}) and gravimetric (C_{sg}), as a function of the starting copolymer structure, were plotted in Figure 3.

As Figure 3 shows, C_{sg} continuously decreased with decreasing AN content because this characteristic is correlated with the content of ionic groups in the dry state of the anion exchanger, which strongly depends on the AN content in the starting copolymer. On the other hand, C_{sv} also depends on the content of ionic groups, but furthermore are correlated with the ability of the anion exchanger to swell in water, usually slightly increases with the increase of the crosslinking degree. The values of C_{sg} in Figure 3 show a slight increase with the increase of DVB from 6 to 12% and than a plateau up to 20% of DVB, and thus confirmed our expectation.

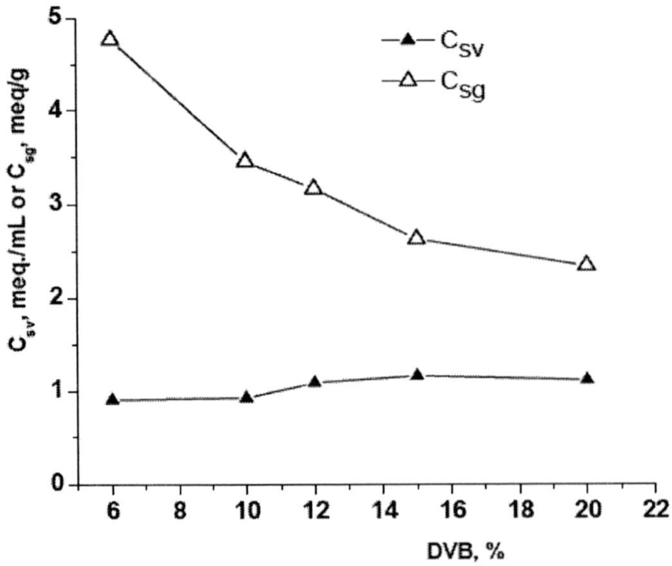

Figure 3. Ion exchange capacities of the anion exchangers derived from AN-DVB as a function of DVB content.

The porosity values of the starting copolymers as a function of the nominal crosslinking degree plotted in Figure 4 show an abrupt decrease with the increase of DVB content from 6 to 10% and a monotonous but slight decrease after that. A different behavior was observed for the corresponding anion exchangers.

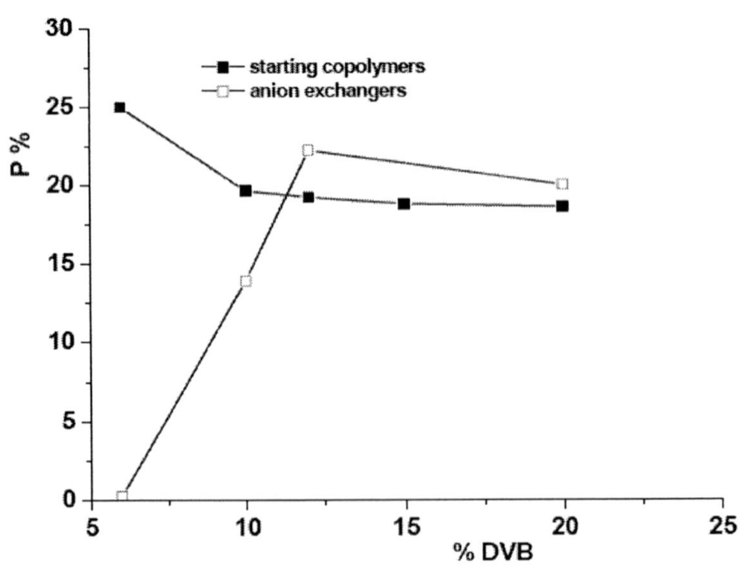

Figure 4. Porosity of the AN-DVB copolymers and the corresponding anion exchangers.

The absence of a permanent porosity for the anion exchanger derived from the copolymer with 6% DVB can be observed, followed by a fast increase up to about 12% DVB and a slight decrease after that. The anion exchanger samples used for the determination of porosity have been dried by heating the samples at 105 °C, up to a constant weight, this thermal treatment being probably responsible for the collapse of the pores in these samples. Water is a good solvent for the anion exchangers and therefore a collapse of their structure occurred at a low content of the crosslinker. Such anion exchangers seem to become permanently porous only after 12% of DVB. The postderivatization of the primary amine groups by the reaction with NaMCA was performed in order to obtain CIE with iminodiacetic groups, according to Scheme 6.

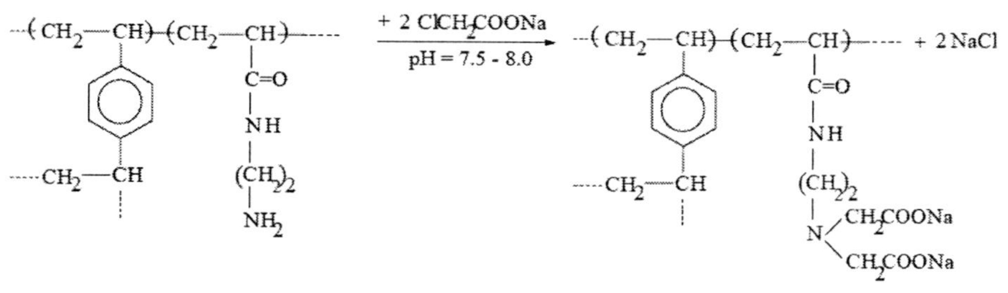

Scheme 6.

The progress of the nucleophilic substitution reaction was followed by the ratio between the ionic chlorine content found at a given moment, and the theoretical value expected for a complete transformation of the primary amine groups into tertiary amine groups (Scheme 6), this being considered the transformation degree (α). The transformation degree (α) of the primary amine groups into iminodiacetic groups, as a function of the reaction duration (h), for the anion exchanger based on the copolymer with 12% DVB and gasoline as porogen, at a monomer dilution of 0.15 mL/mL, has been plotted in Figure 5.

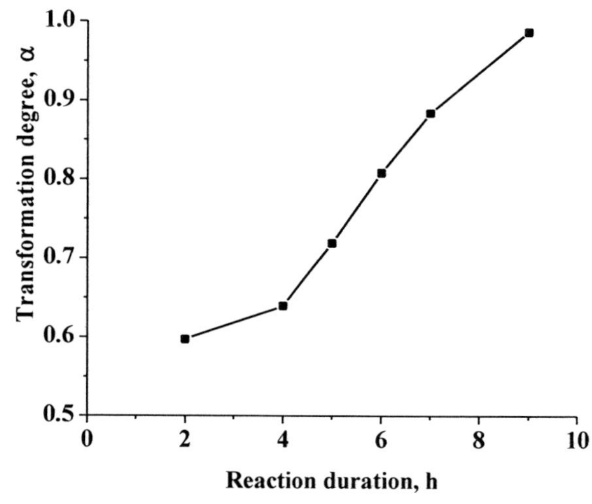

Figure 5. Transformation degree of the primary amine groups into iminodiacetic groups: starting copolymer: 12% DVB and a dilution of 0.15 ml/ml, gasoline with t_f=110–130 °C was the porogen.

By using NaI as a catalyst, the reaction was complete in about 9 h, because an interhalogen exchange occurred in this case. The FTIR spectrum of the CIE was compared with that of the starting anion exchanger with primary amine groups in Figure 6. Besides the characteristic bands of the secondary amide (1640, 1291 cm^{-1}), the new bands at 1452 cm^{-1}

and 1394 cm^{-1}, assigned to carboxylic groups, were observed in the FTIR spectrum of the CIE. Metal complexation capacities of two CIE samples, different by their porosity, are illustrated by Figure 7 as a function of the initial concentration of $CuSO_4$.

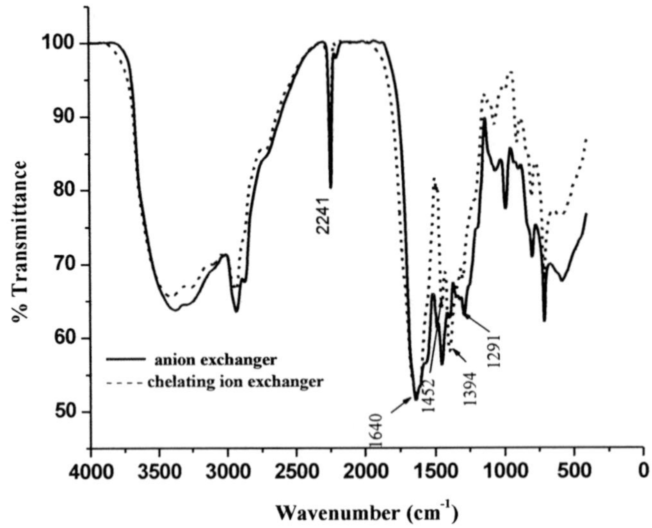

Figure 6. FT-IR spectra of the anion exchanger with primary amine groups and of the corresponding chelating ion exchanger.

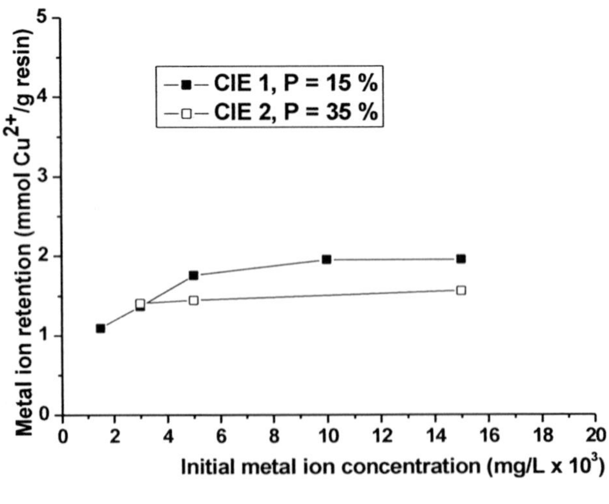

Figure 7. Cu^{2+} retention as a function of the initial concentration of the metal solution for two samples different by their porosities.

A monotonous increase of Cu^{2+} retention was observed for the resin with the higher ion exchange capacity and a lower porosity. On the other hand, the adsorption capacity of the resin with the higher porosity (CIE2) was less dependent on the initial concentration of the metal solution, but increased significantly after 48 h (results not shown here). The Cu^{2+} retention continuosly decreased with the increase of pH from 3 to 6 for both CIE resins [52]. This showed that the optimum adsorption pH of Cu^{2+} was located between 2.0 and 3.0.

3.3. Ion Exchangers Containing Iminodiacetic Groups Based on AA-DVB Copolymers

Anion exchangers with primary amine groups were successfully prepared in our group by the Hofmann degradation of the amide groups contained in acrylamide (AA)-DVB copolymers [134,136]. To obtain chelating ion exchangers with iminodiacetic groups, the postderivatization of the primary amine groups by the reaction with NaMCA was performed, according to Scheme 7.

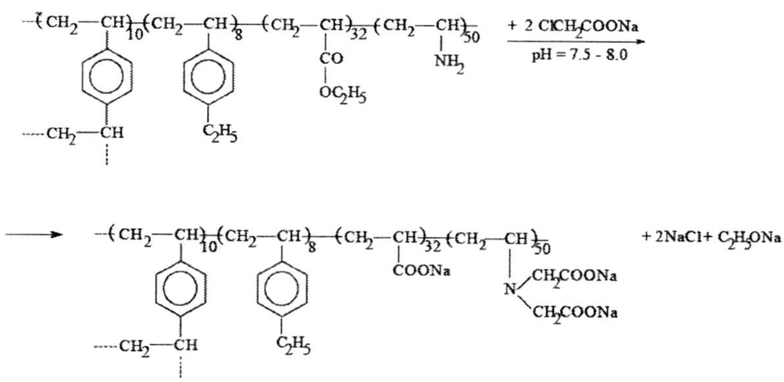

Scheme 7. Postderivatization reaction of the primary amine groups.

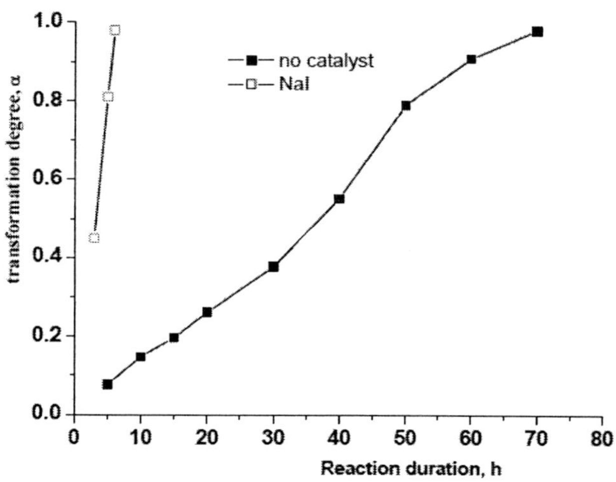

Figure 8. Transformation degree of the primary amine groups into iminodiacetic groups, according to Scheme 7.

The reaction of the primary amine groups with NaMCA was performed in water at 70 °C, in basic medium (pH = 7.5 – 8.0), first without catalyst.

To monitor the progress of the nucleophilic substitution reaction, the transformation degree (α) was followed. As Figure 8 shows, to reach a transformation degree close to 1.0, a very long time of reaction was necessary (about 70 h). By using NaI as a catalyst, the reaction was accelerated for more than 10 times, being complete in about 6 h.

Swelling degree in water (Q_w) and ion exchange capacities (C_{sv} and C_{sg}) of the anion exchanger (AA3H) and of the chelating ion exchanger (AA3HC) are collected in Table 4.

Table 4. Functional characteristics of the ion exchangers derived from AA-DVB copolymers

Sample	Q_w, g/g	C_{sv}, meq/mL		C_{sg}, meq/g	
		Weak basic	Weak acid	Weak basic	Weak acid
AA3[a]	0.6638	-	-	-	-
AA3H[b]	1.1333	0.5950	-	1.8931	-
AA3HC[c]	1.8899	-	2.1948	-	9.1766

[a] AA-DVB copolymer with 10% DVB and 50% AA.
[b] Anion exchanger with primary amine groups derived from AA3 by Hofmann degradation.
[c] Chelating ion exchanger

The high values of the weak acid capacity found for the ion exchanger AA3HC have been explained by the two further contributions, as follows:

- reaction of the primary amine groups with NaMCA at a pH of 7.5 – 8.0, in noncatalyzed conditions, lasted about 70 h and the hydrolysis of the ester groups occurred too, the FT-IR spectra demonstrated this by the absence of the characteristic band for C=O stretching from ester groups;
- the method used for the weak acid capacity determination, did not take into account the presence of tertiary amine groups, which definitely have a contribution by their protonation in the regeneration step of the ion exchanger with HCl 1 N; in the next step, which is the treatment with NaOH, the transformation of carboxylic groups into carboxylate groups occurred, but also the formation of NaCl corresponding to the eliberation of the tertiary amine groups; the titration of the chlorine ions showed that only part of tertiary amine groups were involved in this process.

4. Conclusion

The research in the area of organic chelating sorbents developes very fast, the main interest being focused on their selectivity in order to design novel specialized polymers for specific environmental and industrial applications.The paper presented in brief some recently published information focused on the organic chelating sorbents, in general, and on the chelating sorbents bearing amidoxime and iminodiacetate functional groups, in particular, the own contributions of the authors being also referenced.

The synthesis and characterization of some chelating ion exchangers, by two less used ways, were reported in the paper:

- anion exchangers with 2-aminoethylacrylamide units were obtained by the aminolysis-hydrolysis reaction of the nitrile groups contained in porous AN-DVB copolymers prepared in the presence of toluene, gasoline and 2EH as porogens;

- CIE with iminodiacetic groups were generated by the carboxymethylation of the primary amine groups with NaMCA, in the presence of NaI;
- weak basic anion exchangers with 2-aminoethylacrylamide units were obtained by the aminolysis-hydrolysis reaction of the nitrile groups contained in AN-DVB copolymers prepared in the presence of gasoline as a porogen;
- CIE with iminodiacetic groups were generated by the carboxymethylation of the primary amine groups contained in a vinylamine-ethylacrylate-divinylbenzene copolymer, vinylamine units resulted by a Hofmann degradation of the acrylamide units present in the starting copolymer;
- sorption capacity of Cu^{2+} of the CIE was influenced by the porosity of the resin and solution pH.

Further investigations are necessary especially for the preparation of new hybrid chelating sorbents with tailored properties for special destinations.

ACKNOWLEDGEMENT

The financial support of this research by the Romanian Ministry of Education and Research (Grant Nr. 34670) is gratefully acknowledged.

REFERENCES

[1] Helfferich, H. Ion Exchange, McGraw-Hill, New York, 1962, pp 26-81.

[2] Schmuckler, G. *Talanta* 1965, 12, 281-290.

[3] Schmuckler, G. In: *Encyclopedia of Polymer Science and Technology*, Mark, H. F.; Ed.; Suppl. Vol. 2, John Wiley and Sons: New York, 1977, pp 197-204.

[4] McCrary, A. L.; Howard, W. L. *Encyclopedia of Chemical Technology*, Mark, H. F.; Othmer, D. F.; Overberger, D. F.; Seaborg, G. T.; Eds.; Vol. 5, John Wiley and Sons: New York, 1979, pp 339-368.

[5] Albright, R. L.; Yarnell, P. A. In: *Encyclopedia of Polymer Science and Engineering*, Kroschwitz J. I.; Ed.; Vol. 8, John Wiley and Sons: New York, 1987, pp 341-393.

[6] Alberti, G.; Casciola, M.; Constantino, U. In: *Encyclopedia of Analytical Chemistry,* Townshend, A.; Worsfold, P. J.; Eds.; Vol. 4, Academic Press: London, 1995, pp 2298-2306.

[7] Denizli, A.; Sanli, N.; Garipcan, B.; Patir, S.; Alsancak, G. *Ind. Eng. Chem. Res.* 2004, *43*, 6095-6101.

[8] Sharma, R. K. *Pure Appl.Chem.* 2001, *73*, 181–186.

[9] Luca, C. In: *Encyclopedia of Separation Science*; Academic Press: London, 2000; Vol. 4, pp 1617-1632.

[10] Egawa, H.; Kabay, N.; Shuto, T.; Jyo, A. *Ind. Eng. Chem. Res.* 1993, *32*, 540-547.

[11] Park, I.-H.; Suh, J.-M. *Angew. Makromol.Chem.* 1996, *239*, 121-132.

[12] Bulbul Sonmez, H.; Senkal, B. F.; Bicak, N. *J. Polym. Sci. Part A Polym. Chem.* 2002, *40*, 3068-3078.

[13] Rivas, B. L.; Jara, M.; Pereira, E. D. *J. Appl. Polym. Sci.* 2003, *89*, 2852-2856.
[14] Henry, W. D.; Zhao, D.; SenGupta, A. K.; Lange, C. *React. Funct. Polym.* 2004, *60*, 109-120.
[15] Rivas, B. L.; Quilodran, B.; Quiroz, E. *J. Appl. Polym. Sci.* 2004, *92*, 2908-2916.
[16] Kavakli, P. A.; Güven, O. *J. Appl. Polym. Sci.* 2004, *93*, 1705-1710.
[17] Roy, P. K.; Rawat, A. S.; Choudhary, V.; Rai, P. K. *J. Appl. Polym. Sci.* 2004, *94*, 1771-1779.
[18] Luca, C.; Maruţă, C.; Bunia, I.; Urmenyi, A. M.; Sandu, M.; Raţă, D. *J. Appl. Polym. Sci.* 2005, *97*, 930-938.
[19] Wang, C.-C.; Wang, C.-C. *J. Appl. Polym. Sci.* 2005, *97*, 2457-2468.
[20] Rivas, B. L.; Pereira, E.; Jara, M.; Esparza, C. *J. Appl. Polym. Sci.* 2006, *99*, 706-711.
[21] Bicak, N.; Sherrington, D. C.; Sungar, S.; Tan, N. *React. Funct. Polym.* 2003, *54*, 141-147.
[22] Neagu, V.; Bunia, İ.; Luca C. *Macromol. Symp.* 2006, *235*, 136-142.
[23] Collela, M. B.; Siggia, S.; Barnes, R. M. *Anal. Chem.* 1980, *52*, 967-972.
[24] Collela, M. B.; Siggia, S.; Barnes, R. M. *Anal. Chem.* 1980, *52*, 2347-2350.
[25] Egawa, H.; Nonaka, T.; Matsumoto, S.; Nakayama, M. *Israel J. Chem.* 1985, *26*, 56-59.
[26] Egawa, H.; Nonaka T.; Abe, S.; Nakayama, M. *J. Appl. Polym . Sci.* 1992, *45,* 837-841.
[27] Egawa, H.; Nakayama, M.; Nonaka, T.; Yamamoto, H.; Uemura, K. *J. Appl. Polym. Sci.* 1987, *34*, 1557-1575.
[28] Egawa, H.; Kabay, N.; Saigo, S.; Nonaka, S.; Shuto, T. *Bull. Soc. Sea Water Sci. Jpn.* 1991, *45*, 324-332.
[29] Kabay, N.; Egawa, H. *Tr. J. Chem*. 1993, *17*, 62-73.
[30] Boudakgi, A.; Jezierska, J.; Kolarz, B. N. *Makromol. Chem. Macromol. Symp.* 1992, *59*, 343-352.
[31] Lutfor, M. R.; Silong, S.; Yunus, W. M. Z. W.; Ab Rahman, M. Z.; Ahmad, M.; Haron, J. *Eur. Polym. J.* 2000, *36*, 2105-2113.
[32] Rivas, B. L.; Maturana, H. A.; Villegas, S. *J. Appl. Polym. Sci.* 2000, *77*, 1994-1999.
[33] Pekel, N.; Şahiner, N.; Güven, O. *J. Appl. Polym. Sci.* 2001, *81*, 2324-2329.
[34] Pekel, N.; Güven, O. *Colloids Surf. A* 2003, *212*, 155-161.
[35] Bryant, D. E.; Stewart, D. I.; Kee, T.; Barton, C. S. *Environ. Sci. Technol.* 2003, *37*, 4011-4016.
[36] Zohuriaan-Mehr, M. J.; Pourjavadi, A.; Salehi-Rad, M. *React. Funct. Polym.* 2004, *61*, 23-31.
[37] Zhang, A.; Asakura, T.; Uchiyama, G. *React. Funct. Polym.* 2003, *57*, 67-76.
[38] Zhang, A.; Uchiyama, G.; Asakura, T. *React. Funct. Polym.* 2005, *63*, 143-153.
[39] Seung Lee, T.; Il Hong, S. *J. Appl. Polym. Sci.* 1995, *57*, 311-317.
[40] Seung Lee, T.; Il Hong, S. *J. Polym. Sci. Part A Polym. Chem.* 1995, *33*, 203-210.
[41] Moustafa, A. B.; Faizalla, A. *J. Appl. Polym. Sci.* 1999, *73*, 149-159.
[42] Lutfor, M. R.; Sidik, S.; Wan Yumus, W. M. Z.; Rahman, M. Z. A.; Mansor, A.; Haron, M. J. *J. Appl. Polym. Sci.* 2001, *79*, 1256-1264.
[43] Alakhras, F. A.; Dari, K. A.; Mubarak, M. S. *J. Appl. Polym. Sci.* 2005, *97*, 691-696.
[44] Pande, C. S.; Mehta, I. K.; Ambasta, B. K.; Archana, M. K. *J. Appl. Polym. Sci.* 2000, *77*, 475-483.
[45] Kabay, N.; Katakai, A.; Sugo, T.; Egawa, H. *J. Appl. Polym. Sci.* 1993, *49*, 599-607.
[46] Bicak, N.; Senkal, B. F.; Yarbas T. *Macromol. Chem. Phys.* 1998, *199*, 2731-2735.

[47] Bicak, N.; Senkal, B. F.; Melekaslan, D. *J. Appl. Polym. Sci.* 2000, *77*, 2749-2755.
[48] Wang, C. C.; Chen, C. Y.; Chang, C. Y. *J. Appl. Polym. Sci.* 2002, *84*, 1353-1362.
[49] Dragan, E. S.; Avram, E.; Dinu, M. V. CD-ROM, 2005, ISBN 3-86010-784-4.
[50] Dinu, M. V.; Dragan, E. S.; Vlad, C. D. 2005, ISBN 973-718-284-7, S9_168 - S9_175.
[51] Dinu, M. V.; Dragan, E. S.; Vlad, C. D. 2005, ISBN 973-718-284-7, S9_61 - S9_68.
[52] Dragan, E. S.; Avram, E.; Dinu, M. V. *Polym. Adv. Technol.* 2006, *17*, 1-8.
[53] Maeda, H.; Egawa, H. *J. Appl. Polym. Sci.* 1990, *39*, 1519-1525.
[54] Maeda, H.; Egawa, H. *J. Appl. Polym. Sci.* 1987, *33*, 1275-1279.
[55] Egawa, H.; Nonaka, T.; Ikari, M. *J. Appl. Polym. Sci.* 1984, *29*, 2045-2055.
[56] Yokoyama, T.; Kikuchi, A.; Kimura, T.; Susuki, T. M. *Bull. Chem. Soc. Jpn.* 1983, *56*, 463-466.
[57] Rivas, B. L.; Pooley, S. A.; Aceiton, E. *J. Appl. Polym. Sci.* 2002, *84*, 1251-1256.
[58] Maeda, H.; Egawa, H. *J. Appl. Polym. Sci.* 1984, *29*, 2281-2287.
[59] Trochimczuk, A. W. *Eur. Polym. J.* 1998, *34*, 1657-1662.
[60] Zhu, X.; Alexandratos, S. D. *Ind.. Eng. Chem. Res.* 2005, *44*, 8605-8610.
[61] Reddy, A. R.; Reddy, K. H. *Proc. Indian Acad. Sci. (Chem. Sci.)* 2003, *115*, 155-160.
[62] Barsanescu, A.; Buhaceanu, R.; Dulman, V.; Bunia, I.; Neagu, V. *J. Appl. Polym. Sci.* 2004, *93*, 803-808.
[63] Sun, C.; Qu, R.; Wang, C.; Ji, C.; Cheng, G. *J. Appl. Polym. Sci.* 2005, *95*, 890-896.
[64] Jeragh, B. J. A.; Elassar, A. Z. A.; El-Dissouky, A. *J. Appl. Polym. Sci.* 2005, *96*, 1839-1846.
[65] Qu, R., Wang, C.; Ji, C.; Sun, C.; Sun, X.; Cheng G. *J. Appl. Polym. Sci.* 2005, *95*, 1558-1565.
[66] Rivas, B. L.; Maturana, H. A.; Catalan, R. E. *J. Appl. Polym. Sci.* 1989, *38*, 801-807.
[67] Duru, P. E.; Bektas, S.; Geng, O.; Patir, S.; Denizli, A. *J. Appl. Polym. Sci.* 2001, *81*, 197-205.
[68] Say, R.; Tuncel, A.; Denizli, A. *J. Appl. Polym. Sci.* 2002, *83*, 2467-2473.
[69] Chanda, M.; Rempel, G. L. *React. Funct. Polym.* 1997, *35*, 197-205.
[70] Navarro, R. R.; Sumi, K.; Fujii, N.; Matsumura, M. *Water Res.* 1996, *30*, 2488-2494.
[71] Say, R.; Emir, S.; Garipcan, B.; Patir, S.; Denizli, A. *Adv. Polym. Technol.* 2003, *22*, 355-364.
[72] Tbal, H.; Delporte, M.; Morcellet, J.; Morcellet, M. *Eur. Polym. J.* 1992, *28*, 671-679.
[73] Saegusa, T.; Kobayashi, S.; Hayashi, H.; Yamada, A. *Polym. J.* 1978, *10*, 403-408.
[74] Mathew, B.; Rajasekharan, V. N. *Polym. Bull.* 1991, *26*, 603-610.
[75] Chen, Y. Y.; Gu, Z. M.; Ren, X. M.; Cheng, X. L. *J. Makromol. Sci.Chem.* 1987, *A24*, 319-325.
[76] Kuriakose, S.; Rajasekharan, V. N. *Polym. Int.* 1994, *34*, 225-228.
[77] Lezzi, A.; Cobianco, S. *J Appl. Polym. Sci.* 1994, *54*, 889-897.
[78] Trochimczuk, A. W.; Kolarz, B. N. *Eur. Polym. J.* 2000, *36*, 2359-2363.
[79] Pearson, R. G. *J. Am.. Chem. Soc.* 1963, *85*, 3533-3539.
[80] Zuo, G.; Muhamed, M. *Sep. Sci. Technol.* 1990, *25*, 1785-1802.
[81] Zuo, G.; Muhamed, M. *React. Polym.* 1995, *24*, 165-181.
[82] Trochimczuk, A. W.; Streat, M. *React.. Funct. Polym..* 1999, *40*, 205-213.
[83] Utkelov, B. A.; Ergozhin, E. E.; Ashkuva R. K. *React. Polym.* 1991, *14*, 187-191.
[84] Cogost, M. A.; Salvanerra, D.; Marques, G.; Bourdelande, J. L. *React. Polym.* 1996, *28*, 191-200.

[85] Jerieska, J.; Trochimczuk, A. W.; Kedzierska, J. *Polymer.* 1999, *40*, 3611-3616.

[86] East, G. C.; Mcentyru, J. E.; Shao, J. *Polymer.* 1997, *38*, 3973-3984.

[87] McClain, A; Hsieh, Y- Lo. *J. Appl. Polym. Sci.* 2004, *92*, 218-225.

[88] Chen, Y. Y.; Chai, G. P.; Wang, N. D. *J. Macromol. Sci. Chem.* 1988, *A27*, 1321-1333.

[89] Grote, M.; Schumacher, U. *React. Funct. Polym.* 1997, *35*, 179-196.

[90] Sartore, L.; Penco, M.; Bignotti, F.; Peroni, I.; Gill, M. H.; Ramos, M. A.; D'Amore, A. *J. Appl. Polym. Sci.* 2002, *85*, 1287-1296.

[91] Qu, R.; Wang, C.; Sun, C.; Ji, C.; Cheng, G.; Wang, X.; Xu, G. *J. Appl. Polym. Sci.* 2004, *92*, 1646-1652.

[92] Qu, R.; Sun, C.; Ji, C.; Wang, C.; Zhao, Z.; Yu, D. *Polym. Eng. Sci.* 2005, *45*, 1515-1521.

[93] Wang, Y.; Astilean, S.; Haran, G.; Warshawsky, A. *Anal. Chem.* 2001, *73*, 4096-4103.

[94] Williams, S. M.; Brodbelt, J. S.; Marchand, A. P.; Cal, D.; Majerski, K. M. *Anal. Chem.* 2003, *74*, 4423-4433.

[95] Gailliez-Degremont, E.; Bacquet, M.; Laureyns, J.; Morcellet, M. *J. Appl. Polym. Sci.* 1997, *65*, 871-882.

[96] Kuo, P. L.; Chen, W. F.; Liang, W. J. *J. Polym. Sci. Part A Polym. Chem.* 2005, *15*, 3359-3367.

[97] Okay, O. *Prog. Polym. Sci.* 2000, *25*, 711-779.

[98] Dragan, E. S.; Vlad, C. D. In: *New Trends in Nonionic (Co)Polymers and Hybrids*; Dragan, E. S.; Ed.; Nova Science Publishers: New York, NY, 2000, pp 121-166.

[99] Vernon, F.; Shah, T. *React. Polym.* 1983, *1*, 301-308.

[100] Yamashita, H.; Ozawa, Y.; Nakajima, F.; Murita, T. *Sep. Sci. Technol.* 1981, *16*, 676-681.

[101] Abbasi, W. A.; Streat, M. *Sep. Sci. Technol.* 1994, *29*, 1217-1222.

[102] Choi, S. H.; Nho, Y.C. *Radiat. Phys. Chem.* 2000, *57*, 187-193.

[103] Kawai, T.; Saito, K.; Sugita, K.; Katakai, A.; Seko, N.; Sugo, T.; Kanno, J.; Kawakami, T. *Ind. Eng. Chem. Res.* 2000, *39*, 2910-2915.

[104] Omichi, H.; Katakai, A.; Sugo, T.; Okamoto, J. *Sep. Sci.Technol.* 1986, *21*, 299-311.

[105] Saraydın, D.; Karadag, E.; Guven, O. *Sep. Sci. Technol.* 1995, *30*, 3287-3291.

[106] Karadag E.; Saraydın, D.; Guven, O. *Sep. Sci. Technol.* 1995, *30*, 3747-3760.

[107] Akkas, P.; Guven, O. *J. Appl. Polym. Sci.* 2000, *78*, 284-289.

[108] Rivas, B. L.; Pooley, S. A.; Maturana, H. A.; Villegas, S. *Macromol. Chem. Phys.* 2001, *202*, 443-447.

[109] Guler, H.; Sahiner, N.; Aycik, G. A.; Güven, O. *J. Appl. Polym. Sci.* 1997, *66*, 2475-2480.

[110] Lin, W. P.; Lu, Y.; Zeng, H. M. *J. Appl. Polym. Sci.* 1993, *47*, 45-52

[111] Shiraishi, T.; Tamada, M.; Saito, K.; Sugo, T. *Radiat. Phys. Chem.* 2003, *66*, 43-47.

[112] Kabay, N.; Katakai, A.; Sugo, T. *Radiat. Phys. Chem.* 1995, *46*, 833-836.

[113] Kavaklı Akkas, P.; Uzun, C.; Güven, O. *React. Funct. Polym.* 2004, *61*, 245-254.

[114] Kavaklı Akkas, P.; Seko, N.; Tamada, M.; Güven, O. *Sep. Sci. Technol.* 2004, *39*, 1631-1643.

[115] Kavaklı Akkas, P.; Güven, O. *Adsorption* 2005, *10*, 309-315.

[116] Pekel, N.; Sahiner, N.; Güven, O. *J. Polym. Sci. Part B Polym. Phys.* 2004, *42*, 986–993.

[117] Verraest, D. L.; Peters, J. A.; Kuzee, H. C.; Raaijmakers, H. W. C.; Van Bekkum, H. *Carbohydr. Polym.* 1998, *37*, 209-214.
[118] Ferreira, L. M.; Loureiro, J. M.; Rodrigues, A. *Sep. Sci..Technol.* 1998, *33*, 1585-1604.
[119] Kang, D. W.; Choi, H. R.; Keon, D. *J. Appl. Polym. Sci.* 1999, *73*, 469–476.
[120] Kubota, H.; Shigehisha, Y. *J. Appl. Polym. Sci.* 1995, *56*, 147-151.
[121] Chauhan, G.S.; Mahajan, S.; *J. Appl. Polym. Sci.* 2002, *86*, 667-671.
[122] Cao, Z.; Ge, H.; Lai, S. *Eur. Polym. J.* 2001, *37*, 2141-2143.
[123] Kasgoz, H.; Ozgumus, S.; Orbay, M. *Polymer* 2003, *44*, 1785-1793.
[124] Dragan, S.; Cristea, M.; Airinei, A.; Poinescu, Ig.; Luca, C. *J. Appl. Polym. Sci.* 1995, *55*, 421-430.
[125] Svec, F. *Acta Polym.* 1980, *31*, 68-73.
[126] Blasius, E.; Bock, I. *J. Chromatogr.* 1964, *14*, 244-254.
[127] Barnes, J. H.; Esslemont, G. F. *Makromol. Chem.* 1976, *177*, 307-310.
[128] Kaliyappan, T.; Kannan, P. *Prog. Polym. Sci.* 2000, *25*, 343-370.
[129] Beauvais, R. A.; Alexandratos, S. D. *React. Funct. Polym.* 1998, *36*, 113-123.
[130] Lecat-Tillier, C.; Lafuma, F.; Quivoron, C. *Eur. Polym. J.* 1980, *16*, 467-470.
[131] Sherman, J. H.; Danielson, N. D.; Taylor, R. T.; Marsh, J. R.; Esterline, D.T. *Environ. Technol.* 1993, *14*, 1097-1100.
[132] Senkal, B. F.; Bicak, N. *React. Funct. Polym.* 2001, *49*, 151–157.
[133] Dragan, S.; Grigoriu, G. *Angew. Makromol. Chem.* 1992, *200*, 27–36.
[134] Dragan, E. S.; C. D. *Macromol. Symp.* 2002, *181*, 47-55.
[135] El-Hamshary, H.; El-Garawany, M.; Assubaie, F. N.; Al-Eed, M. *J. Appl. Polym. Sci.* 2003, *89*, 2522–2526.
[136] Dragan, E. S.; Dinu, M. V.; Vlad, C. D. *J. Appl. Polym. Sci.* 2003, *89*, 2701–2707.

In: New Trends in Ionic (Co) Polymers and Hybrids
Editor: Ecaterina Stela Dragan, pp. 91-108 ISBN 978-1-60021-611-4

Chapter 4

NANOCOMPOSITES FOR POLYMER ELECTROLYTES

Han Ping Zhang, Peng Zhang, Yu Ping Wu **[*]** ***and Hao Qing Wu***

Department of Chemistry, Fudan University,
Shanghai 200433, China

ABSTRACT

Polymer electrolyte has been studied since 1973, and its ionic conductivity was still not high enough. Recently, nanotechnology provided some new insights, and some marked progress was achieved, indicating great promise for its practical application. After introduction of nano fillers, the formed nanocomposites with poly(ethylene oxide) as polymer electrolytes for lithium ions present evidently an enhancement in ionic conductivity, which is due to several factors such as decreasing the content of crystal phase and producing new pathways for the movement of lithium ions along the surface of the nanoparticles. The behavior of the interphase between polymer electrolytes and electrodes is also ameliorated. Of course, different species of nano fillers and their amount and surface groups will lead to different effects. Prior to its practical applications, there are still some challenging questions to solve.

LIST OF ABBREVIATIONS

Content of crystal phase	χ_c
Ionic conductivity	σ
Layered double hydroxide	LDH
Organic modified MMT hybrid	OMMT
Poly(ethylene oxide)	PEO
Polymer electrolyte:	PE
Temperature	T

[*] Email: wuyp@fudan.edu.cn

1. Introduction

In 1973, it was found that a composite of poly(ethylene oxide) (PEO) and alkaline salt presented high ionic conductivity [1,2]. Since then a new research area, polymer electrolytes, came into birth. They present promising application in energy storage systems, displayers, sensors and so on. Polymer electrolytes can be divided into solid polymer electrolyte and gelled polymer electrolyte. If there is no specification, it usually refers to the former one. There are many kinds of ions that are movable in the polymer bases. Here only polymer electrolytes for lithium ions are reviewed. To polymer electrolytes for other ions such as Na^+, K^+, Mg^{2+}, and Cl^-, some literatures can be referred [3].

So far, many kinds of polymer-based electrolytes including PEO, polyacrylonitrile, poly(methyl acrylate) and poly(vinylidenedifluoride) have been studied. Among them, PEO has been widely explored since it has strong polarity and includes flexible ether units to dissolve a lot of lithium salts therein. Oxygen atoms in the polymer chains can form coordination complexes with lithium ions. Due to the movement of the polymer segments and coordination and dissociation of the complex, lithium ions can be transported. There are mainly two ways of transportation of lithium ions and ionic clusters: inter chains and intrachain, which are shown in Figure 1 [3]. The transportation mainly happens in the amorphous area, and the ionic conductivity is about 2-3 orders of magnitudes higher than that in the crystal phase. However, PEO can easily crystallize due to its polarity. So far, a lot of methods have been used to decrease the amount of crystal phase such as blending, copolymerizing including random, graft, block and comb-shape one, crosslinking, and adding inorganic fillers [2]. However, the ionic conductivity of these kinds of polymer electrolytes is still not high enough. As a result, practical application is not available.

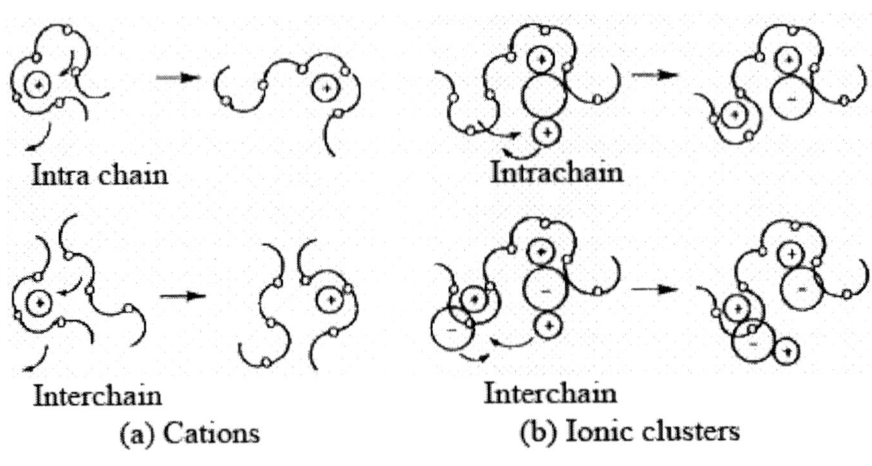

Figure 1. Schematic model for the transportation of (a) cations and (b) ionic clusters in PEO *via* the segmental motion (from reference [3]).

Recently, with the development of nanotechnology, it was found that nano fillers could increase the ionic conductivity of polymer electrolytes [4]. Consequently, nanocomposites for polymer lithium ion electrolytes have attracted much attention. The studied nanofillers mainly include MgO, Al_2O_3, SiO_2, TiO_2 and Fe_2O_3.

2. Nanocomposites of PEO with MgO Nanoparticles

The size effect of MgO fillers on the ionic conductivity of PEO/$LiBF_4$ is shown in Figure 2 [5]. After the addition of MgO fillers, the ionic conductivity of the composites is evidently increased at room temperature. The added MgO filler depresses the PEO melting temperature, interacts with the PEO phase and retards the kinetics of its crystallization, providing a stable amorphous phase. A resistivity relaxation or conductivity enhancement below the melting temperature of PEO (68 °C) occurs, which appears to be a characteristic of these electrolytes and related to an interaction of dipoles associated with polymer chains and MgO [6]. In the case of the nano size filler, it provides higher ionic conductivity at room temperature, about 10^{-4} S/cm, one order of magnitude higher than that of the PEO/$LiBF_4$ composite added with micro size MgO. This enhancement is mainly due to an enhanced interaction. For example, if the weight percent of MgO is maintained constant and the particle size is reduced from 5 μm to 20 nm, over 15 million additional polymer-filler interaction sites, which are shown in Figure 3, are created.

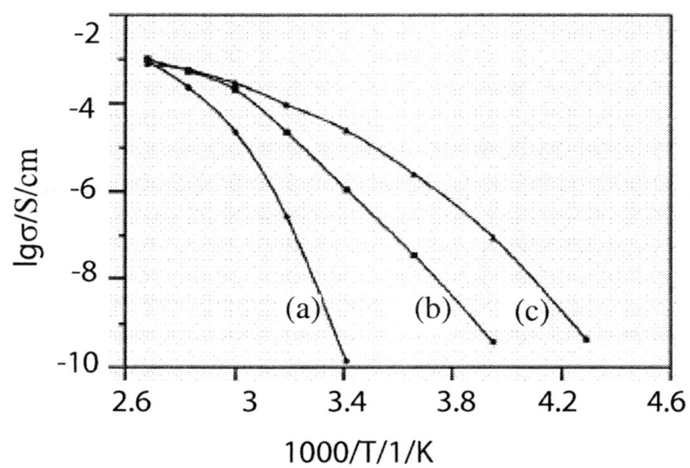

Figure 2. Effect of fillers on ionic conductivity of PEO/$LiBF_4$ (8:1): (a) without filler; (b) with 10 wt % 5 μm MgO; (c) with 10 wt % 20 nm MgO (from reference [5], reprinted with permission from the Electrochemical Society).

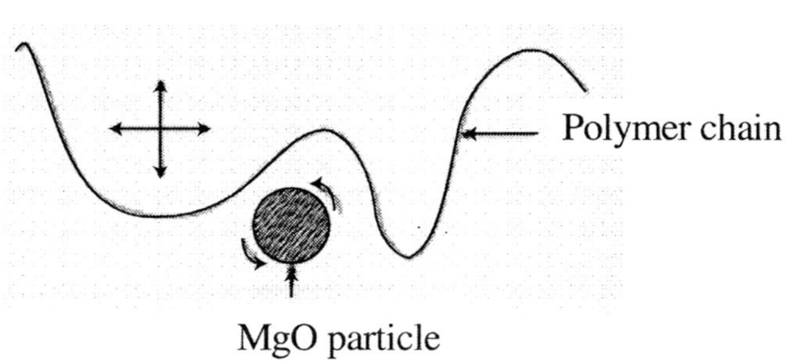

Figure 3. Schematic representation of a polymer chain segment and MgO particle interaction (from reference [5], reprinted with permission from the Electrochemical Society).

The temperature dependence of ionic conductivity diminishes as the MgO particle is reduced from micro- to nanosize. At low temperature (about 20 °C), the dipole-dipole interaction is favored, whereas at high temperature (about 100 °C) a reverse trend is observed [5]. When the added amount of nano MgO is high (> 30%), it segregates and PEO crystallizes, leading to a decrease of ionic conductivity [6].

3. NANOCOMPOSITES OF PEO WITH Al_2O_3 NANOPARTICLES

There are several crystal phases of Al_2O_3 such as α, β, γ and δ. These nanoparticles can present favorable enhancement on ionic conductivity [4,7-9].

When δ-Al_2O_3 nanoparticles (diameter about 10 nm) was added into PEO with molecular weight of 2000 (also called PEG), the glass transition temperature, T_g, and melting temperature do not change evidently and the effective mobility of the lithium ions is markedly increased. As shown in Figure 4, with the increase of the amount of the nano fillers, the content of crystal phase decreases [7,10]. When the amount arrives at about 10% (molar ratio), the activation energy of transportation of lithium ions becomes lowest and the ionic conductivity highest, about 4.5×10^{-6} S/cm, an order of magnitude higher than that of the virginal composite of PEO/$LiClO_4$ [7].

Concentration and surface area of nano fillers present different effect on the enhancement of ionic conductivity, which is shown in Figure 5. The conductivity enhancement is attributed to Lewis acid-base type surface interactions of ionic species with O/OH groups at the filler surface, with an additional contribution below 60 °C coming from the retention of an increased fraction of the amorphous phase due to the presence of the filler. The curves of ionic conductivity *versus* filler concentration exhibit two conductivity maxima. This can be explained in terms of the surface interactions, blocking effect and grain consolidation. The geometrical restrictions on the long polymer chains of the bulk polymer at moderately higher filler concentrations would give rise to the first conductivity drop. The subsequent conductivity increase is quite likely to come from the interconnected high conducting pathways due to the close proximity of the filler grains. At still higher filler concentrations, high volume fraction of inert fillers reduces the conductivity, giving rise to the observed second conductivity maximum. Below 60 °C, and for certain filler concentrations, an additional contribution to reduce the ionic conductivity may come from the formation of crystallites. The melting temperature of the $(PEO)_9LiCF_3SO_3$ system changes appreciably due to the addition of the nano filler and passes through a minimum with increasing filler concentration with the exception for the porous grains. This minimum generally corresponds to the filler concentration range of the second conductivity maximum. The conductivity enhancement appears to saturate beyond 100 m^2/g grain surface area [11].

Perhaps, the main contribution to enhancement of ionic conductivity is due to the existence of terminal hydroxyl groups at the surface of Al_2O_3 nano particles. They interact with the lithium ions and promote the movement of lithium ions in the composite. When the hydroxyl groups are methyl-capped, it presents no evident effects on its nanocomposite with PEO as polymer electrolyte whether the surface of alumina is neutral, acidic or basic [12]. The degree of enhancement of ionic conductivity depends on the nature of the filler surface group and decreases in the following series: acidic > basic > neutral > weakly acidic [13].

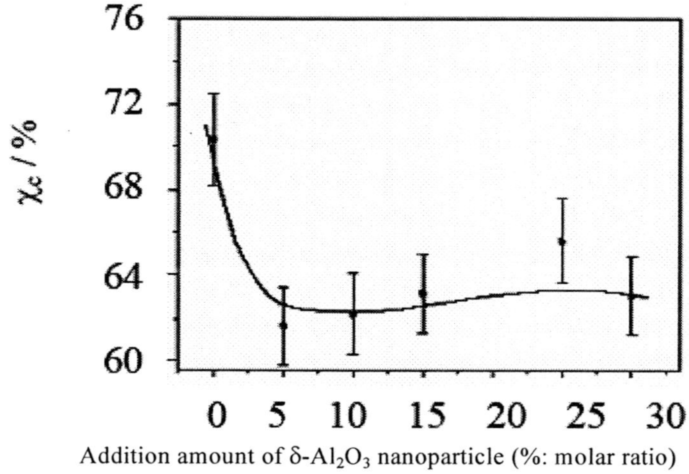

Figure 4. The relationship between the content of crystal phase and the added amount of δ-Al_2O_3 nano particles (from reference [5], reprinted with permission from the Electrochemical Society).

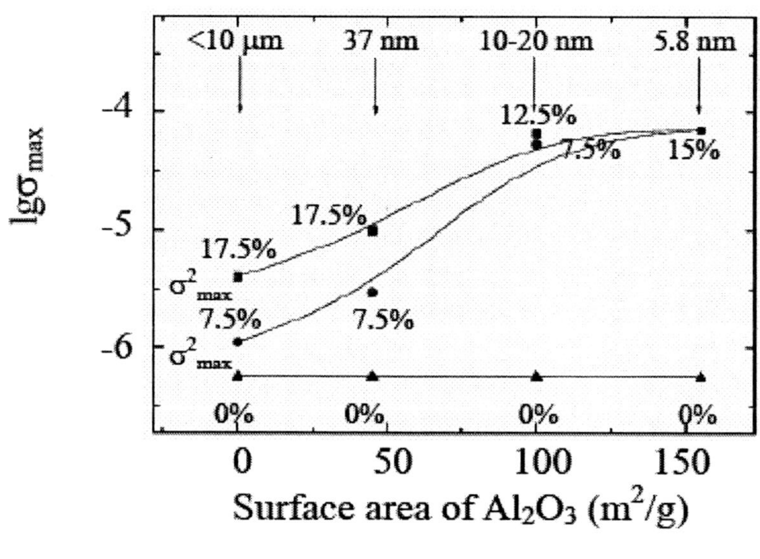

Figure 5. Variation of ionic conductivity at 30 °C with specific surface area of alumina grains for the composite polymer electrolyte $(PEO)_9LiCF_3SO_3$+Al_2O_3 (from reference [11], reprinted with permission from the Elsevier Ltd.).

4. Nanocomposites of PEO with SiO_2 Nanoparticles

Nano silica can be added into the polymer electrolyte by mechanical milling, which lowers the glass transition temperature of the composite polymers and increases the ionic conductivity, which is one order of magnitude greater than that of the un-milled samples [10].

There is an interaction between SiO_2 and polymer component, which is proven by *in situ* synthesis of nanocomposite polymer electrolytes of PEO with SiO_2. As shown in Figure 6, the hydroxyl groups at the surface of SiO_2 interact with the ether-type oxygen atoms on the PEO backbone. Of course, they can react with lithium ions and anions. Consequently, the transportation number of lithium ions increases markedly to 0.56, and that for anions decreases [14].

Silica is usually fumed. The fumed silica has several distinct advantages [15], i.e. (1) its "starburst" particulate shape helps to reduce the crystallinity of the host polymer at low solid loading; (2) its large surface area leads to an open network structure supporting high ionic mobility; (3) its surface groups can be modified to meet a specific need.

When the temperature is below 80 °C, the transportation coefficient of lithium ions is changed after the addition of nano silica. The lower the temperature, the more evident the change [16]. It has been illustrated that the ionic conductivity of polymer electrolytes of PEO with lithium salts is related to the formation history since the formation process greatly affects the content of crystal phases in the polymer electrolytes [2,17]. It will also act in this case. When the temperature is below its melting point, nano silica can improve the mechanical strength and prevent the re-crystallization of segments, leading to an increase of ionic conductivity [18].

As to PEO or PEG with ethered terminal hydroxyl groups, a network can be formed after adding lithium salt, nano silica and methylacrylates monomer. When the substitution alkyl group is longer, an increase in ionic conductivity and tensile modulus is favorable.

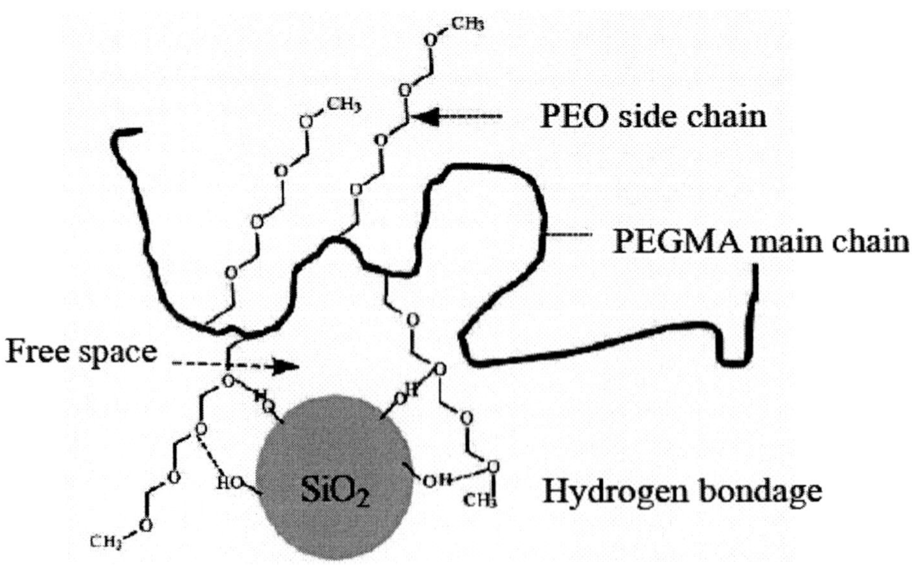

Figure 6. Schematic representation of interaction between polymer chains and surface groups of SiO_2 particles (from reference [14], reprinted with permission from the Elsevier Ltd.).

However, when the amount of the monomer increases, the tensile modulus increases accompanying the decrease of ionic conductivity. When butyl and n-hexyl methylacrylates are added, the stability of interphase between the polymer electrolytes and lithium metal electrode is better than the other methylacrylates [19].

SiO_2 nano filler can also be added into the copolymer of PEO. For example, when it is added into comb-shaped PEO, the melting temperature of amorphous polymer is increased and the ionic conductivity can be up to 1.6×10^{-4} S/cm at 30°C [20].

The surface of the nano silica can also be modified. When the surface of nano silica was modified by a surfactant such as n-dodecylsulfate lithium and the modified nano silica was added into PEO/$LiCF_3SO_3$ polymer electrolyte, the ionic conductivity at room temperature can be up to 5×10^{-5} S/cm at the level where the amount is enough to cover a layer at the surface of the nano silica [21]. When the surface of 7 nm SiO_2 particles is covered by trimethylsilyl, the dispersion of nano fillers is greatly improved though it presents no influence on ionic conductivity. The optimal amount was 12 wt.% [22]. However, some groups introduced on the surface of SiO_2 nanoparticles are favorable for the decrease of crystallization degree and the increase of ionic conductivity [15].

Whether the surface of silica is hydrophilic or hydrophobic, they can increase ionic conductivity. However, the detailed mechanism is different [23]. As for hydrophilic SiO_2, the increase in ionic conductivity is attributed to the enhancement in the amount of the amorphous phase (favoring the increase of the ion mobility) as well as to the Lewis acid-base type ceramic-electrolyte interactions (favoring the increase of the number of charge carriers). While in the case of hydrophobic SiO_2, the increase in ionic conductivity may be mostly attributed to the former. However, the degree is different. At 30°C, the ionic conductivity of the former is lower than that of the latter, which seems to indicate that the ion mobility (related to the degree of the amorphous phase in the composite system) plays a predominant role in influencing the ionic conductivity. At higher temperature (above 40°C), the ionic conductivity of the former is higher than that of the latter, and the number of charge carriers plays a more dominant role in influencing the ionic conductivity [23].

Silica can also be added in the form of mesoporous filler. For example, when mesoporous SBA-15 is added into the polymer electrolyte of PEO-$LiClO_4$, it favors the dissociation of lithium salts, the increase of ionic conductivity and the inter-dissolution of organic and inorganic phases. The main reason is that both the surface hydroxyl groups and the inner channels can provide transportation routes for lithium ions. Figure 7 shows the ^7Li nuclear magnetic resonance spectra of the polymer electrolyte of PEO/$LiClO_4$ containing different amount of SBA-15. It presents that there are three kinds of lithium ions except ion pairs in the virginal polymer electrolyte. Li^{Out} represents the lithium species on the outer surface (0.5–0.7 ppm); Li^{In} represents the lithium species inside the pore channels of SBA-15, which is situated at higher field (1.1–1.5 ppm) and corresponds to those interact with the surface groups of mesoporous SBA-15; Li^{PEO} represents lithium species in the amorphous PEO phase (0.8–1.0 ppm). When the amount of mesoporous SBA-15 increases, the relative contents of Li^{Out} and Li^{In} increase and those of Li^{PEO} decrease. This clearly shows that the added SBA-15 favors the "hopping" of lithium ions in and at the surface of the filler. The best amount of mesoporous SBA-15 is about 10 wt.%. If PEO is heated above its melting point, a large degree of disorder is produced. After cooling, the disorder phase is kept and ionic conductivity is increased [24]. The inner of the mesoporous silica can be modified. For example, as schematically shown in Figure 8, after adding room temperature melt salt, the channel for transportation of lithium ions is increased and ionic conductivity is further increased [25].

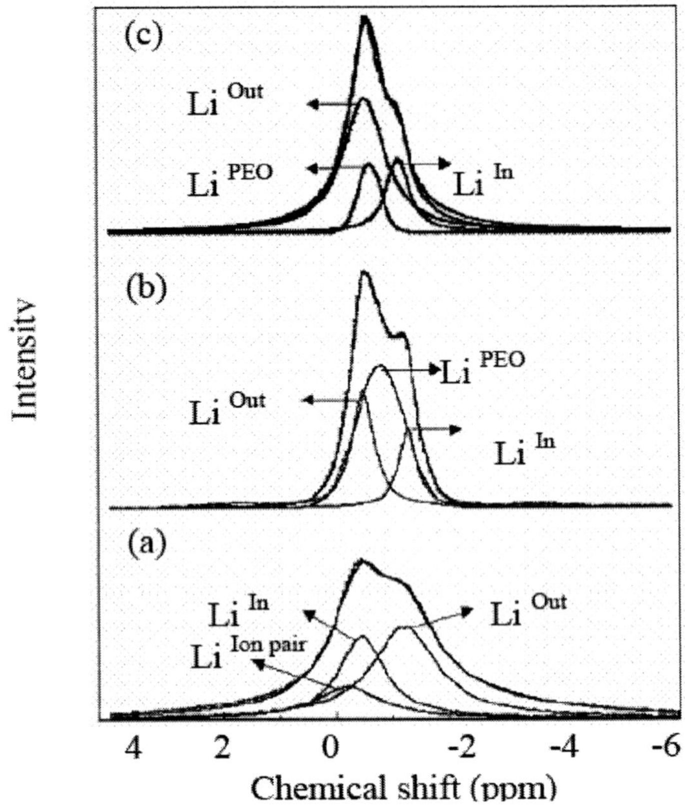

Figure 7. Least square peak fits of ^{7}Li NMR spectra of SBA-15/$LiClO_4$ composite: (a) 0 wt.%, (b) 8 wt% and (c) 15 wt.% of SBA-15 (from reference [24], reprinted with permission from the Elsevier Ltd.).

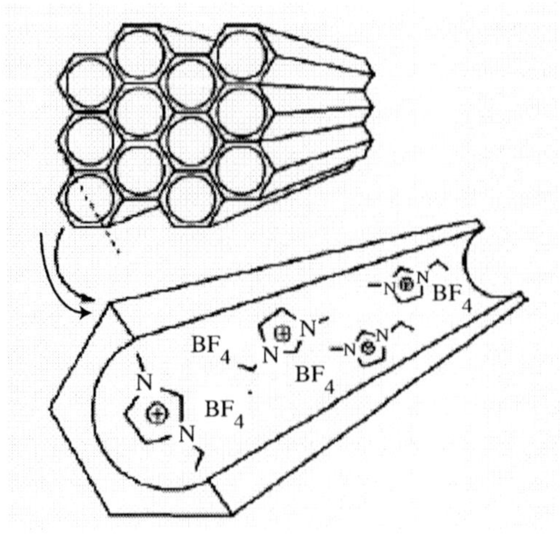

Figure 8. Structural images of neat mesoporous silica domain (up) and $EMImBF_4$ phase in the mesopore (down) (from reference [25], reprinted with permission from the Elsevier Ltd.).

Thus, when 5 wt % 3-methyl-1-ethyl-imidazolyl tetrafluorate has been added, its ionic conductivity increased up to 3 times [26]. From the action mechanism, it is possible that the surface of the filler and mesopores favor transportation and exchange of ions. When the surface of mesoporous silica (MCM-41) is treated to be hydrophilic, its addition of 5-10wt.% into PEO/$LiClO_4$ polymer electrolyte results in an evident increase in ionic conductivity. As shown in Figure 9, lithium ions can move at the surface and in the inner of silicas [27].

When mesoporous ZSM-5 or zeolite is added into PEO-based polymer electrolyte, the ionic conductivity is also increased. The increase is mainly due to the following three reasons: (1) Interaction of Lewis acid-base leads to a higher amount of "free" lithium ions. (2) As shown in Figure 10, since there are some pores with fixed size in ZSM-5, lithium ions can selectively pass through, but anions like ClO_4^-, whose size is large, could not pass through due to electrostatic repulsion of the negative charge in the channels. (3) There are ionic exchange centers in ZSM-5 frame, lithium ions can exchange with these centers and balance the charge in the channels, providing a special transportation route for lithium ions. Consequently, the ionic conductivity at room temperature is increased by 2 orders of magnitudes with large transportation coefficient of lithium ions [28].

Silica can also be doped by alumina, and aluminosilica is thus obtained. After adding aluminosilica nano filler into PEO or PEG of different molecular weight, the filler exists in two phases: nano particles with 20-60 nm and platlet particles of larger size. In the case of low molecular weight PEO, the amount of platlet particles increases with the amount of nano filler. The nano particles do not present pores and the platlet particles show mesopores. In contrast, when the molecular weight of PEO is large such as 100000 g.mol^{-1}, the main phase is platlet particles without pores. The pores in the filler are closed and do not participate the ionic conduction of lithium ions. In this case, when the amount of nano filler increases, ionic conductivity also increases due to an enlarged surface area of the interphase [29].

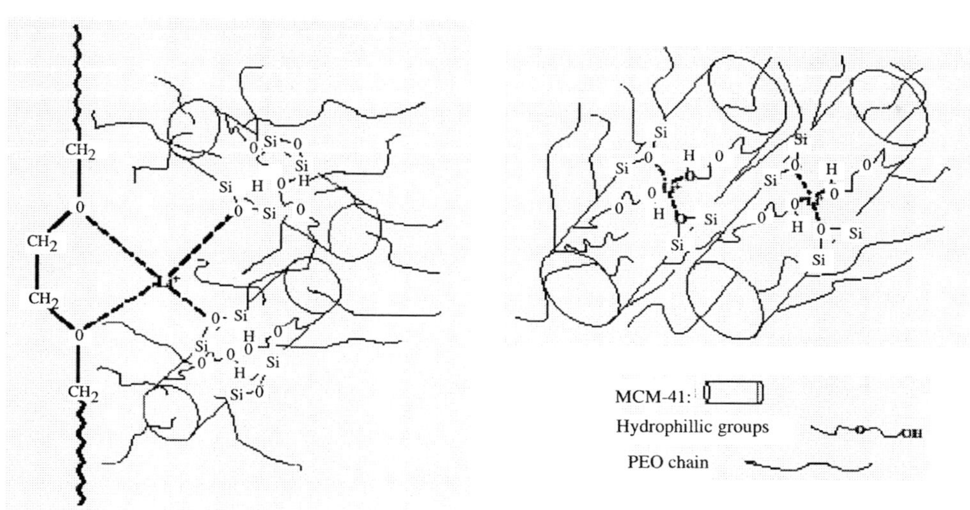

a b

Figure 9. Schematic structures of possible complexes formed by lithium ion with (a) polyether and MCM-41 surface, and (c) MCM-41 mesoporous channels (from reference [27], reprinted with permission from the Elsevier Ltd.).

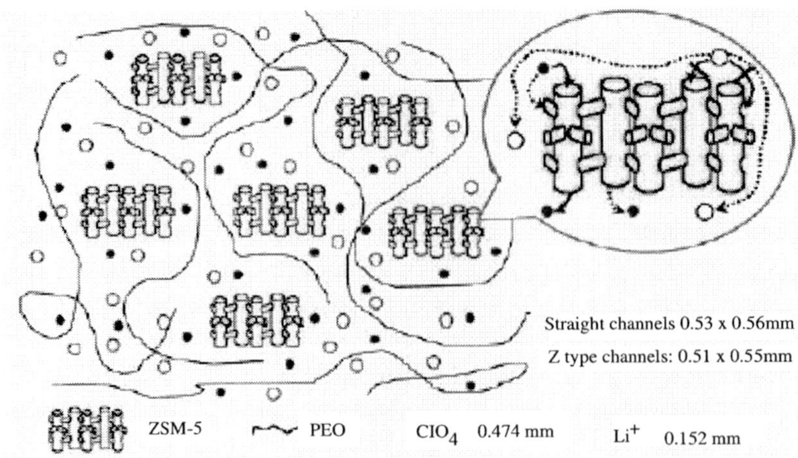

Figure 10. Schematic representation of selective passing of Li^+ ion by ZSM-5 in PEO-$LiClO_4$-ZSM5 composite polymer electrolyte (from reference [28], reprinted with permission from the American Chemical Society).

5. Nanocomposite of PEO with TiO_2 Nanoparticles

The added nano TiO_2 can also decrease the glass transition temperature of PEO, leading to an increase of ionic conductivity [9,10].

Calculation based on ab initio Hartree-Fock (HF) and B3LYP type of density functional theory (DFT) shows that anions can be easily absorbed by the surface of TiO_2 nano particles and bi-dentate complex is formed. This shows clearly that the addition of TiO_2 nano particles may create new diffusion pathways for the lithium ion at the nano-particle surfaces [30]. For example, when nano filler of TiO_2 (21 nm) is added into the amorphous copolymer of EO-PO, there is an interaction between nano filler and polymer, leading to transportation of lithium ions at the interphase between the polymer and the filler. The added nano filler increases the free volume, providing large space for the movement of lithium ions. In the case of the addition of 3.7 nm TiO_2 (5wt.%) into PEO/10wt.% $LiClO_4$, the ionic conductivity increases from 1.03×10^{-5} S/cm to 1.40×10^{-4} S/cm, and the transportation number of lithium ions increases from 0.21 to 0.51[31]. When 40 nm TiO_2 (anatase) is added into low molecular PEO, the structure of PEO does not change basically, which is consistent with the calculation results from DFT and inelastic neutron scattering (INS) [32].

In order to remove the complication of multiple phases presented by the semi-crystal nature of PEO, a completely amorphous polyether-based system has been used to study the effects of nano TiO_2. Multinuclear magnetic resonance spectroscopy shows that the lithium ion environment is changed by the addition of the filler. Vibrational spectroscopy shows that the filler influences the disordered longitudinal acoustic modes in the case of an amorphous polyether and suggests an interaction between the filler surface and the polymer. Positron annihilation lifetime spectroscopy indicates an increase in free volume upon addition of filler to an amorphous polyether-salt complex, coinciding with an apparent increase in polymer mobility as determined from ^1H T_2 NMR measurements [33].

6. Nanocomposite of PEO with Fe_2O_3 Nanoparticles

Nano particles of Fe_2O_3 are also added into the polymer electrolytes as filler. Since Fe_2O_3 has magnetic property, ordered arrangement of polymer structure can be formed if the polymer electrolyte is processed under magnetic field. The order degree of the helic structure for the transportation of lithium ions is markedly improved. Both intra- and interchain ion mobility is enhanced by about one order of magnitude. The magnetic field effect is even more pronounced in polymer electrolytes with incorporated diamagnetic and ferromagnetic nanofillers. The lithium transference number increases from 0.2 to 0.6 in magnetic field oriented polymer electrolytes [34].

7. Nanocomposite of PEO with Layered Compounds

Layered compounds include double hydroxide and montmorillonite (MMT). $Mg_2Al(OH)_6$ is a layered double hydroxide, and the anion OH^- is exchangeable. When oligo(ethylene oxide) is modified by introducing phosphate anion (OLDH), it is used as a nanoscale nucleating agent to add into PEO to form PEO/OLDH nanocomposite electrolytes. Exfoliated nanocomposites are formed due to the surface modification of LDH layers with PEO matrix, which is a compatible oligo(ethylene oxide)s. The nanoscale dispersed OLDH layers inhibit the crystal growth of PEO crystallites and result in a plenty amount of intercrystal grain boundary within PEO/OLDH nanocomposites. The ionic conductivity of nanocomposite electrolytes is enhanced by three orders of magnitude compared to the pure PEO polymer electrolytes at room temperature. It can be attributed to the ease transport of Li^+ along the intercrystal amorphous phase [35].

Due to the existence of abundant anionic silicates on the layers of MMT, a polymer can intercalate into its layers to form nanocomposite. To eliminate the trapping effect of anionic silicates, organic modified MMT hybrids (OMMTs) are prepared as a nanoscale dispersing agent and added into PEO matrix to enhance ionic conductivity [36-39]. Of course, it can also be added into PEO-based copolymers such as crosslinked PEG [40]. In general, the dispersion of layer particles in a polymer matrix can result in the formation of three types of phases as illustrated in Figure 11 [40]. The environment of the lithium ions in the nanocomposite changes with the amount of OMMT.

For example, in the case of a poly(ether-urethane) network polymer (PUN) from crosslinking reaction of PEG with $-(CH_2CH_2O)-$ unit, four complexes are suggested in the composite polymer electrolyte [38]:

Complex 1: OMMT ----$Li^+(ClO_4^-)_n$---- PUN---- PUN ($n \geq 0$)
Weak interaction connecting the PUN matrix and silicate layers (mostly present in phase A).
Complex 2: OMMT ----$Li^+(ClO_4^-)_n$---- PUN---- OMMT ($n \geq 0$)
Strong interaction connecting the PUN matrix and silicate layers (mostly present in phase B).

Complex 3: OMMT----$Li^+(ClO_4^-)_n$----OMMT ($n \geq 0$)
Interaction near the surface of silicate layers (mostly present in phase C).
Complex 4: PUN----$Li^+(ClO_4^-)_n$----PUN ($n \geq 0$)
Coordination within the polymer matrix (mostly present in polymer phase).

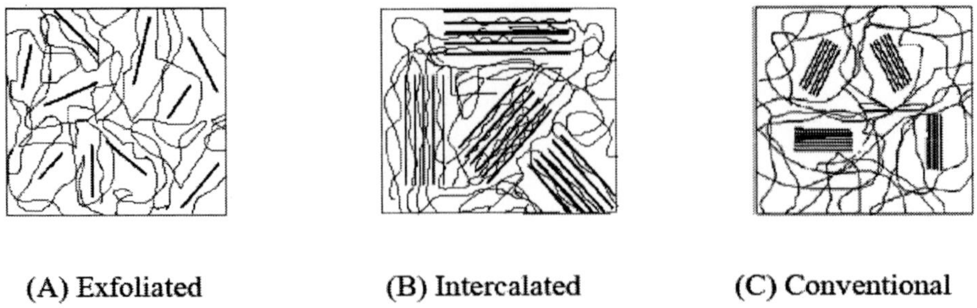

Figure 11. Schematic illustrations of the three phases in the dispersion of OMMT in a polymer matrix (from reference [40], reprinted with permission from the Elsevier Ltd.).

When the content of OMMT is low, OMMT is completely exfoliated in the PUN matrix, and complex 1 mainly exists in the phase A. It acts as a retarder of crystallization, leading to lower crystallinity of PUN. In addition, the formation of the complex 1 also provides extra sites for ionic transport at the surface of the silicate layers. Furthermore, it is known that each silicate layer possesses numerous negative charges, and great numbers of the well-dispersed clay layers can produce huge numbers of dipoles [41]. The negative charges in the silicate layers are able to effectively disperse the salt and impede the ionic aggregation process, and play the same role as the polar group in the polymer matrix [38]. Therefore, the charge environment in the system would contribute to the dissociation of lithium salt and weaken interaction between lithium cation and oxygen atoms in the polymer matrix [42], and leads to an increase in the number of charge carriers.

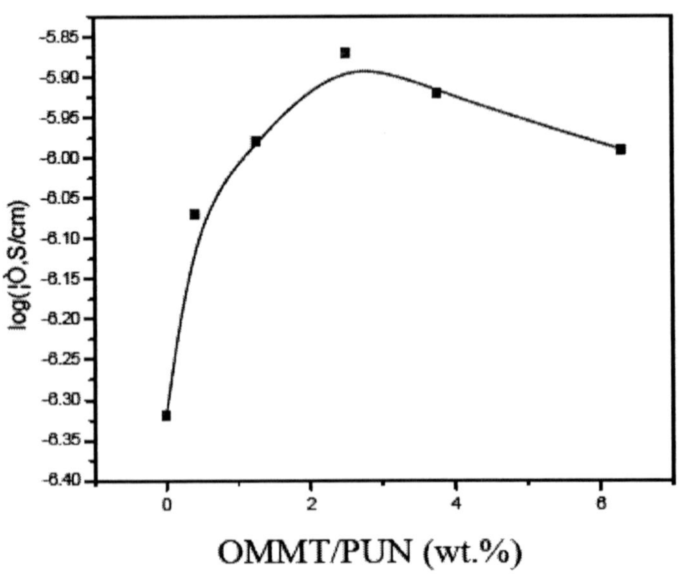

Figure 12. Dependence of the ionic conductivity for PUN/MMMT/$LiClO_4$ composite electrolytes at 30°C on the added amount of MMMT based on PUN, and the ratio of [EO]/[Li] is 12.5 (from reference [40], reprinted with permission from the Elsevier Ltd.).

The overall actions (reduction of crystallinity of the PUN, the extra sites for ionic transport and the increase in the number of charge carriers) lead to the increase in the ionic conductivity. However, an excess of clay tends to form complex 2 in phase B and complex 3 in phase C due to the increasing aggregation tendency. Comparing with the lithium ions in the complex 1, the lithium ions in complexes 2 and 3 are surrounded by more negative charges provided by more silicate layers, and the interactions between lithium ions and silicate layers are stronger. Thus, complexes 2 and 3 clutch the lithium ions tightly, and restrict the mobility of the cations, leading to a decrease in ionic conductivity of the polymer electrolyte. As a result, as shown in Figure 12, the ionic conductivity increases at first with the amount of OMMT. After arriving at a peak, it decreases [40]. The temperature dependence of the ionic conductivity of the nanocomposite polymer electrolyte follows the Vogel-Tamman-Fulcher (VTF) equation.

8. Effects of Nano Fillers on Nanocomposite of PEO for Polymer Lithium Ion Electrolytes

The above results clearly show that the addition of nano particles can present great influence on the nanocomposite as polymer electrolytes. The main actions of the nano fillers can be ascribed to the following aspects: (1) Generally, the degree of crystallization is decreased and the amount of amorphous phase is increased, leading to the corresponding increase of ionic conductivity. (2) Lithium ions can transport *via* the surface of nano fillers, but this kind of transportation depends on the species of nano fillers, host polymers and doped salts. Ionic conductivity can be increased with good stability. The behavior of ionic transportation can be expounded by effective medium theory.

Mechanism of ionic conduction is related to temperature [43]. When temperature is higher or lower than T_g, it will be different. At temperature lower than T_g, mechanism of ionic conduction transits from normal coupling mechanism and polymer segmental movement into decoupling of ions and ionic hopping *via* thermal stimulation, i.e. chain-assisted ionic movement mechanism and ion hopping mechanism [44].

The optimal amount of nano particles in PEO is dependent on the species of nano fillers, size of the nano particles, and their surface structure. By the way, the preparation method of nano fillers affects their surface structures, which present different influences on polymer electrolytes. This is one main reason why there are different reports on the same kind of nano fillers such as Al_2O_3 and SiO_2 [8]. Of course, the data on the ionic conductivity are also related to the species of the polymers, their molecular weight and measurement condition. When the temperature is above the melting point of the polymers, ionic conductivity decreases since the added nano particles stiffen the segments of the polymers and increase the T_g. When temperature is below melting point of the polymers, the movement of segment is not increased, but lithium ions can move *via* the surface of nano particles, leading to the increase of the ionic conductivity [45].

9. Effects of Nanocomposite on Interphase between Polymer Electrolytes and Electrodes

In the case of electrodes, there are two kinds: anode and cathode. As for application, it has been realized that the interphase between polymer electrolyte and electrode is very important. The added nanoparticles mainly influence the interphase for anode, which refers to lithium metal in this chapter.

When polymer electrolytes contact lithium metal, interphase passivation phenomena will happen or an interphase between electrode and polymer electrolyte will be produced. For example, in the case of Li/PEO_8+$LiCF_3SO_3$/Li system, upon increasing of time, the interphase film between electrolyte and lithium metal will continuously grow up [2]. Research results of electrochemical impedance spectra (EIS) show that the existence of impurities such as residual water in the polymer electrolytes will easily form SEI (solid-electrolyte interface) film. However, the SEI film is mainly related to the chemical and electrochemical reactions between polymer electrolytes and lithium metal. Different polymer host will lead to different SEI film and resistance. Of course, the species of ions in the polymer electrolytes and chemical potential of salts also play an important role. The activation energy of the interphase is usually higher than that in polymer electrolyte, and in most cases, the resistance between Li and polymer electrolyte is 2-3 orders of magnitude higher than that of the polymer electrolyte [2].

For dry systems, with no moisture, the surfaces of nanoparticle fillers inhibit the segmental motion and reduce the transportation of lithium ions. The problem of reduced polymer mobility is compounded by the generation of salt concentration gradients within the composite electrode. Highly concentrated polymer electrolytes have reduced transport properties due to the increased ionic cross-linking. Combined with the interfacial interactions this leads to the generation of low mobility electrolyte layers within the electrode and to loss of capacity and power capability. Even with planar lithium metal electrodes, the concentration gradients can significantly impact the interfacial impedance. The interfacial impedance of lithium/PEO–$LiCF_3SO_3$ cells varies depending upon the elapsed time. The behavior is consistent with relaxation of the salt concentration gradients and indicates that a portion of the interfacial impedance, which is usually attributed to the SEI layer, is due to concentrated salt solutions next to the electrode surfaces that are very resistive. These resistive layers may undergo actual phase changes in a non-uniform manner, and reduce the mobility of polymer layers and the initiation and growth of lithium dendrites [46].

After adding nano particles, the interphase and its resistance will be changed. For example, in the case of nano Al_2O_3, compared with the virginal polymer electrolyte, the resistance of the composite polymer electrolyte is much lower than that of the virginal one with increase of storage time at 70 °C, which is shown in Figure 13 [4].

In the case of Fe_2O_3 nano particles, the resistance of SEI in the lithium cells comprising of the oriented polymer electrolytes under magnetic field is much smaller than that of virginal one and that without magnetic field, which is shown in Figure 14 [34].

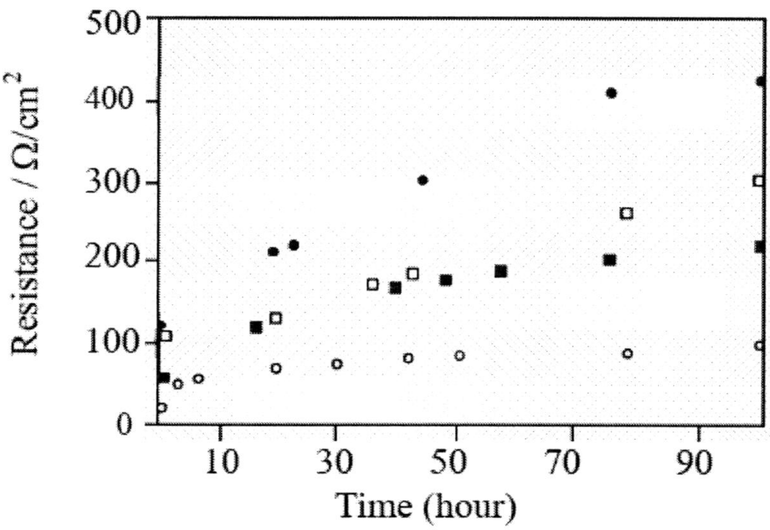

Figure 13. Resistance *vs.* storage time of Li/Polymer electrolyte/Li at 70°C for the $P(EO)_8$-$LiBF_4$ system before and after addition of Al_2O_3 nano particles: (•) $P(EO)_8 \cdot LiBF_4$, (■) $P(EO)_8 \cdot LiBF_4$+10wt.% Al_2O_3 nano particles, (o) $P(EO)_8 \cdot LiBF_4$+20wt.% Al_2O_3 nano particles, and (□) $P(EO)_8 \cdot LiBF_4$+20wt.% Al_2O_3 micro particles (from reference [4]).

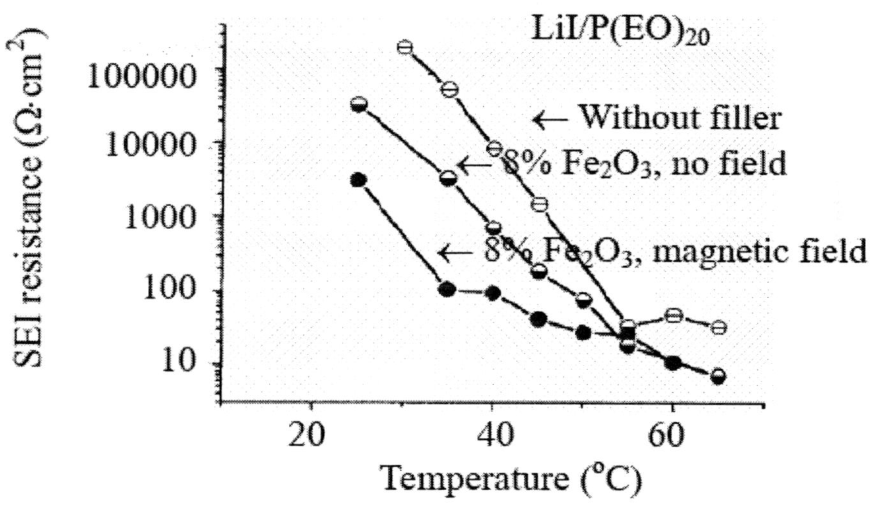

Figure 14. SEI *vs.* temperature plot of the Li/PE/Li cells with $LiI/P(EO)_{20}$ 9%(v/v) Fe_2O_3 polymer electrolytes (from reference [34], reprinted with permission from the Electrochemical Society).

However, the added nano particles do not always improve the interphase performance. As shown in Figure 15, the presence of nano-sized silica in $P(EO)_{20}LiN(SO_2CF_2CF_3)_2$ polymer electrolytes results in no substantial improvement on the electrochemical and interfacial properties of the electrolytes after storage at 90 °C. On contrary, the added nano silica causes higher interface resistance. For the galvanostatic Li stripping/deposition at 0.2 mA/cm^2 at 90 °C, the silica-containing electrolyte shows higher initial overvoltage due to the high interface resistance. But in the case of composite electrolyte with nano $LiAlO_2$, it is favorable for the interphase as shown in Figure 15 [47].

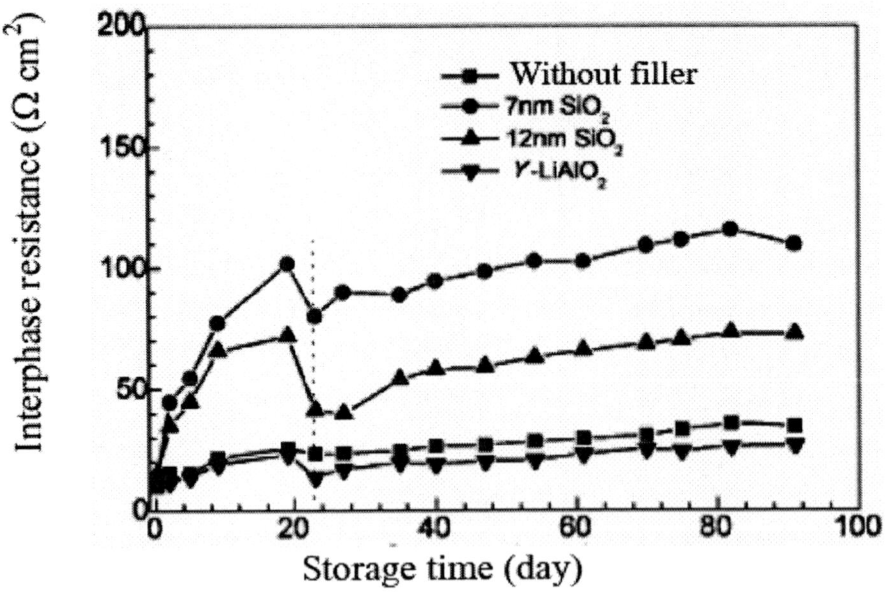

Figure 15. Interphase resistance of $Li/P(EO)_{20}LiN(SO_2CF_2CF_3)_2/Li$, with or without nano particles, *vs.* storage time, at 90 °C (from reference [47], reprinted with permission from the Elsevier Ltd.).

10. CONCLUSION

Nano fillers can surely improve the performance of the polymer electrolytes such as electrochemical stability, ionic conductivity and mechanical strength. Different species, surface structure and particle size present different influences. However, there is still much work to do. For example, in the case of nano TiO_2, its effect is very difficult to model because little is known of which crystal planes are involved or even how the surface is terminated [32]. In the case of fumed nano silica, there is a report different from the above results, i.e. no influence was observed including ionic conductivity, transportation coefficient and diffusion coefficient of lithium ions [48,49]. This needs further investigation for a clear explanation.

At present, PEO and ethylene oxide-based polymers are less than ideal with respect to the interfacial behavior [46]. This implies that polyelectrolyte single-ion conductors will perform better since they do not generate concentration gradients. The question then is whether the polyelectrolytes remain mobile at the surface and do not self-assemble into a non-conducting layer on the electrode surface [2].

Of course, there is an interphase between polymer electrolyte and cathode material, and further study on this aspect is necessary [50].

The practical application of nanocomposite polymer electrolytes in lithium ion batteries and other areas is very promising in the near future. Before that day comes, further investigation, such as rate capability except ionic conductivity and interphase property, is necessary.

ACKNOWLEDGMENT

Financial supports from China Natural Science Foundation (20333040) and Shanghai Science and Technology Committee (04QMX1406) are greatly appreciated.

REFERENCES

[1] Fenton, D. E.; Parker, J. M.; Wright, P. V. *Polymer*. 1973, *14*, 589-590.

[2] Wu, Y. P.; Dai, X. B.; Ma, J. Q.; Chen, Y. J. *Lithium ion batteries: Practice and applications*, Chemical Industry Press, Beijing, 2004, pp 268-298.

[3] Gray, F. M. *Polymer electrolytes*; The Royal Society of Chemistry: Cambridge, UK, 1997, pp 47-78.

[4] Croce, F.; Appetecchi, B. B.; Persi, L.; Scrosati, B. *Nature*. 1998, *394*, 456–458.

[5] Kumar, B.; Rodrigues, S. J.; Scanlon, L.G. *J. Electrochem. Soc.* 2001, *148*, A1191-A1195.

[6] Kumar, B.; Scanlon, L.; Marsh, R.; Mason, R.; Higgins, R.; Baldwin, R. *Electrochim. Acta* 2001, *46*, 1515-1521.

[7] Singh, T.; Bhat, S. V. *J. Power Sources* 2004, *129*, 280–287.

[8] Tambelli, C.; Bloise, A.; Rosario, A.; Pereira, E.; Magon, C.; Donoso, J. *Electrochim. Acta* 2002, *47*, 1677-1682.

[9] Croce, F.; Curini, R.; A. Martinelli, Persi, L.; Ronci, F.; Scrosati, B. *J. Phys. Chem. B* 1999, *103*, 10632–10638.

[10] Ahn, J.; Wang, G.; Liu, H.; Dou, S. *J. Power Sources* 2003, *119-121*, 422– 426.

[11] Dissanayake, M.; Jayathilaka, P.; Bokalawala, R.; Albinsson, I.; Mellander, B. *J. Power Sources* 2003, *119-121*, 409–414.

[12] Marcinek, M.; Ciosek, M.; Zukowska, G.; Wieczorek, W.; Jeffrey, K.; Stevens, J. *Solid State Ionics* 2004, *171*, 69-80.

[13] Jayathilaka, P.; Dissanayake, M.; Albinsson, I.; Mellander, B. E. *Electrochim. Acta* 2002, *47*, 3257-3268.

[14] Liu, Y.; Lee, J. Y.; Hong L. *J. Power Sources* 2004, *129*, 303-311.

[15] Liu, Y.; Lee, J. Y.; Hong, L. *J. Power Sources* 2002, *109*, 507-514.

[16] Suareza, S.; Abbrenta, S.; Greenbaum, S. G.; Shin, J. H.; Passerini, S. *Solid State Ionics* 2004, *166*, 407-415.

[17] Volel, M.; Armand, M.; Gorecki, W. *Macromolecules* 2004, *37*, 8373-8380.

[18] Xie, J. B.; Duan, R.; Han, Y. B.; Ke, J. *Solid State Ionics* 2004, *175*, 755-758.

[19] Yerian, J. A.; Khan, S. A.; Fedkiw, P. S. *J. Power Sources* 2004, *135*, 232-239.

[20] Chung, J.; Sohn, H. *J. Power Sources* 2002, *112*, 671-675.

[21] Matsuo, Y.; Kuwano, J. *Solid State Ionics* 1995, *79*, 295-299.

[22] Kim, J.; Ji, K.; Lee, J.; Park, J. *J. Power Sources* 2003, *119-121*, 415-421.

[23] Wang, X. J.; Zhang, H. P.; Kang, J. J.; Wu, Y. P.; Fang, S. B. *J. Solid State Electrochem.* 2006, in press.

[24] Reddy, M.; Chu, P. P. *J. Power Sources* 2004, *135*, 1-8.

[25] Tominaga, Y.; Asai, S.; Sumita, M.; Panero, S.; Scrosati, B. *J. Power Sources* 2005, *146*, 402-406.

[26] Tominaga, Y.; Asai, S.; Sumita, M.; Panero, S.; Scrosati, B. *Electrochem. Solid-State Lett.* 2005, *8*, A22-A25.
[27] Kao, H. M.; Tsai, Y. Y.; Chao, S. W. *Solid State Ionics* 2005, *176*, 1261-1270.
[28] Xi, J. Y.; Miao, S. J.; Tang, X. Z. *Macromolecules* 2004, *37*, 8592-8598.
[29] Bronstein, L. M.; Ashcraft, E.; DeSanto, P.; Karlinsey, R. L.; Zwanziger, J. W. *J. Phys. Chem. B.* 2004, *108*, 5851-5858.
[30] Johansson, P.; Jacobsson, P. *Solid State Ionics* 2004, *170*, 73-78.
[31] Lin, C. W.; Hung, C. L.; Venkateswarlu, M.; Hwang, B. J. *J. Power Sources* 2005, *146*, 397-401.
[32] Eijck, L.; Best, A. S.; Kearley, G. J. *Macromolecules* 2004, *37*, 9591-9595.
[33] Forsyth, M.; MacFarlane, D. R.; Best, A.; Adebahr, J.; Jacobsson, P.; Hill, A. J. *Solid State Ionics* 2002, *147*: 203-211.
[34] Golodnitsky, D.; Livshits, E.; Kovarsky, R.; Peled, E.; Chung, S. H.; Suarez, S.; Greenbaum, S. G. *Electrochem. Solid-State Lett.* 2004, *7*, A412-A415.
[35] Liao, C. S.; Ye,W. B. *Electrochim. Acta* 2004, *49*, 4993-4998.
[36] Chen, H. W.; Chiu, C. Y.; Wu, H. D.; Shen, I. W.; Chang, F. C. *Polymer* 2002, *43*, 5011-5016.
[37] Chen, H. W.; Chiu, C. Y.; Chang, F. C. *J. Polym. Sci. Part B Polym. Phys*. 2002, *40*, 1342-1353.
[38] Fan, L.; Nan, C. W.; Dang, Z. *Electrochim. Acta* 2002, *47*, 3541-3544.
[39] Fan, L. ; Nan, C. W. ; Li, M. *Chem. Phys. Lett*. 2003, *369*, 698-702.
[40] Wang, X. J.; Kang, J. J.; Wu, Y. P.; Fang, S. B. *Electrochem. Commun*. 2003, *5*, 1025-1029.
[41] Pinnavaia, T. J. *Science* 1983, *220*, 365-366.
[42] Lan, T.; Kaviratna, P. D.; Pinnavaia, T. J. *Chem. Mater*. 1995, *7*, 2144-2150.
[43] Wieczorek, W.; Chug, S. H.; Stevens, J.R. *J. Polym. Sci. Part B Polym. Phys*. 1996, *34*, 2911-2917.
[44] Kumar, B.; Weissman, P. T.; Marsh, R. A. *J. Electrochem. Soc*. 1993, *140*, 320-323.
[45] Zhou, J.; Fedkiw, P. S. *Solid State Ionics* 2004, *166*, 275-293.
[46] Ke, J. B.; Han, Y. B.; Liu, G.; Reeder, C.; Xie, J. B.; Sun, X. G. *Electrochim. Acta* 2004, *50*, 235-242.
[47] Shin, J. H.; Passerini, S. *Electrochim. Acta* 2004, *49*, 1605-1612.
[48] Shin, J. H.; Jeong, S. S.; Kim, K. W.; Passerini, S. *Solid State Ionics* 2005, *176*, 571-577.
[49] Shin, J. H.; Passerini, S. *J. Electrochem. Soc*. 2004, *151*, A238-A245.
[50] Seki, S.; Tabata, S.; Matsui, S.; Watanabe, M. *Electrochim. Acta* 2004, *50*, 379-383.

In: New Trends in Ionic (Co) Polymers and Hybrids
Editor: Ecaterina Stela Dragan, pp. 109-143

ISBN 978-1-60021-611-4

Chapter 5

NEW APPLICATIONS OF FUNCTIONALIZED SOLID SURFACES

Florin Bucatariu[1], Frank Simon[2] and Ecaterina Stela Dragan[1*]

[1] "Petru Poni" Institute of Macromolecular Chemistry,
Aleea Grigore Ghica Voda 41 A, RO-700487 Iasi, Romania
[2] Leibniz Institute of Polymer Research, Hohe Strasse 6,
D-01069 Dresden, Germany

ABSTRACT

Controlling the surface chemical composition and mastering its modification at the nanometer scale are critical issues for high-added value applications. Three main directions of this topic have been followed in the chapter: modification of membrane properties by sequential adsorption of polyelectrolytes, multicomponent stimuli-responsive systems based on chemically or physically modification of solid substrates with appropriate stimuli-responsive polymers, and generation and uses of polyelectrolyte-colloids reactive hybrids. The original contribution of the authors concerning the preparation of reactive hybrids by the adsorption of poly(vinylamine-*co*-vinylformamide) onto silica particles as a function of polycation concentration and the ionic strength, and the post-reactions of crosslinking with epichlorohydrin, and the preparation of oligopeptides on the resulted hybrid were discussed in more detail. The reactive hybrid and the peptide formation on this new substrate were followed by total carbon analysis, electrokinetic measurements and X-ray photoelectron spectroscopy.

Keywords: modified membranes, layer-by-layer self-assembly, pervaporation, transport of ions, pH-responsive systems, thermoresponsive systems, photoresponsive systems, polymer brushes, silica, poly(vinylamine-co-vinylformamide), epichlorohydrin, zeta-potential, S-benzyl-L-cysteine, XPS.

* E-mail: sdragan@icmpp.ro

LIST OF ABBREVIATIONS

AFM	atomic force microscopy
BY	brilliant yellow dye
CHI	chitosan
CPS	counts per second
DMF	N,N-dimethylformamide
DSHA	4,4'-di(6-sulfatohexyloxy)azobenzene
ECH	epichlorohydrin
FMI	fluorescence microscopy image
GOD	glucose oxidase
HSA	human serum albumine
iep	isoelectric point
IgG	immunoglobulin G
LB	Langmuir-Blodgett
LbL	layer-by-layer
LCST	low critical solution temperature
P_R	permeation rate
P2VP	poly(2-vinylpyridine)
PAA	poly(acrylic acid)
PAH	poly(allylamine hydrochloride)
PAN	poly(acrylonitrile)
PCA_5	poly(N,N-dimethyl-2-hydroxypropyleneammonium chloride)
PAZO	poly{1-4[4-(3-carboxy-4-hydroxyphenylazo)benzene-sulfonamido]-1,2-ethanediyl sodium salt}
PDADMAC	poly(diallyldimethylammonium chloride)
PEI	poly(ethylene imine)
PET	poly(ethylene terephthalate)
PLGA	poly(L-glutamic acid)
PLL	poly(L-lysine)
PMAA	poly(methacrylic acid)
PMDA-PDA	poly(pyromellitic dianhydride phenylenediamine)
PMP	poly(4-methyl-1-pentene)
PNIPAAm	poly(N-isopropylacrylamide)
PP	polypropylene
PS	polystyrene
PS-*b*-PAA	poly(styrene-*b*-acrylic acid)
PSS	poly(sodium styrenesulfonate)
PTFE	poly(tetrafluoroethylene)
PVA	poly(vinyl amine)
P(VFA-*co*-VAm)	poly(vinyl formamide-*co*-vinyl amine)
PVI	poly(vinyl imidazole)
PVS	poly(vinyl sulfate)
pzc	point of zero charge
SAM	self-assembled monolayer

SEM	scanning electron microscopy
TEM	transmision electron microscopy
TFA	trifluoroacetic acid
TOC	total organic carbon
XPS	X-ray photoelectron spectroscopy

1. INTRODUCTION

The modification of the surface of a material allows the production of far superior products in terms of increased corrosion resistance, better biocompatibility, improved optical properties and altered electrical/electronic properties. Controlling the surface chemical composition and mastering its modification at the nano-meter scale are critical issues for high-added value applications. The basic applications of surface functionalization range from altering the wetting or adhesion characteristics and improving the nano-particles dispersion in matrices to enhancing the catalytic properties and ordering the interfacial region. The creation of specific surface sites on solid surfaces for selective molecular attachment is considered a promising approach for their applications in nano-fabrication, nano-patterning, self-assembly, nano-sensors, bioprobes, drug delivery, pigments, photocatalysis, LEDs, etc [1].

Organic–inorganic hybrids with well-defined morphology and structure controlled at the nano-metric scale represent a very interesting class of materials. Their unique features and their role envisaged as components of electronic and optoelectronic devices can be exploited in controlled release and bioencapsulation, as active substrates for chromatographic separation and catalysis, as nano-fillers for composite films in packaging and coating, in nano-writing and nano-lithography, etc. A synergistic combination or totally new properties with respect to the two components of the hybrid can arise from nano-structuration, achieved by surface modifications, self-assembling or simply heterophase dispersion. In fact, owing to the extremely large total surface area associated with the resulting morphologies, the interfacial interactions can deeply modify the bulk properties of each component. A wide range of starting materials and production processes have been studied in recent years for the controlled synthesis and characterization of hybrid nano-structures, from nano-particle or lamellar dispersions to mesoporous materials obtained from templating nano-particle dispersions in a continuous, e.g. ceramic precursor, matrix [2].

Self-Assembled Monolayer (SAM) films can provide surface modification that is both uniform and stable. The direct deposition of functionalized self-assembled monolayers is a simple route to get surfaces with tailored chemical and physical properties. SAM modification can also fine-tune monolayer reactivity and provide useful approaches to pattern surfaces which can then be used as templates for further, site specific, surface elaboration. The effective implementation of in situ monolayer chemistry requires reactions that do not compromise monolayer integrity as well as an awareness of the constraints imposed by the organization of the monolayer environment.

Due to their outstanding properties, surface-functionalized polymer latexes proved to offer suitable colloidal supports in many biotechnological applications. A broad selection of functionalities can be installed to the particle interface due to both the availability of a large amount of functional molecular and macromolecular species and versatility of manufacturing protocols. Surface functionalization with appropriate reactive groups has now become a prerequisite in handling polymer particles for biotechnological applications. The immobilization of a hydrophilic and sometimes stimuli-responsive layer as well as the possibility of coating particles with specific ligands (saccharide moieties, lipids, biomolecules) open very attractive potentialities [3].

In this chapter our attention will be focused on three directions: (1) modified membranes by sequential adsorption of polyelectrolytes; (2) multicomponent stimuli-responsive systems and (3) polyelectrolyte-colloids reactive systems.

2. Modified Membranes by Sequential Adsorption of Polyelectrolytes

A membrane is a natural or artificial interface that separates two fluid phases or volumes of different composition, but enables a selective mass transfer between these two phases. The driving forces, which enable mass transfer through the membranes are: the pressure difference Δp, the concentration difference Δc, the difference in chemical potential $\Delta\mu$, the electrical potential difference ΔE (depending on the separation process) [4]. Two different types of membranes are used: porous membranes (with the mean pore diameter larger than 2 nm), and dense membranes (with the mean pore diameter less than 2 nm). A good membrane is characterized by a high flux and a good separation capability. Composite membranes consisting of a very thin, dense separating layer and a thick, highly porous supporting membrane can combine both properties in an optimum manner. The Langmuir-Blodgett (LB) technique and the electrostatic self-assembly of oppositely charged polyelectrolytes have been used to prepare composite membranes with dense separating layers with a thickness on the nano-meter scale. LB separation layers did not provide high permeability and selectivity because they essentially consist of densely packed aliphatic chains with low molecular mobility and flexibility. In the early 1990s, a new preparation route to organize polymer films was reported by Decher et al. [5,7] which is based on the electrostatic layer-by-layer (LbL) assembly of cationic and anionic compounds on a solid substrate. In the LbL technique a charged substrate is dipped into a dilute solution of an oppositely charged polyelectrolyte so that the polymer is adsorbed at the substrate surface and the surface charge is reversed. After washing, the coated substrate is dipped into aqueous solution of an oppositely charged polyelectrolyte and the surface charge is reversed again. A multi-layer assembly is obtained after repeating the adsorption steps. The thickness of the films is controlled by the number of dipping cycles, pH and ionic strength of the polyelectrolyte solutions. The charge of the polyelectrolyte of the outer layer determines the sign of the membrane surface potential. That means, positively or negatively charged membrane surfaces can be produced employing the same polyelectrolytes by variation the number of dipping processes (odd or even cycles). Numerous studies [1-52] and among these several comprehensive reviews have been published [7-9].

The LbL technique has been often applied to the preparation of composite membranes. In this way, supporting membranes hydrophilized by plasma treatment or wet chemical methods were alternately dipped into aqueous solutions of oppositely charged polyelectrolytes. This new type of composite membrane was extensively investigated in, pervaporation separation of alcohol/water mixtures [10-22], separation of gases [23-27], separation of ions in aqueous solutions [28-48], micro-filtration of protein solutions [49-51].

2.1. Pervaporation Separation of Alcohol/Water Mixtures

The performance of a porous membrane is defined by the selectivity (separation factor α) and the permeate flux, *J*. The separation factor α in pervaporation is given by [4]:

$$\alpha = \frac{c_{\mathrm{A}}^{\mathrm{Permeate}} \cdot c_{\mathrm{B}}^{\mathrm{Feed}}}{c_{\mathrm{A}}^{\mathrm{Feed}} \cdot c_{\mathrm{B}}^{\mathrm{Permeate}}} \qquad (1)$$

where A is the faster permeating species. The permeate flux can be calculated using the Hagen-Poiseuille equation [4]:

$$J_V = \frac{V}{F \cdot t} = \frac{r_P^2 \cdot \varepsilon \cdot \Delta P_{tr}}{8 \cdot \eta \cdot d} \qquad (2)$$

where: J_V – the volume flux; V – the volume; F – the membrane area; t – time; ε – the porosity; r_p – the mean pore radius; ΔP_{tr} – the transmembrane pressure; d – the effective capillary length, and η – the dynamic viscosity of the permeate.

Studies on alcohol/water separation and dehydration of organic solvents have been reported by the research groups of Tieke et al. [10-17] and Meier-Haack et al. [18-20].

Krasemann and Tieke [11] investigated a composite membrane with a separation layer formed by poly(allylamine hydrochloride) (PAH) and the sodium salt of poly(styrene sulfonate) (PSS) in the separation of an ethanol/water mixture. After the annealing of composite membrane at 60 °C, the separation capability increased strongly, while the flux decreased. A maximum separation factor α of 70 was found for a feed solution containing 6.2 wt% water. If the feed mixture contained more than 24 wt% water, a rearrangement in the multi-layer assembly took place and the separation capability strongly decreased. Krasemann and Tieke [13-15] studied the separation of a variety of polyelectrolyte multilayer membranes with different charge density ρ_c. The total flux was found to be inversely proportional to ρ_c, while the separation increased with ρ_c. The water flux and the separation capability of the membranes are mainly controlled by the charge density ρ_c, a membrane consisting of poly(ethylene imine) (PEI) and poly(vinyl sulfate) (PVS) having an improved separation factor for the ethanol/water mixture compared with the PAH/PSS membrane [13]. For 20% (w/w) water in the feed solution, a water content of 99.6% in the permeate was reached, which corresponds to an α-value of nearly 10^3 (the PEI/PVS membrane was annealed at 90 °C for 1 h). To improve the separation factor, PEI was replaced by poly(vinyl amine) (PVA) with the same charge density [14,16]. The resulting PVA/PVS membranes showed improved separation characteristics compared with PEI/PVS membranes and the subsequent thermal treatment was no longer necessary (the water content of the permeate being 99.9% or higher). The best separation factor was found for *t*-butanol/water mixture, followed by propanol/water and ethanol/water mixtures. Toutianoush and Tieke [17] found that the PVA/PVS, PVA/polyacrylate (PAA) and PVA/poly(vinyl sulfonate) (PVSu) composite membranes are suited for alcohol/water separation, but at low water content in the feed (< 20 wt%). At high water content, the less hydrophilic PVA/PAA membrane showed the best results. They also studied the influence of pH and neutral salts on the formation of composite membranes. Membranes prepared from polyelectrolyte solutions which contain salt showed a separation factor higher than the membranes prepared from salt-free polyelectrolyte solutions.

Meier-Haack et al. [18] used as support an asymmetric polyamide membrane and, after the adsorption of 4-6 PEI/PAA double layers, they found that the separation factor was considerably improved while the permeate flux decreased. For an ethanol/water mixture containing 10 wt% water, the separation factor was 1400. In agreement with the studies of Tieke et al. [12-14], Meier-Haack et al. [19] found that membranes modified with two weak polyelectrolytes with high charge density (such as PEI/PAA) gave the best separation properties, while the membranes modified with strong polyelectrolytes of low charge density (PSS, poly(diallyldimethylammonium chloride) (PDADMAC), chitosan (CHI)) led to poorer separation. The influence of preparation conditions on the pervaporation separation properties of polyelectrolyte multilayer composite membranes was also studied. Variation of polyelectrolyte concentration in the dipping solution and adsorption time had only a minor effect on the membrane performance. On the other hand, polyelectrolytes with a high molar mass had a high impact on membrane performance. A separation factor α of 2500 was observed for a membrane having four double layers. Best results were obtained when the adsorption temperature was raised from 25 °C to 80 °C. A membrane with two double layers showed a separation factor α of 900 [20].

Budd et al. [21] prepared multilayer membranes by alternating adsorption of CHI and the sodium salt of poly(4-styrene sulfonate) onto zeolites. The multilayer membranes exhibited high selectivity for water over alcohols. The membranes were successfully applied to the selective removal of water by pervaporation in the esterification of lactic acid with ethanol catalyzed by *p*-toluene sulfonic acid.

Sullivan and Bruening [22] obtained polyimide composite membranes by alternating electrostatic adsorption of poly(amic acid) salts and polycations onto porous alumina supports followed by a heat-induced imidization (250 °C for 2 h) of the film. The pervaporation was investigated as a function of crosslinking by varying either the polycation, i.e., PAH, PEI, or PDADMAC, or the number of crosslinkable groups in the poly(amic acid). Maximum crosslinkage and selectivities occured with imidized films prepared from PAH and a poly(amic acid) that contains diaminobenzoic acid in each repeat unit. Such membranes exhibit water/alcohol selectivities of 1100 and 6100 for solutions containing 10 and 90% isopropanol, respectively, and the minimal thickness of the multilayer films still allows these selectivities to occur at fluxes of 11 and 2 kg m^{-2} h^{-1}, respectively. Fluxes were essentially the same for water/ethanol solutions, but selectivities decreased to 100 and 500 for 10 and 90% ethanol.

2.2. Dense Polyelectrolyte Multi-Layer Membranes for Gas Separation

Stroeve et al. [23] alternately adsorbed PAH and PSS on porous polypropylene (Celgard 2400) supports and on nonporous poly(dimethyl siloxane) membranes. They studied the permeabilities of pure CO_2 and N_2 through the coated Celgard and silicone membranes. Permeabilities in coated silicone membrane indicated higher CO_2/N_2 selectivity than the bare membrane at elevated temperatures. Leväsalmi and McCarthy [24] adsorbed PAH and PSS onto poly(4-methyl-1-pentene) (PMP). The composite membranes were 18.000 times better barries to nitrogen than PMP alone. The coated PMP membranes showed significantly higher O_2/N_2, H_2/N_2, and H_2/O_2 selectivities in gas transport than the bare PMP membrane. Van Ackern et al. [25] used three different porous substrates (Celgard 2400, a poly(ethylene terephthalate) (PET) fabric coated with a thin layer of poly(acrylonitrile) (PAN) and a polycarbonate membrane) and adsorbed PAH and PSS, as well as PAH and anionic bipolar amphiphiles at the surface. The PAN/PET membranes with 20 double layers of PAH/PSS reduced the argon flow to 7 % of the initial value, while 60 double layers reduced the flow to 0.1%. If 60 bilayers were deposited on polycarbonate membrane, the flow rate of CO_2 was to 2.4 times higher than that for the other gases, while for PAN/PET membranes coated with 60 bi-layers of PAH/PSS a selectivity of 1.8 was found. Blumstein et al. [26] studied the permeation of oxygen and nitrogen through a Nafion membrane coated with poly(N-octadecyl-2-ethylpyridinium bromide) and PSS. The membrane showed a high selectivity for oxygen, the separation factor being 3.5. Sullivan and Bruening [27] obtained a gas-selective polyimide membrane on porous alumina supports by electrostatic deposition of poly(amic acid) salts and PAH followed by heat-induced imidization of the multi-layer film. FTIR spectroscopy showed that heating of poly(amic acid)/PAH membranes at 250 °C for 2 h completely converts the poly(amic acid) into the corresponding polyimide. Scanning electron microscopy (SEM) revealed uniform films with minimal deposition in substrate pores. The permeability and selectivity coefficients of the tested membranes are the same as those of cast, pure polyimide membranes with micrometer thicknesses. The O_2/N_2 and CO_2/CH_4 selectivity decreased from 6.9 to 1.9, and from 68 to 4.1, respectively, when polyimide thickness was reduced from 50 nm to 25-30 nm by reducing the number of bilayers from 9.5 to 5.5.

2.3. Transport of Ions

Numerous authors investigated the ion permeation through membranes modified with polyelectrolyte multilayers such as: PAH/PSS [12,28-35]; PVA/PVS [36-39]; PSS/PDADMAC [40,41], and PAA/PAH [42].

Several years ago, Tsuru et al. [28,29] investigated the transport of ions through bipolar polyelectrolyte membranes. It was found that bipolar membranes represent barriers for divalent ions in general and can be used for ion separation. In a first study, Krasemann et al. [12] investigated the permeation of aqueous NaCl and $MgCl_2$ solutions through PAH/PSS membranes. It was found that permeation rate for NaCl was 15 times higher than that for $MgCl_2$. In subsequent studies, the transport of NaCl, $MgCl_2$ and Na_2SO_4 was investigated [30]. The separation factor α for Na^+/Mg^{2+} and Cl^-/SO_4^{2-} was 112.5, and 45 respectively. The separation factor α and the permeation rates were influenced by the number of adsorbed layer pairs, the charge density of the polyelectrolytes, pH and the presence of NaCl in the polyelectrolyte solution used for membrane preparation. When the number of adsorbed layer pairs increased, the α values increased too while the permeation rates decreased.

Harris et al. [31] coated the porous alumina supports with two, four, and five bilayers of PAH/PSS. Deposition of five bilayers fully covered the 0.02 μm diameter pores in the support. After diffusion dialysis experiments, the five bi-layers PAH/PSS membranes showed Cl^-/SO_4^{2-} and $Cl^-/Fe(CN)_6^{3-}$ selectivity values of 7 and 310, respectively. Stanton et al. [32] obtained nano-filtration membranes that allow high water flux along with selective ion transport. The membranes composed of 4.5-5 layer pairs of PSS/PAH on porous alumina allow water fluxes of 1-2 $m^3.m^{-2}.day^{-1}$ at 4.8 bar. Rejections of $CaCl_2$ and Na_2SO_4 depended on polyelectrolyte deposition conditions and the composition of the outer layer of the membrane. The rejection of divalent ion increased when the charge of the outer layer of the membrane had the same sign as the divalent ion being rejected. Nano-filtration with mixed solutions of NaCl and Na_2SO_4 yields Cl^-/SO_4^{2-} selectivities of about 30 when the top layer of PSS/PAH membranes was deposited from a solution of high ionic strength. The Cl^-/SO_4^{2-} selectivities in mixed solutions were higher than those determined from single-salt measurements.

Bruening et al. [33-35] studied the anion-transport through the PSS/PAH membranes capped with few layer pairs of polyelectrolytes of high charge density such as PAA/PAH. After the deposition of eight double-layers onto support membrane, the Cl^-/SO_4^{2-} and $Cl^-/Fe(CN)_6^{3-}$ selectivities were 150 and 3000, respectively. If the last adsorbed layer was PAH instead of PAA the Cl^-/SO_4^{2-} selectivity was greatly reduced. This demonstrates that in capped multilayers the high selectivity is largely due to ion rejection at the membrane surface. The crosslinking of pure PAA/PAH membranes through heat-induced amidation increased Cl^-/SO_4^{2-} selectivity up to a factor of 4. The ion rejections have been improved by capping the multilayers with one to three bilayers of poly(phenylene pyromellitamic acid) and PAH. Heating at 150 °C led to partial imidization of the poly(amic acid) and formation of an ultrathin polyimide membrane. The sulfate rejection by this membrane was greater than 99%. Sullivan and Bruening [35] reported the formation of selective, ultrathin (4-9 nm) polyimide skins on porous alumina by alternating polyelectrolyte (PSS, PAH) deposition and subsequent imidization. To fully cover the porous substrate, authors deposited poly(pyromellitic dianhydride phenylenediamine) (PMDA-PDA)/PAH bilayers on the top of five bilayers of PSS/PAH. After heating PMDA-PDA under N_2, a polyimide membrane was obtained. The extent of imidization was controlled to alter ion-transport selectivity. The membranes obtained showed remarkably selective transport of singly over doubly charged ions (Cl^-/SO_4^{2-} selectivities reach values as high as 1000).

Toutianoush et al. [36,37] investigated the ion permeation through PVA/PVS membranes. The composite membranes exhibited very good separation properties because of the high charge density of the polyelectrolytes. They found the permeation rates (P_R) of the four electrolytes decreased in the sequence $P_R(NaCl) > P_R(Na_2SO_4) > P_R(MgCl_2) > P_R(MgSO_4)$, i.e. in the direction of increasing charge density of the ions [36]. Toutianoush and Tieke [37] studied the transport of $CuCl_2$, $LaCl_3$, $BaCl_2$, and $K_4[Fe(CN)_6]$ through PVA/PVS membranes. After the permeation of aqueous $CuCl_2$ through PVA/PVS membrane, the copper ions were incorporated into membrane. The complex formed between copper and amino groups of PVA increased the network density of the membrane and consequently the permeability was generally decreased. Upon repeated alternating permeation of aqueous NaCl and $CuCl_2$ solutions through the PVA/PVS membrane, the P_R values decreased until a constant rate was obtained, when the membrane was saturated with copper ions. In the case of incorporation of ions in the composite membranes, the permeation rates of salts (NaCl, KCl, and $MgCl_2$) decreased, because the incorporated ions further crosslink the membrane [37]. The multilayered PVA/PVS membranes have been used for desalination of aqueous salt solutions, diluted and non-diluted artificial seawater under reverse osmosis conditions [38]. Using 60 layer pairs of PVA/PVS as separating membrane, it was possible to completely reject $MgCl_2$ and $MgSO_4$ from feed solutions. The rejection of NaCl and Na_2SO_4 increased from 84 and 96%, at 5 bar, to 93.5 and 98.5%, at 40 bar, respectively. The permeation flux J increased linearly with the pressure applied. For a membrane of 60 PVA/PVS layer pairs, a flux value of 4 ± 0.2 L m^{-2} h^{-1} was found at 40 bar.

Jin et al. [39] studied the ion-transport of aqueous electrolyte solutions containing NaCl, Na_2SO_4, $MgCl_2$, and $MgSO_4$, and water flux across ultrathin multilayered membranes of polyelectrolytes (PVA and PVS) carried out under nano-filtration and reverse osmosis conditions. The salt rejection increased in the series 1,1- < 1,2- < 2,1- < 2,2-electrolyte, because the permeating divalent ions exhibit stronger electrostatic interactions with the membrane-bound charged groups than the monovalent ones. For $MgCl_2$ and $MgSO_4$, a complete rejection independently from the concentration of the feed solution and the applied pressure was observed, while for NaCl and Na_2SO_4, the rejections were 84 and 96%, at 5 bar, and 93.5 and 98.5%, at 40 bar, respectively.

Farhat and Schlenoff [40] studied the transport of redox-active probe ions through multilayers made from highly charged polyelectrolytes (PSS, PDADMAC). The authors believed that membranes behave as "reluctant" ion exchange membranes: at low solution salt concentration, there were no bulk ion exchange sites in the interior of the film but increasing the salt concentration opened up internal exchange sites. Equilibrium considerations led to the conclusion that, under conditions of high salt concentration and strong polymer ion pairing, the concentration in ion exchange sites is proportional to the solution salt concentration, while the population of redox ions within the multilayer remains constant. As a consequence, the permeating ions are attracted by the ion exchange sites and their mobility is decreased. Significant transport selectivity was observed because the highly charged ions are more strongly affected than the monovalent ones.

Miller and Bruening [41,42] used a variety of polyelectrolyte multilayers as selective films in composite membranes for nano-filtration and diffusion dialysis. The polyelectrolytes used in these studies were PDADMAC, PSS, hyaluronic acid, CHI. The use of polycations and polyanions with lower charge densities allows separation of larger analytes. To separate sugars, Miller and Bruening [41] employed PSS/PDADMAC films, which allow 42% passage of glucose along with a 98% rejection of raffinose and a pure water flux of 2.4 m^3/(m^2 day). PSS/PDADMAC membranes are also capable of separating NaCl and sucrose (selectivity of ca. 10). In nanofiltration experiments [42], the HA/CHI membranes permit a 250-fold greater fractional passage of sucrose. The films prepared from polyelectrolytes with a high charge density showed low swelling and, therefore, a slow solute transport.

Balachandra et al. [43,44] developed two methods to introduce charged groups into multilayer polyelectrolyte membranes. The first method involves partial complexation of carboxylate groups with Cu^{2+} ions during PAA/PAH membrane formation and subsequent removal of these ions after film formation [43]. Cu^{2+}-templated PAA/PAH membranes showed a four-fold increase in Cl^-/SO_4^{2-} transport selectivity compared to pure PAA/PAH membranes deposited under similar conditions. Alternating deposition of PAA partially complexed with Cu^{2+} and PAH on porous alumina supports, followed by removal of Cu^{2+} and deprotonation, resulted in fixed $-COO^-$ sites in these films. To limit film swelling, authors crosslinked PAA-Cu/PAH films by heating under N_2 to form amide bonds through the reaction between the ammonium groups of PAH and the uncomplexed carboxylate groups of PAA. The Cl^-/SO_4^{2-} selectivity of these crosslinked membranes was approximately 610. Changing the capping polymer from negatively charged PAA to positively charged PAH, greatly reduced Cl^-/SO_4^{2-} selectivity, showing that selectivity is highly dependent on surface charge [43]. The second method employed the partial derivatization of polyelectrolytes with photocleavable groups prior to film formation [44]. Thus, PAA was previously modified with 2-nitrobenzyl bromide. Diffusion dialysis experiments showed that PAA/PAH membranes prepared with 23%, 50%, and 65% 2-nitrobenzyl esterified PAA exhibit Cl^-/SO_4^{2-}selectivities of 100, 150, and 170, respectively (after photolysis). The cation selectivity (Na^+/Mg^{2+}) has been improved by modifying PAH with photo-labile 2-nitrobenzyl-oxycarbonyl groups [44].

Hollman et al. [45] prepared highly permeable ion-selective membranes via immobilization of polyelectrolyte multilayer networks within the inner pore structure of a microporous (pore size = 0.2 μm) support. The first layer consisted of covalently bound charged poly(L-lysine) (PLL), establishing a charged support for subsequent adsorption. Multilayer assemblies have been fabricated using poly(L-glutamic acid) (PLGA) and PLL. For comparison of adsorption and separation properties of the resulting materials, the PSS and PVA were used. Constriction of the pore size induced by multilayer formation was monitored through permeability measurements and dextran rejection studies at each stage of the deposition process. Multilayer assemblies prepared at high ionic strength provide excellent separation of divalent co-ions, even in the ultradilute regime.

Lajimi et al. [46] obtained nano-filtration cellulose acetate membranes by LbL deposition of acidic CHI and sodium alginate as the cationic and anionic polyelectrolyte. They found that the permeation rate of salted solutions was higher than that of pure water. The rejection of monovalent salt (NaCl) decreased whereas that of divalent salt ($MgSO_4$) remained constant, so that the retention ratio increased. Increasing the concentration of feed solutions enhanced the selectivity.

Toutianoush et al. [47,48] presented the first studies of ion transport across self-assembled multilayered films of *p*-sulfonato-calix[*n*]arenes and polyelectrolytes. The films were prepared as LbL assemblies of alternating layers of the anionic calixarenes (tetra-*p*-sulfonato-calix[4]arene (calix4), hexa-*p*-sulfonato-calix[6]arene (calix6), and octa-*p*-sulfonato-calix[8]arene (calix8)) and cationic PVA onto porous PAN supports. The permeation rates, P_R, of monovalent alkali-metal chlorides (Li, Na, K, Cs), magnesium chloride, divalent transition-metal chlorides (Ni, Cu, Zn), trivalent lanthanide chlorides (La, Ce, Pr, Sm), and sodium sulfate across the calix4/PVA, calix6/PVA, and calix8/PVA membranes were studied and compared with the corresponding P_R values across a PSS/PVA multilayer membrane prepared under the same conditions. The P_R values of the alkali-metal salts were between 4 and 17×10^{-6} cm s^{-1}, those of $MgCl_2$ and the transition-metal salts were 0.2-1.3 $\times 10^{-6}$ cm s^{-1}, and those of the lanthanide salts were about 0.1×10^{-6} cm s^{-1}. Divalent sulfate ions replaced the calixarene polyanions in the membrane.

2.4. Separation of Proteins

Rieser et al. [49] studied the effect of self-assembled polyelectrolyte complex layers on protein filtration and fouling behavior of modified Celgard membranes. PAA was grafted on the CO_2 plasma-treated Celgard membranes. Subsequently, a single layer pair of PDADMAC/PAA was adsorbed. The composite membranes showed a lower tendency for protein adsorption (rejection > 90%) and fouling than the unmodified membrane.

Müller et al. [50,51] used *in-situ* attenuated total reflection (ATR)-FTIR spectroscopy to study CO_2 plasma-treated polypropylene (PP) films modified with PEI/PAA multilayers. Human serum albumine (HSA) adsorption experiments revealed a strong decrease of fouling for the PP films, which were modified with polyelectrolyte multilayers, in comparison to those unmodified. The interactions of HSA, lysozyme, and immunoglobulin G (IgG) with multilayer assemblies were then investigated [51]. Using ATR-FTIR spectroscopy, low amounts of adsorbed protein were found if the top polyelectrolyte layers and the protein were equally charged and this showed the important role of electrostatic interaction as the driving force when proteins adsorb at charged surfaces.

3. Multicomponent Stimuli-Responsive Systems

The stimuli-responsive polymers can be chemically or physically deposited onto solid supports to obtain smart surfaces or interfaces. The behavior of a solid surface with respect to surrounding medium is essential for many applications in industry, biology, medicine and environmental science (permeation switches, protein-resistant bio-surfaces, chromatographic devices, etc.) [53-93]. Polymer brushes are the systems in which the chains of polymer molecules are attached with one end through an anchor group to the surface, in such a way that the graft density of the chains is high enough to be stretched away from the surface in a brush-like conformation. In most cases, the interactions between substrate and polymer are mainly van der Waals forces, hydrogen bonds and electrostatic interactions. Hence, such polymer brushes exhibit thermal and solvolytic instability. Covalent bonds formed between the surface and the polymer chain make the brushes resistant to environmental conditions. According to Zhao and Brittain [53], the covalent attachment can be classified into "grafting from" and "grafting to" approaches. The "grafting from" approach consists in an *in-situ* polymerization of precursor monomers onto solid surfaces (surface initiated polymerization). The "grafting to" process uses specifically synthesized end-functionalized macromolecules that are chemically attached to appropriate reactive sites on the substrate. This part of the chapter is focused on the multicomponent stimuli-responsive systems generated by either polymer brushes attached on the surface or LbL multilayers.

3.1. pH-Responsive Systems

Sevick and Williams [54] described theoretically the operation and fabrication of a pressure-sensitive switch or valve using (non-electrolytic) polymer brushes. The micro-valve is controlled by the swelling behavior of the polymer brush. For brushes which partially fill the gap, the brush pair acts as a constant discharge micro-valve. For brushes which nearly fill the gap, the brush pair behaves as a cutoff micro-valve. In addition, a theoretical study predicted controlled transportation through a polyelectrolyte-grafted porous membrane [55]. Kontturi et al. [56] proposed a simple theoretical model describing the effects of pH on the permeability and validated it experimentally for the case of a poly(vinylidene fluoride) membrane grafted with a small amount of PAA.

Ito et al. [57] grafted poly(methacrylic acid) (PMAA) onto a porous polycarbonate membrane (Nuclepore membrane: 200 nm average pore diameter, DuPont). The rate of water permeation through the grafted membrane was found to be less than that of the non-grafted membrane, and was dependent on pH. The rate was high at low pH, and nearly zero at neutral pH. PMAA chains were contracted at low pH (opened pores) and extended at higher pH (covered pores). The authors showed that thus the water transport through the membrane was regulated by pH. In subsequent studies, Park et al. [58] grafted PAA onto a porous glass filter in order to construct a pH-dependent system for the control of liquid permeation. The water transport through the PAA-grafted glass filter was high at low pH, and low at high pH. The permeation rate of a water/acetone mixture through the grafted glass filter was also determined by pH. Zhang and Ito [59] studied the filtration of water solutions of ionic (oligodeoxyribonucleotide) and non-ionic (poly(ethylene glycol)) trough a gold-coated nano-porous membrane grafted with cysteamine-modified PAA. It was found that water permeability through the membrane was reversibly regulated by pH and ionic strength. The permeability was high at low pH and low at high pH, and an increase in ionic strength increased the permeability at high pH. The permeability of the solutes depended on their molecular weight.

Ito et al. [60-62] fixed polypeptide brushes onto different porous membrane. The PLGA was grafted onto porous poly(tetrafluoroethylene) (PTFE) [60] and onto a gold-coated polycarbonate membrane surface [61,62]. The authors studied the effects of pH and ionic strength on water permeation rate. The water permeation rate through grafted polymer membrane depended on pH (slow at high pH and fast at low pH), on the length and density of the grafted chains (decreased with increasing degree of polycondenzation), and on the ionic strength (strong dependence at high pH and small effect at low pH). Sui and Schlenoff [63] coated fused silica capillaries with a diameter of 50 μm with multilayers made from PSS, PDADMAC, poly(vinyl imidazole) (PVI), quaternized PVI, and a random copolymer of PDADMAC-*co*-PAA. Using multi-layers with pH-responsive polyelectrolytes, they controlled precisely the rate and direction of electro-osmotic flow within fused silica capillaries.

Hollman et al. [64] investigated the pressure-driven transport of dilute electrolytes in microporous membranes containing terminally-anchored charged polypeptides. The membrane pore structure was modified with PLGA, poly(L-arginine) (PLA) and PLL. The authors examined the effects of solute type, concentration, pH, polypeptide loading and pore coverage by the attached macromolecules, on the solute rejection and hydraulic permeability. The solute rejection of environmentally-toxic species, such as divalent oxyanions of As(V) and Cr(VI), using a PLGA functionalized silica support (pore size ca. 100 nm), was higher than 80% at a pressure of 0.7 bar.

3.2. Thermoresponsive Systems

Iwata et al. [65,66] grafted poly(N-isopropylacrylamide) (PNIPAAm) onto a porous membrane. It was found that water permeation is high above the lower critical solution temperature (LCST) of the grafted polymer and low below the LCST. Below the LCST the

PNIPAAm chains extended and closed the pores and above LCST the chains shrank and opened the pores. Temperature-sensitive micro-filtration membranes exhibited reversible temperature-dependent permeability to calcein and fluorescein isothiocyanate-dextran in phosphate buffer solutions (pH 7.4), with the most drastic change in permeability being observed in the temperature range between 27 and 32 °C. Yamaguchi et al. [67] made a molecular recognition ion gating membrane using PNIPAAm with a crown ether receptor. They studied the transport through grafted membrane of the water containing Ca^{2+} and Ba^{2+}. A smart gating system has been achieved at 40 °C, which can perform "on" and "off" functions in response to specific chemicals.

Baldi et al. [68] obtained a new hydrogel-based micro-valve. They used a silicon membrane having an array of orifices (140 μm diameter) on which two different hydrogels, one based on *N*-isopropylacrylamide (temperature-sensitive), and another based on phenylboronic acid (pH and glucose-sensitive), have been polymerized. In the swollen state, the hydrogels completely occupied the void space of the orifice, completely blocking the fluid flow. In the shrunken state, the hydrogel is contracted and allows the fluid to flow through the opened annular gap. The measured response times were 10 s (for temperature change), 4 min (for pH change), and 10 min (for changing the glucose concentration).

3.3. Photoresponsive Systems

The macromolecular backbones, containing photochromic moieties, can change their conformational state after photoisomerization. Park et al. [69] grafted PMAA containing spiropyran on a glass filter to control the permeation of organic liquids by photoirradiation. Toluene permeation through this grafted glass filter can be increased by ultraviolet-light irradiation, and decreased by visible-light irradiation. *N,N*-dimethylformamide (DMF) permeation did not depend on the irradiation. The authors demonstrated that the permeation of organic liquids through a glass filter can be controlled by the extension/contraction of polymer chains grafted onto the surface.

Wu et al. [70] produced uniform photoresponsive multilayer films using azo group containing polyelectrolytes by the LbL technique. They used a series of azo-polyelectrolytes derivated from the poly(acrylic acid) having different types of azo-chromophores and degrees of functionalization. The self-assembled multilayers showed a photochromic effect upon 365 nm UV light irradiation. The isomerization rates depended on both the azo chromophore structure and the local environments of the condensation states. The photochromic studies showed that the photoisomerization of the azopolyelectrolytes reached the photo-stationary states in less than 100 s. Pyrash and Tieke [71] prepared Prussian blue/PAH multilayer films onto a quartz and glass supports. The photoresponse of a film prepared upon 16 dipping cycles at an applied potential of 1 V showed that switching on and off the white light from a 500 W lamp is accompanied by repeated generation and disappearance of a photo-current.

Zucolotto et al. [72-74] have investigated the nano-scale interactions between adjacent layers of LbL PAH/azodye brilliant yellow (BY) and PAH/azopolymer films. The films were used for optical storage and formation of surface-relief gratings. It was found that the electrostatic interactions were responsible for the slow kinetics of writing in the optical storage experiments, due to a tendency to hinder photoisomerization and the subsequent reorientation of the azochromophores. The presence the electrostatic interactions prevented thermal relaxation of the chromophores after the writing laser was switched off, leading to a very stable written pattern.

Dragan and Schwarz [75,76] studied the formation and stability of some polycation/azo dye multilayers formed by LbL adsorption of a polycation of integral type (N,N-dimethyl-2-hydroxypropyleneammonium chloride) (PCA_5) and four azo dyes (Crocein Scarlet MOO, Ponceau SS, Direct Blue 1, Direct Red 80). The multilayer growth was followed by UV-vis spectroscopy and by streaming potential measurements. In these studies, the influence of the dye structure on the polycation/azo dye multilayer growth has been clearly observed. From UV-vis spectroscopy and streaming potential measurements it was concluded that the dyes were mainly bound by the electrostatic interactions, the hydrophobic interactions being also possible as a driving force in the multilayer growth. The streaming potential measurements demonstrated the high stability of these assemblies.

Saremi et al. [77,78] studied multilayers of azobenzene bipolar amphiphiles and polyelectrolytes. Self-assembled multilayers were prepared by alternately dipping a quartz substrate, silanized with (3-aminopropyl)trimethoxysilane, into aqueous solutions of the disodium salt of 4,4'-di(6-sulfatohexyloxy)azobenzene (DSHA) and a cationic polyelectrolyte. Samples containing DSHA and PAH or PEI had a strong and narrow absorption band with a maximum at 314 nm, while samples of DSHA and PDADMAC or CHI showed a broad absorption with a maximum at 340 nm. The isomerization was reversed upon irradiation with visible light with $\lambda > 450$ nm or annealing of the multilayers at 95 °C.

Ziegler et al. [79] prepared ultrathin polyelectrolyte multilayers with azobenzene-containing bipolar amphiphiles or ions. In the initial films, the distribution of the azobenzene units was azimuthally isotropic. The photoorientation process of the photochromic units was studied by irradiation with linearly polarized visible light. This procedure led to a significant dichroism due to the photoorientation of the photochromic units, where the long axes become perpendicularly oriented to the electric field vector of the incident light. A dichroism of about 0.34 was induced for a sample of PDADMAC with DSHA. The results demonstrated a restricted mobility of the azobenzene units in the LbL assemblies caused by aggregation, molecular orientation and/or electrostatic interactions of the dye's charged groups with the polyelectrolyte matrix.

Dante at al. [80] investigated polyionic films containing photoactive azobenzene groups. As photoactive polyanion, authors used poly{1-4[4-(3-carboxy-4-hydroxyphenylazo)benzene-sulfonamido]-1,2-ethanediyl sodium salt} (PAZO). PDADMAC and PEI were used as polycations. To monitor the assembly of multilayers, the ellipsometry, X-ray reflectivity, quartz crystal microbalance and UV-vis spectroscopy were used. The photoreaction in the multilayers, induced by UV-light (340 nm), has been followed by the decrease in the intensity of the π-π^* band which was associated with the transition of the trans form of the azo molecules to the cis form.

3.4. Hydrophilic and Hydrophobic Polymer Brushes

It was shown that wetting behavior of solid surfaces can be tuned by composition of random polymers grafted to the solid substrates. Mixed polymer brushes have been intensively investigated theoretically and experimentally [81–93]. The switching mechanism of the polymer brushes can change the morphology, surface energetic state, and functionality upon exposure to controlled environments [84]. For instance, an initially hydrophobic surface will gradually change to hydrophilic in water, responding in this way to the changes of the environment and optimizing its wetting properties. The surface becomes hydrophobic again in contact with a non-polar organic solvent. This behavior resemble with an artificial chameleon, which can adapt its appearance to the changes of surroundings. Husemann et al. [85] obtained continuous films with discrete areas of hydrophilic and hydrophobic polymer brushes. The PAA domains exhibited a high wettability for water, which can be controllable by external agents (e.g., by shifting the pH).

Minko et al. [86-91] synthesized polymer brushes from polymers with different polarity (PS, poly(2-vinylpyridine) (P2VP), PMAA, etc.) using the two approaches, "grafting from" where the polymer has been obtained by polymerization direct on the solid surface [89,90], and "grafting to", where an end-functionalized polymer has been directly grafted to the substrate [86-88,91]. The brushes were sensitive to the surrounding medium, and their morphology and composition have been switched by exposure to different solvents. Thus, surface energetic states and roughness of the film were precisely tuned.

Retsos et al. [92] investigated the adhesive properties of binary heterogeneous polymer brushes made from end-functionalized PS and P2VP chains. The molecular organization of the mixed brush varied reversibly by exposure to selective solvents, toluene for PS, and acidic water for P2VP. The adhesion switching has been tuned by the composition of mixed brushes. It was found a higher sensitivity of the hydrophobic/hydrophilic brushes to polar molecules than to non-polar molecules. Xu et al. [93] assembled perpendicular cylinders by converting an asymmetric diblock copolymer from poly(styrene-*b*-tert-butylacrylate) to poly(styrene-*b*-acrylic acid) (PS-*b*-PAA). After exposure of the PS-b-PAA films to water, PAA cylinders constrained by the continuous, glassy PS phase and swelled laterally to form mushroom caps (entire surface hydrophilic). Upon annealing, the original nano-structures have been reformed, demonstrating reversibility of swelling.

4. Polyelectrolyte-Colloids Reactive Systems

4.1. Short Literature Overview on Some Applications

The high surface area of colloids and the possibility to introduce multifunctional character to the particles, make them suitable candidates for catalysis applications. The activity of enzyme multilayer-coated colloids can be tailored designing the film architecture formed on the colloid surfaces. The total amount of enzyme deposited on colloid particles and the architecture of the enzyme film can be controlled using the LbL technique [94-97].

Caruso and Schüler [94,95] assembled glucose oxidase (GOD), horseradish peroxidase in alternation with oppositely charged polyelectrolytes onto polystyrene (PS) particles. The PS particles coated with the enzyme multi-layers were used as specific enzyme reactors. GOD immobilized on particles was more stable against higher temperatures and pH than free GOD in solution. The enzymatic activity increased with the number of immobilized enzyme layers. Caruso et al. [95-97] used the enzyme containing multilayers also for coating premodified particles with magnetic nano-particles (magnetite). These particles were rapidly and easily separated with a magnet after the enzymatic reactions.

Liang et al. [98] elaborated a core-shell colloid with multilayers for biocatalysis using the LbL technique. Urease was adsorbed in alternation with the oppositely charged polyelectrolytes onto PS colloidal nano-particles. The nano-particles coated with negatively charged urease were found to be more stable than those coated with positively charged urease. The catalytic activity of the immobilized urease was 23.67% of that for the free urease in aqueous solution. The addition of 0.05 M NaCl increased the activity of the immobilized urease up to 65%.

Colloids coated with optical functional materials such as dye-labeled polymers and metal nano-particles can be used as colloids with interesting optical properties [99-101]. Yang et al. [99] constructed biofunctional fluorescent micro-particles for application in immunoassays using the LbL method to deposit multiple layers of fluorescently labeled polyelectrolytes onto colloidal particles, followed by deposition of a protein (IgG) layer. Fluorescence microscopy image (FMI) analysis provided direct measurement of the fluorescence intensity of single micro-particles. FMI analysis showed an increase of the fluorescence intensity of individual micro-particles with increasing polyelectrolyte layer number. The assembly of IgG and fluorescein isothiocyanate-labeled IgG onto polyelectrolyte multilayer coated PS micro-particles and their potential use was ultimately confirmed by a solid phase immunotest. Caruso et al. [100] prepared PS-Au/SiO_2/PDADMAC core shell particles by LbL technique. It was shown that the optical properties of this particles can be modulated via altering the thickness of the multilayer shell.

Rogach et al. [102] obtained 3D colloidal photonic crystals by the self-organization of PS latex spheres covered by the consecutive electrostatic adsorption of charged polyelectrolytes and luminescent CdTe and CdTe(S) semiconductor nano-crystals. The CdTe nano-particle/polyelectrolyte multilayers displayed luminescence properties due to the nano-particles embedded in the polymer coatings. The luminescence maximum was red-shifted from 590 nm, for CdTe(S) nano-particles in aqueous solution, to 595 nm for the nano-particles in the colloidal crystal. They attributed this difference to the close packing of the CdTe(S) nano-particles in the polyelectrolyte layer and the change in environment from water to polymer.

Metal nano-particles are attractive for catalysis because their large surface area-to-volume ratio allows effective utilization of expensive metals. Kidambi et al [103] prepared colloidal particle catalysts by alternating adsorption of PAA and a complex PEI-Pd(II) onto 150-μm diameter alumina particles. The reduction of Pd(II) by $NaBH_4$ yielded catalytic Pd nano-particles embedded in multilayer polyelectrolyte film. They used alumina particles because particles are small enough to be suspended in solution by vigorous stirring and large enough for simple filtration, allowing catalyst recovery. Shi et al. [104] fabricated cobalt oxide nano-tubes (length 3-4 μm, width 200-400 nm) by a colloidal templating approach. In this approach, the Co(III)-cysteinato complex was employed as precursor of cobalt oxide. The precursor was incorporated into polyelectrolyte multilayers deposited onto PS latex particles via LbL deposition. In the yielded composite the colloidal particles became threaded in a pearl necklace-like structure. After calcination, the cobalt oxide nano-tubes were analyzed by transmission electron microscopy (TEM) and atomic force microscopy (AFM).

4.2. Adsorption of Polyelectrolytes onto Silica and Polymer Analogous Reactions on the Adsorbed Layers

4.2.1. Introduction

The adsorption of polyelectrolytes onto various kinds of solid materials [105-108] is a rapid growing field of tremendous importance for colloidal science, biophysics, nano-electronics, etc. Indeed, many practical systems for industrial applications utilize adsorption properties of polyelectrolytes to prepare stabilizing agents, emulsifiers, rheology modifiers, and agents for nano-fabrication. As it has been already mentioned in the previous section, Decher et al. [7] developed a new technique for synthesizing LbL multilayers. The authors used oppositely charged polyelectrolytes that have been adsorbed consecutively from the diluted aqueous solutions onto flat surfaces. Möhwald et al. [109,110] used similar techniques for synthesis of novel polymeric hollow spheres. The strong polyelectrolytes such as PSS and quaternary polyammonium salts, often used in the multilayer construction, are not suitable for further functionalization reactions under mild conditions.

Spange et al. [111-115] obtained stable poly(vinyl formamide-*co*-vinyl amine) [P(VFA-*co*-VAm)]/inorganic oxide hybrid particles. Various procedures such as grafting from, cross-linking surface polymerization and subsequent functionalization of adsorbed P(VFA-*co*-VAm) chains have been used. The P(VFA-*co*-VAm) from aqueous solutions was adsorbed onto silica and titanium dioxide particles [114,115]. The authors used fullerene [114] and (4,4'-diisocyanate)diphenyl methane [115] to irreversibly fix the polyelectrolyte layer onto the particle surfaces. It was shown that P(VFA-*co*-VAm) is a suitable polyelectrolyte for surface functionalization of inorganic particles because after the surface modification, a large number of primary amino groups remain reactive for further functionalization reactions.

In this chapter we discuss in detail the adsorption of P(VFA-*co*-VAm) onto silica particles, the stability of the adsorbed layer under mechanical stress, the cross-linking reaction of the adsorbed P(VFA-*co*-VAm) with epichlorohydrin (ECH) and the coupling reaction of Boc-S-benzyl-L-cysteine with the free amino groups from the surface of the P(VFA-*co*-VAm)/silica hybrid material.

4.2.2. Adsorption of P(VFA-co-VAm) onto Silica Particles

Silica particles suspended in water or in aqueous electrolyte solutions may be considered as a polyelectrolyte. The surface charge of the silica particles is either the result of dissociation processes of the Brønsted-acidic silanol groups (Si–OH), forming negatively charged silanolate ions (Si–O$^-$), or proton adsorption yielding Si–OH_2^+ species. In a wide pH range, the two charge generation mechanisms may take place simultaneously on the silica surface because the acidity of surface silanol groups can be quite different. P(VFA-*co*-VAm) chains also have discrete charges; therefore, an important component of the driving force of P(VFA-*co*-VAm) adsorption onto silica surfaces should be the Coulomb force between the charge centers of the solid surfaces and the polycation. Potentiometric titration in water shows that the point of zero charge (pzc = $pH|_{\Psi=0}$, where Ψ is the potential) is reached at pH = 2.5 for bare silica and at pH = 10.2 for P(VFA-*co*-VAm) (Figure 1).

As mentioned above, the dissociation reactions form negatively charged silanolate ions onto the silica surface. The formation of ammonium groups leads to positively charged P(VFA-*co*-VAm) over a wide pH range (pH < pzc). P(VFA-*co*-VAm) is a weak polyelectrolyte containing a controllable number of reactive and accessible primary amino groups on the polymer backbone. It is an interesting candidate for silica surface functionalization in aqueous solutions [111-118], because it has a high chemical potential for subsequent derivatization reactions with electrophilic agents.

To observe the adsorption process of P(VFA-*co*-VAm) onto silica particles we used electrokinetic measurements, polyelectrolyte and potentiometric titrations as well as the Total Organic Carbon analyzer (TOC). The adsorption of P(VFA-*co*-VAm) and its adhesion are driven by electrostatic forces between the oppositely charged silica surface and the P(VFA-*co*-VAm)'s amino groups. However, the charge degree of P(VFA-*co*-VAm) depends on the protonation/deprotonation equilibrium of the amino groups. The protonation/deprotonation equilibrium can be easily influenced by the pH value of the surrounding aqueous medium. For the adsorption experiments, an aqueous stock solution containing 11.4% P(VFA-*co*-VAm) was diluted to different concentrations. Two series of six samples were prepared. In the first series, the adsorption process was carried out from salt-free aqueous solution and in the second from KCl 10^{-1} mol·L^{-1}. The concentration of the P(VFA-*co*-VAm) varied between 0.97 x 10^{-4} – 2.81 x 10^{-3} mol·L^{-1}.

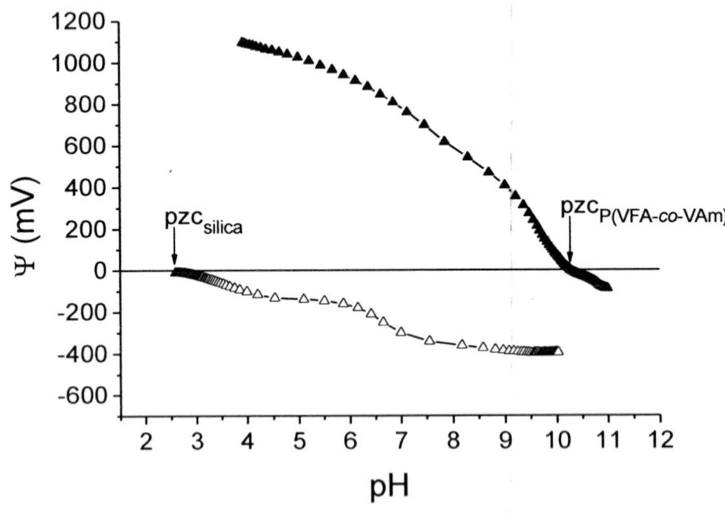

Figure 1. Potentiometric titration of bare silica (8) and P(VFA-*co*-VAm) (7).

Samples of 0.5 g silica were suspended in 50 ml of the diluted P(VFA-*co*-VAm) solutions. During the adsorption process, the suspension was gently shaken at room temperature. The modified particles were filtered off by slight suction. The hybrid materials were carefully washed with distilled water and dried in vacuum at 40 °C for 24 h.

Polyelectrolyte and potentiometric titration were performed with the particle charge detector PCD 02 (Mütek, Germany). For polyelectrolyte titration, a 10^{-3} mol·L^{-1} solution of poly(sodium ethylene sulphonate) was used. Potentiometric titration was carried out between pH 3.5 and 10, using 0.1 mol·L^{-1} KOH and HCl, respectively. TOC analysis is a method of indirectly measure of the polymer amount adsorbed by measuring the amount of carbon in solution. TOC analysis was performed using the TOC 5000-A made by Shimadzu of Kyoto, Japan. TOC oxidizes the sample and measure the peak of CO_2 in the infrared region. TOC analysis is effective in measuring carbon concentrations of over 15 ppm (equivalent to about

0.005% of polymer per sample). The electrokinetic measurements were performed as streaming potential experiments employing an Electrokinetic Analyzer (EKA, Anton Paar, Austria). In a specially designed powder-measuring cell [118], the hybrid materials were packed as diaphragm, which was flown through by an aqueous KCl solution ($c = 0.001$ mol·L^{-1}). The pH dependent measurements start from pH = 6.8. The addition of 0.1 mol·L^{-1} HCl lowered the pH values, while the addition of 0.1 mol·L^{-1} KOH increased the pH values. After recording the streaming potential values of the acidic pH range, the sample was exchanged for measuring the streaming potential values of the basic pH range. The values of the electrokinetic potential (zeta-potential, ζ) were calculated from the measured streaming potential values according to the Smoluchowski's equation [118]:

$$\zeta = (\eta/\varepsilon_0\varepsilon_r)\cdot(\Delta U/\Delta P)\cdot\kappa \tag{3}$$

where: ζ is the apparent zeta-potential; ΔU - the streaming potential; Δp - the pressure; κ - conductivity, η dynamic viscosity of the test liquid; ε_r - relative permittivity of the solution and ε_0 dielectric constant.

Figure 2 shows the influence of the solution pH on the adsorbed amount of P(VFA-*co*-VAm). At low pH, the macromolecules have a high charge density, and as a consequence the chains are stiff and stretched. Hence, the P(VFA-*co*-VAm) amount adsorbed onto silica surface is low. As can be seen in Figure 2, an increased pH value decreased the charge density of the P(VFA-*co*-VAm) molecules. The macromolecules were more flexible and no electrostatic repulsion between the P(VFA-*co*-VAm) molecules could prevent the adsorption process. A higher amount of P(VFA-*co*-VAm) was adsorbed onto the silica surface.

After 48 hours of P(VFA-*co*-VAm) adsorption, the adsorbed amount is nearly the same like after 1 hour (Figure 3), i.e., the majority of pores are inaccessible to the polyelectrolyte molecules.

To study the effect of the P(VFA-*co*-VAm) concentration on its adsorption onto silica, the concentration range was selected in the range ($0.97 \times 10^{-4} - 2.81 \times 10^{-3}$) mol·L^{-1}. The adsorption from salt-free aqueous solution and from KCl 10^{-1} mol·L^{-1} increases with the increase of the P(VFA-*co*-VAm) concentration (Figure 4). When the polyelectrolyte was adsorbed from KCl 10^{-1} mol·L^{-1}, the adsorbed amount was higher than the amount obtained in the case of adsorption from salt-free aqueous solution. KCl is an electrolyte which screens the chain charges, and, as a result, the electrostatic repulsions between P(VFA-*co*-VAm) molecules decreases and the chains are more flexible. In this case, more macromolecules are able to occupy the same unit area on the silica particle surface and, as a consequence, the adsorbed amount is higher.

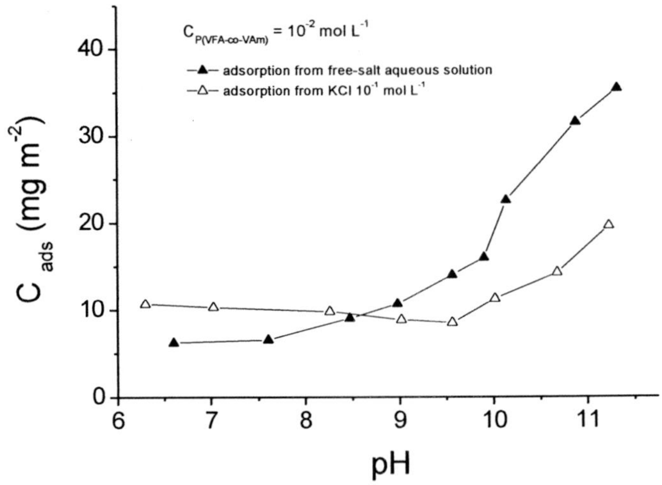

Figure 2. Influence of the stock solution's pH value on the P(VFA-*co*-VAm) amount adsorbed onto silica ($C_{P(VFA\text{-}co\text{-}Vam)} = 10^{-2}$ mol·L^{-1}, adsorption time 3h).

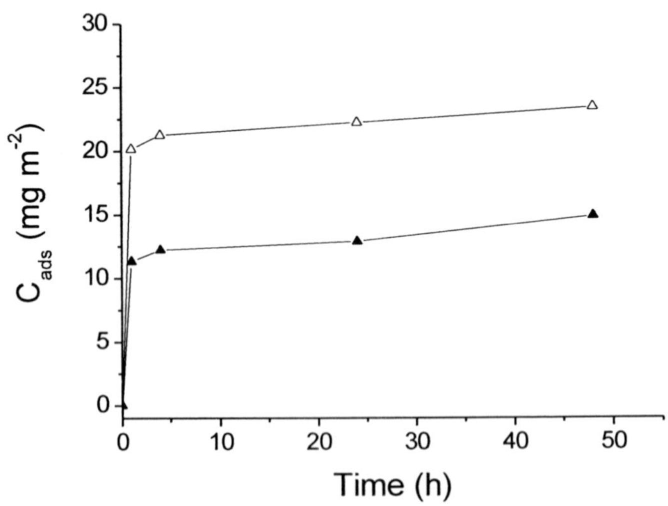

Figure 3. Influence of adsorption time on the P(VFA-*co*-VAm) adsorbed amount onto silica ($C_{P(VFA\text{-}co\text{-}Vam)} = 10^{-2}$ mol·L^{-1}). Adsorption was performed from salt-free aqueous solution (8) and from KCl 10^{-1} mol L^{-1} (7).

The results also reveal that, beyond the polyelectrolyte concentration of 10^{-3} mol·L^{-1}, the adsorption reaches the saturation and evidently a plateau is obtained (adsorption from KCl 10^{-1} mol·L^{-1}).

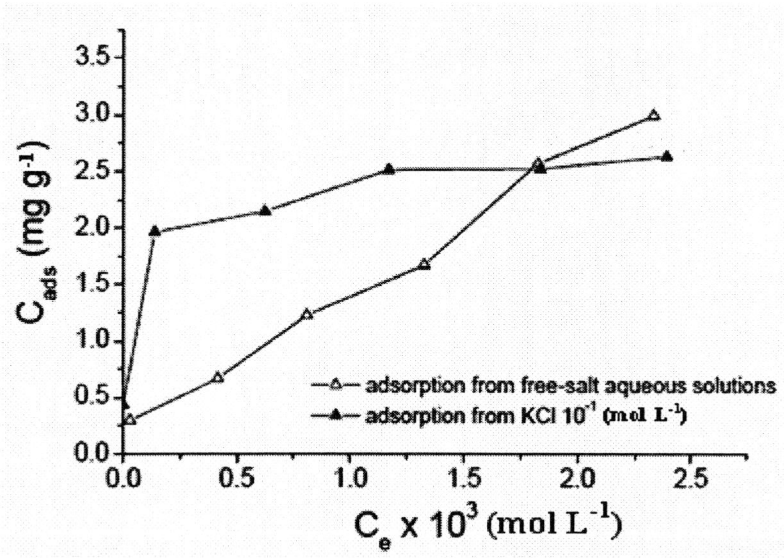

Figure 4. The variation of adsorbed amount C_{ads} of P(VFA-*co*-VAm) with equilibrium concentration C_e of P(VFA-*co*-VAm); silica = 0.5 g; pH ≈ 8; temperature = 23 ± 0.3 °C.

Adsorption data are often presented in adsorption isotherms [119], where, at a constant temperature, the adsorbed amount is plotted against the polyelectrolyte concentration in solution after the adsorption has occurred. The Freundlich and Langmuir equations were applied to the P(VFA-*co*-VAm) adsorption isotherms.

The Freundlich isotherm has the form:

$$C_{ads} = K_f C_e^N \text{ or } \log C_{ads} = \log K_f + N \log C_e \tag{4}$$

where: C_{ads} is the adsorbed amount on the adsorbents; K_f, Freundlich constant which predicts the quantity of polycation per gram of silica at the unit equilibrium concentration, and N, a measure of the nature and strength of the adsorption process and of the distribution of active sites. If $N < 1$, bond energies increase with the surface density; if $N > 1$, bond energies decrease with the surface density, and when $N = 1$, all surface sites are equivalent [119].

The Freundlich equation fits better the experimental data according to the R-squared test ($R^2 = 0.9975$) if the P(VFA-*co*-VAm) is adsorbed from salt-free aqueous solution (Figure 5). The coefficients of the Freundlich isotherm equation are N = 0,69 and K_f = 117.48 (mL·g^{-1}).

The Langmuir isotherm has the form:

$$C_{ads} = \frac{C_{max} K_L C_e}{1 + K_L C_e} \text{ or } \frac{1}{C_{ads}} = \frac{1}{C_{max} K_L C_e} + \frac{1}{C_{max}} \tag{5}$$

where: C_{ads} is the adsorbed amount of P(VFA-*co*-VAm) (mg·g^{-1}); C_e is the equilibrium concentration of the P(VFA-*co*-VAm) in solution; C_{max}, the adsorbed amount of P(VFA-*co*-VAm) (mg·g^{-1}) at saturation (adsorption capacity) and K_L is a constant known as the Langmuir parameter.

If the P(VFA-*co*-VAm) is adsorbed from KCl 10^{-1} mol·L^{-1}, the Langmuir equation fits better the experimental data according to the R-squared test (R^2 = 0.9876) (Figure 6). The coefficients of the Langmuir isotherm equation are C_{max} = 2.51 mg·g^{-1} and K_L = 2.26 x 10^{-5} L·mol^{-1}.

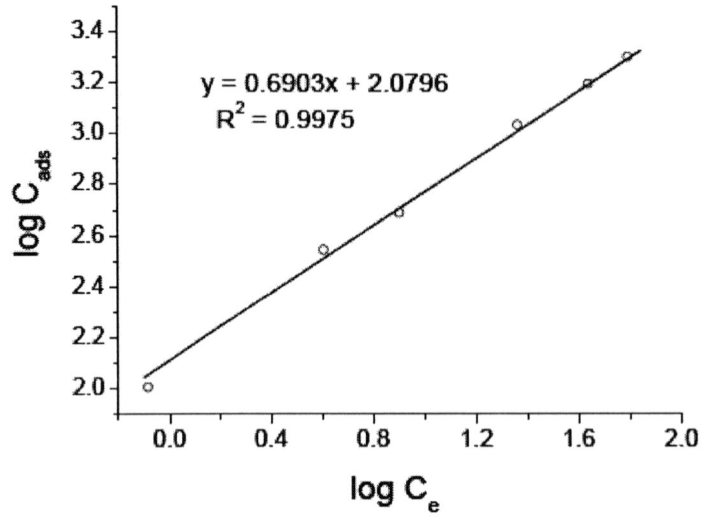

Figure 5. Freundlich plot of the P(VFA-*co*-VAm) adsorption onto silica particles from salt-free aqueous solution.

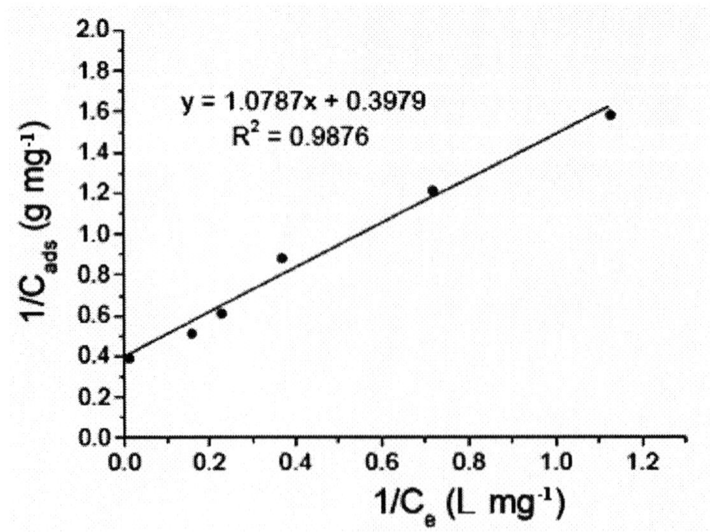

Figure 6. Langmuir plot of the P(VFA-*co*-VAm) adsorption onto silica particles from KCl 10^{-1} mol·L^{-1}.

The three main thermodynamic parameters includes enthalpy of adsorption (ΔH), free energy change (ΔG) due to the transfer of unit mole of solute from solution to the solid-liquid interface, and entropy (ΔS) of adsorption. The thermodynamic parameters of the P(VFA-*co*-VAm) adsorption from salt-free aqueous solutions onto silica particles were determined using the TOC analysis. The standard free energy change ΔG^0 ($kJ \cdot mol^{-1}$) was calculated by equation (6):

$$\Delta G^0 = - RT \ln K \qquad (6)$$

where: K is the equilibrium constant of the adsorption process. The values of ΔG^0 were calculated for two different initial concentrations (C_i) of P(VFA-*co*-VAm) at two temperatures, these values being collected in Table 1.

Table 1. The values of ΔG^0 at two different initial concentrations of the P(VFA-*co*-VAm) at two temperatures

C_i ($\mu g \cdot mL^{-1}$)	C_e ($\mu g \cdot mL^{-1}$)		K		ΔG^0 ($kJ \cdot mol^{-1}$)	
	23 °C	43 °C	23 °C	43 °C	23 °C	43 °C
16.0	7.5	12.5	1.13	0.28	-0.300	3.345
58.5	34,7	45.7	0.68	0.28	0.949	3.345

The standard enthalpy change (ΔH^0) was calculated using equation (7):

$$\ln \frac{K_2}{K_1} = \frac{\Delta H^0}{R} \left[\frac{1}{T_1} - \frac{1}{T_2} \right] \qquad (7)$$

where K_1 and K_2 are the equilibrium constants of the adsorption process at two different temperatures (T_1 and T_2). The values of ΔH^0 were calculated to be –54.41 $kJ \cdot mol^{-1}$, when the initial concentration of P(VFA-*co*-VAm) was C_i = 16.0 $\mu g \cdot mL^{-1}$, and –42.14 $kJ \cdot mol^{-1}$, when C_i = 58.5 $\mu g \cdot mL^{-1}$. The negative value of ΔH^0 indicates that the process is exothermic and the sorption behavior is rather physical in nature. The standard entropy change (ΔS^0) was calculated by equation (8):

$$\Delta G^0 = \Delta H^0 - T\Delta S^0 \qquad (8)$$

The values of ΔS^0 were –182.69 J mol^{-1} K^{-1}, when C_i = 16.0 $\mu g \cdot mL^{-1}$, and –145.43 J mol^{-1} K^{-1}, when C_i = 58.5 $\mu g \cdot mL^{-1}$. It is evident that the enthalpic factor favors the adsorption process ($\Delta H^0 < 0$), but the entropic factor did not ($\Delta S^0 < 0$). Entropy loss shows that the water is not removed from silica surface (water is in the adsorbed layer).

The stability of the adsorbed polyelectrolyte layer can be also affected by the environmental conditions. Investigations of the polyelectrolyte layer stability show that weakly bound polyelectrolyte molecules can be easily removed, but it seems impossible to completely remove the P(VFA-*co*-VAm) layer adsorbed to the silica particles [120]. Figure 7 shows the apparent zeta-potential of the P(VFA-*co*-VAm)/silica hybrid particles in dependence on the pH value of the streaming aqueous KCl solution. At pH > 7 - 8, the ammonium species are gradually deprotonated by the excess of OH^- ions in the aqueous solution and, as a consequence, the zeta-potential value decreased.

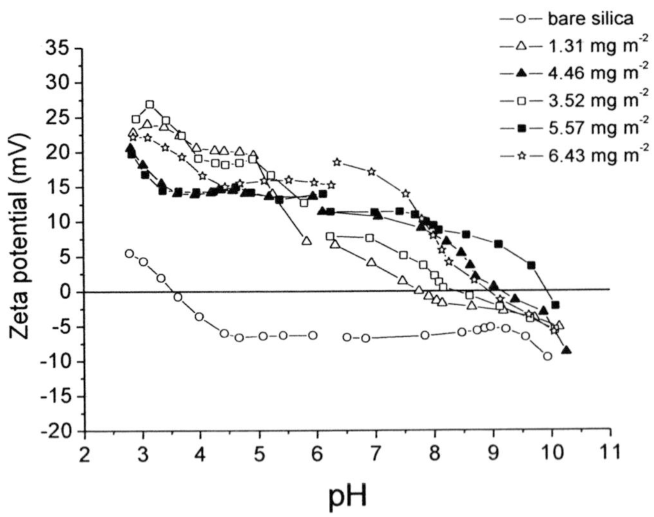

Figure 7. Zeta-potential values determined from streaming potential measurements as a function of pH of an aqueous 10^{-3} mol·L^{-1} KCl solution for different P(VFA-*co*-VAm)/silica hybrid particles [obtained from salt-free aqueous solutions (open symbols) and from KCl 10^{-1} mol·L^{-1} (filled symbols)].

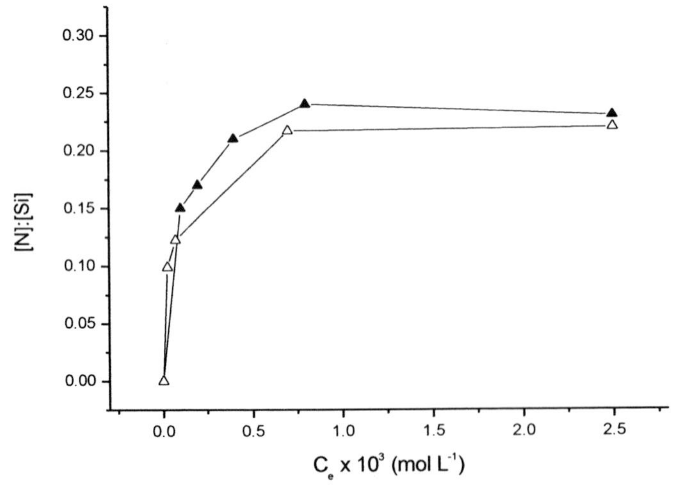

Figure 8. Relative amount of P(VFA-*co*-VAm) adsorbed onto silica surfaces determined by XPS. The amount of the polyelectrolyte is expressed by the nitrogen content [N], and the number of unoccupied silica surface sites is indicated by the silicon value [Si]. Samples were obtained in salt-free aqueous solution (8) and KCl 10^{-1} mol·L^{-1} (7).

With the increasing of the adsorbed amount of P(VFA-*co*-VAm) onto silica particles the isoelectric point (iep = $pH|_{\zeta=0}$) is shifted to a higher value of pH. The iep corresponds to the number of Brønsted base sites on the sample surface. Adsorption isotherms determined from XPS measurements (Figure 8) of the P(VFA-*co*-VAm)/silica hybrid particles showed a clear plateau region of adsorption.

In the case of XPS it can be assumed that the N 1s peak appears only from the polyelectrolyte layer and Si 2p is a specific signal of the substrate material. Then, the [N]:[Si] atomic ratio should be an accurate measure of the adsorbed polyelectrolyte amount.

Nevertheless, for practical applications and subsequent functionalization reactions it is recommended that the P(VFA-*co*-VAm) layer to be stabilized by crosslinking reactions [111]. To irreversibly fix the adsorbed P(VFA-*co*-VAm) layers onto silica, without changing the aqueous medium, subsequent reaction with ECH was carried out. It is expected that this reaction, which occurs between adsorbed P(VFA-*co*-VAm) chains near the silica particle surface, to transform the flexible polyelectrolyte layer into a less flexible one.

4.2.3. Chemical Cross-Linking of Adsorbed P(VFA-co-VAm)

The partial removal of the polyelectrolyte layer from the surface of silica particles was observed in streaming potential experiments. It is expected that, during the experiment, the streaming liquid to generate an additional mechanical stress, which should enhance the abrasion of the polyelectrolyte layer. Figure 9 shows the zeta-potential of P(VFA-*co*-VAm)/silica hybrid material and P(VFA-*co*-VAm)/silica crosslinked with ECH, in dependence on the pH value of the streaming aqueous KCl solution. In the pH range $9 \geq pH \geq 6$, the positive zeta-potential of P(VFA-*co*-VAm)/silica hybrid material remains nearly constant (plateau, region *B*, Figure 9). Here, all amino groups that are able to form positively charged ammonium salt species are protonated. At pH > 9, the ammonium species are gradually deprotonated by the excess of OH^- ions in the aqueous solution and, as a consequence, the zeta-potential value decreased. The isoelectric point (iep = $pH|_{\zeta = 0}$) of the P(VFA-*co*-VAm)/silica hybrid material is 9.8. This indicates that the P(VFA-*co*-VAm) fully covers the silica surface and determines the charging behaviour of the hybrid material. In the acidic range (pH < 6, region *A*, Figure 9), a high content of protonated amino groups is expected and the measured zeta-potential should agree with the potential of the plateau phase. However, at pH ≈ 4 the zeta-potential shows a minimum. This minimum can be explained by the instability of the adsorbed polyelectrolyte layer in an acidic environment. The stepwise addition of HCl results in an increased solvatation of macromolecules, which is the first step of their dissolution. The streaming liquid removes the dissolved polymers, and supports so the dissolving process by changing the adsorption equilibrium to the side of the solved species. In order to proof the assumption mentioned above, the P(VFA-*co*-VAm)/silica hybrid material was measured again in the acidic range (region *A*, the curve with Δ). In this case no minimum of zeta-potential is observed. The same sample, measured two times in the acidic range, was measured in the basic range (region *B*, curve with Δ) and the iep was found to be 8.7, lower than the initial one found at 9.8. This shift of the iep to a lower value shows a partially removal of the polyelectrolyte layer from the silica surface.

To crosslink P(VFA-*co*-VAm) chains, the commercially available reagent ECH was used. It was expected that the adsorbed P(VFA-*co*-VAm) chains would become irreversibly linked to the particle surface. The formation of a two-dimensional network (Figure 10) is desired to effectivelly cover each single particle. The crosslinking reaction was carried out in a two-step synthesis. In the first step P(VFA-*co*-VAm) was adsorbed from a salt-free aqueous solution with the concentration of 10^{-2} mol·L^{-1}. In the second step, the separated particles react with ECH dissolved in acetone:water = 1 : 1. Principally, the chlorine and epoxy groups of ECH can react with amino groups of the P(VFA-*co*-VAm).

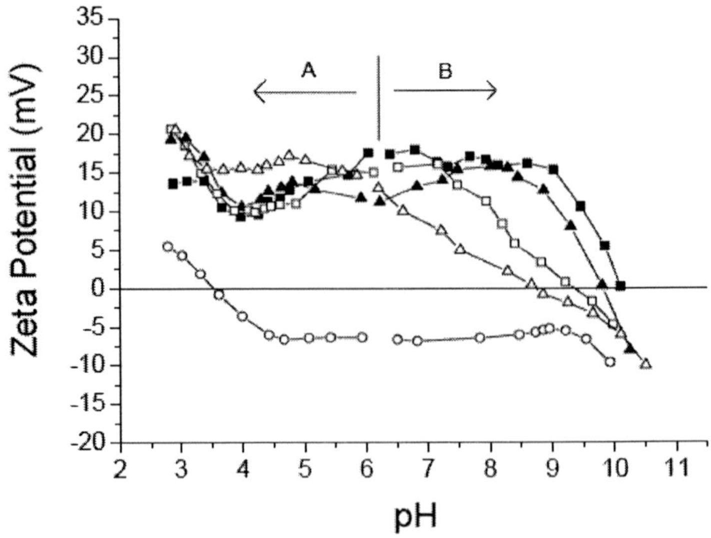

Figure 9. Zeta-potential values determined from streaming potential measurements in dependence on pH of an aqueous 10^{-3} mol·L^{-1} KCl solution. The measurements were carried out with bare silica (O), P(VFA-*co*-VAm)/silica hybrid material (7); P(VFA-*co*-VAm)/silica hybrid material after measurement in the acidic range (Δ), P(VFA-*co*-VAm)/silica hybrid material crosslinked with ECH (■) and P(VFA-*co*-VAm)/silica hybrid material crosslinked with ECH after measurement in the acidic range (□).

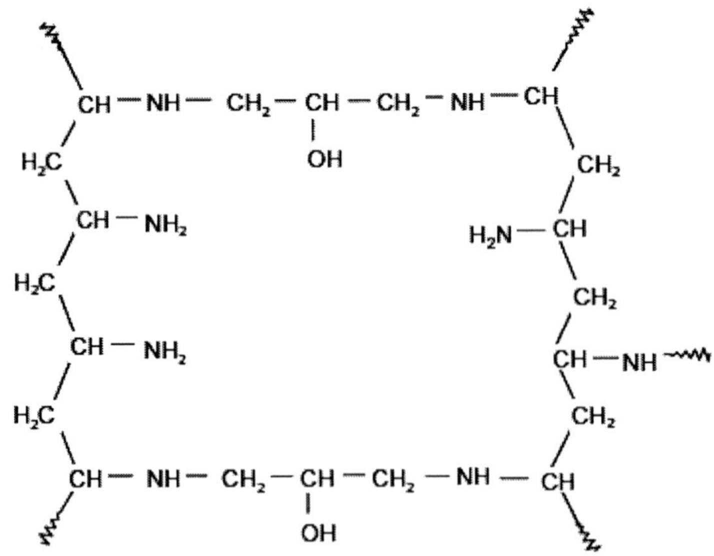

Figure 10. Idealized structure of ECH functionalized P(VFA-*co*-VAm) network.

The same electrokinetic measurements were carried out with P(VFA-*co*-VAm)/silica hybrid particles crosslinked with ECH. In the acidic range (pH < 6, region *A* in Figure 9, the curve with ■) we can see at pH ≈ 4.0 a minimum of the zeta-potential, that means a partially removal of the polyelectrolyte from de silica surface. When the sample, already measured in the acidic range, is measured in the basic range (region *B,* curve with □), the iep is 9.3, higher than the iep of the non-crosslinked sample (iep = 8.7). The polyelectrolyte amount partially removed from de silica surface is proportional with the shift of iep to a lower value. Hence, the polyelectrolyte amount partially removed from the P(VFA-*co*-VAm)/silica hybrid particles crosslinked with ECH, is lower than the polyelectrolyte amount partially removed from the P(VFA-*co*-VAm)/silica hybrid particles which indicates an enhance of the polyelectrolyte stability by crosslinking reaction with ECH.

4.2.4. Reaction of Boc-S-Benzyl-L-Cysteine with the Free Amino Groups from the Surfaces of Silica Hybrid Particles

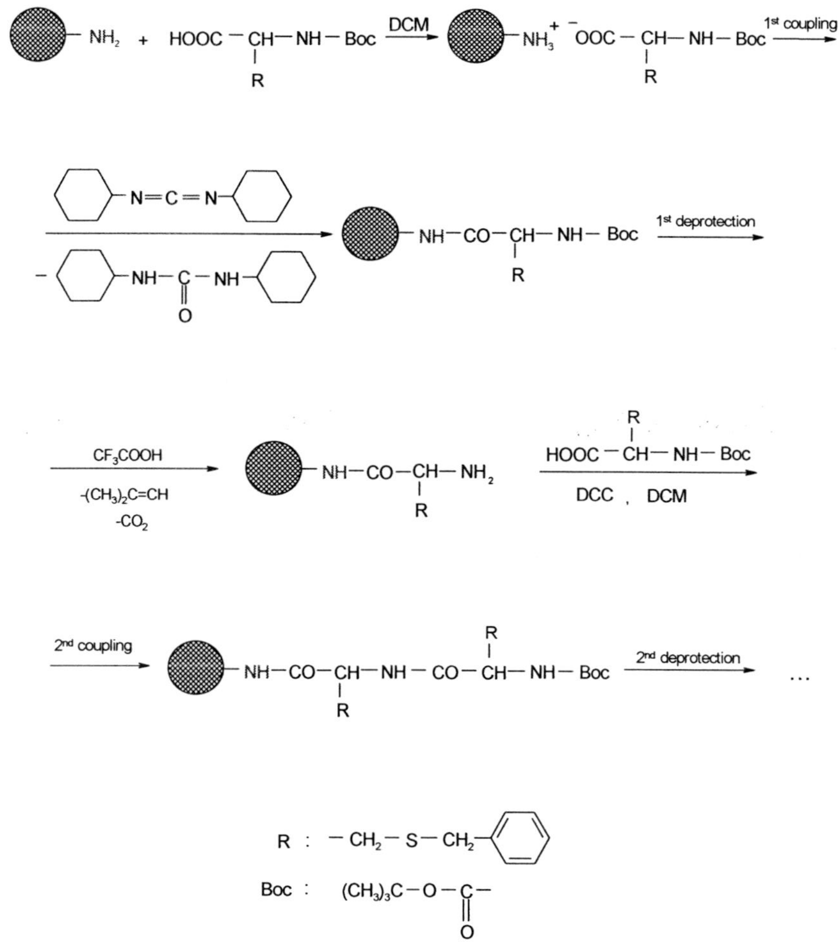

Scheme 1. The step-by-step coupling of Boc-S-benzyl-L-cysteine onto the surface of P(VFA-*co*-VAm)/silica hybrid particles.

After crosslinking, the functionalization with S-benzyl-L-cysteine of the formed hybrid material was carried out. XPS was employed to obtain information about the amount of amino acid S-benzyl-L-cysteine which reacts with the free amino groups on the silica particle surfaces. In the XPS spectra, the main important peak is the sulfur 2s, which indicates qualitatively and quantitatively the presence of the amino acid on the hybrid material surface.

After each coupling and deprotection reaction (Scheme 1), the hybrid materials were characterized using an Axis Ultra photoelectron spectrometer.

Figure 11 shows the elements on the silica surface before (a) and after (b) adsorption of P(VFA-*co*-VAm) and after the first coupling of the Boc-S-benzyl-L-cysteine to the hybrid particle (c). Two new elements, N and C, appeared in the XPS spectrum of P(VFA-*co*-VAm)/silica (Figure 11b), showing the presence of P(VFA-*co*-VAm) on the silica surface. A new element, S, appeared in the XPS spectrum after the first coupling of Boc-S-benzyl-L-cysteine, which indicates the presence of the amino acid on the surface, because the aminoacid molecule contains S (Figura 11c).

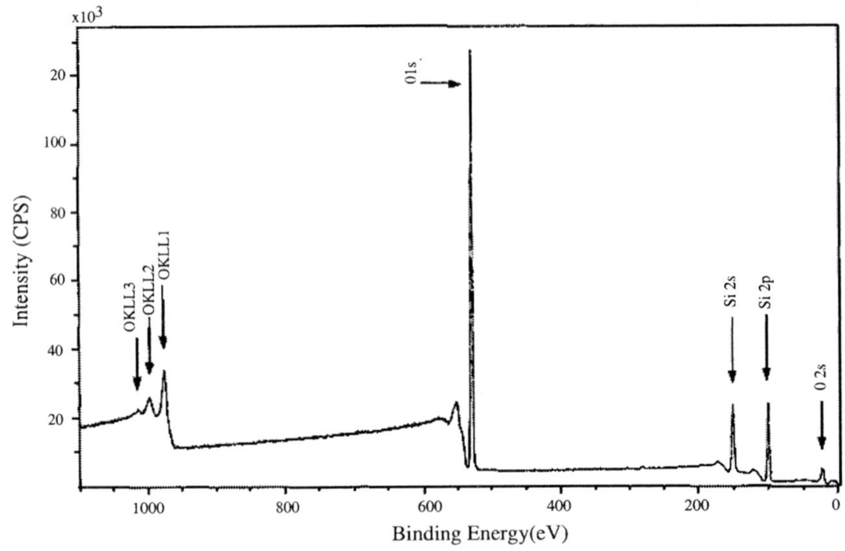

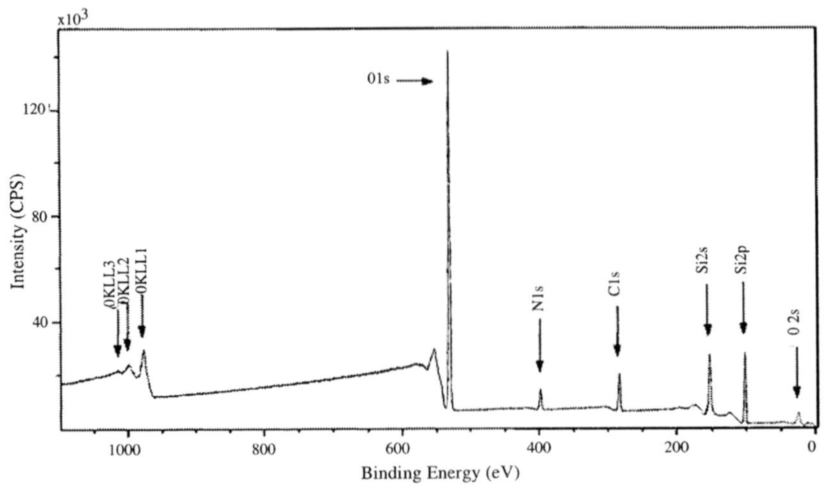

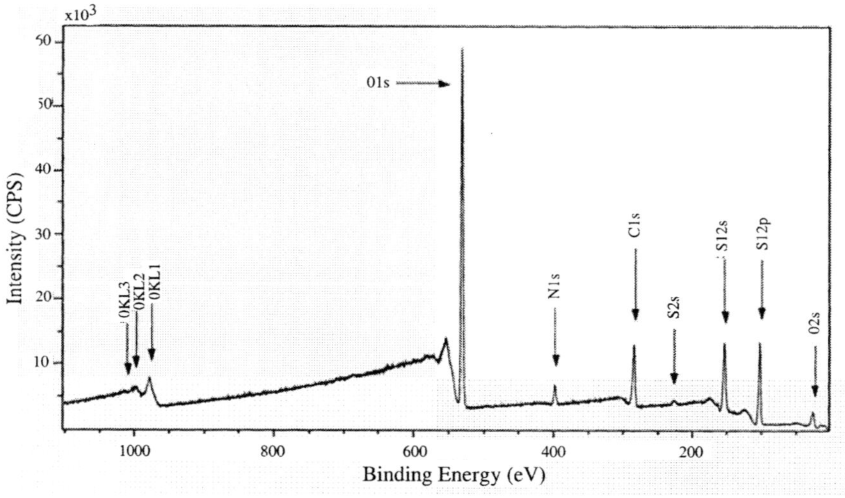

Figure 11. Wide-scan XPS spectra of bare silica (a), P(VFA-*co*-VAm)/silica (b), P(VFA-*co*-VAm)-ECH/silica after first coupling of Boc-S-benzyl-L-cysteine (c).

Besides the wide-scan XPS spectra, the sulfur peaks (S 2s) were followed to see if the step-by-step reaction between aminoacid and the free amino groups from the surface of silica particles takes place or not. Each coupling reaction is followed by a deprotection reaction to form free new amino groups onto the hybrid particles. After the first aminoacid coupling, the raw area of the S 2s peak was 1905 counts per second (CPS), the photoelectrons counted over a second (Figure 12b).

The raw area of S 2s decreased to 1026 CPS after the first deprotection reaction, because the polyelectrolyte macromolecules (together with bound aminoacid molecules), which are not strongly fixed to the surface, were removed in trifluoroacetic acid (TFA), used for deprotection. After this, the second coupling reaction was performed, the raw area of S 2s being 1950 CPS.

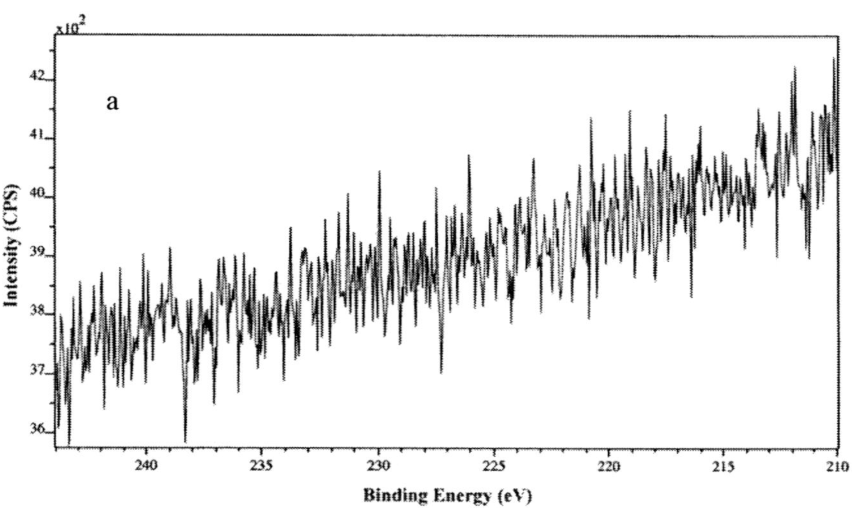

Figure 12. (Continued)

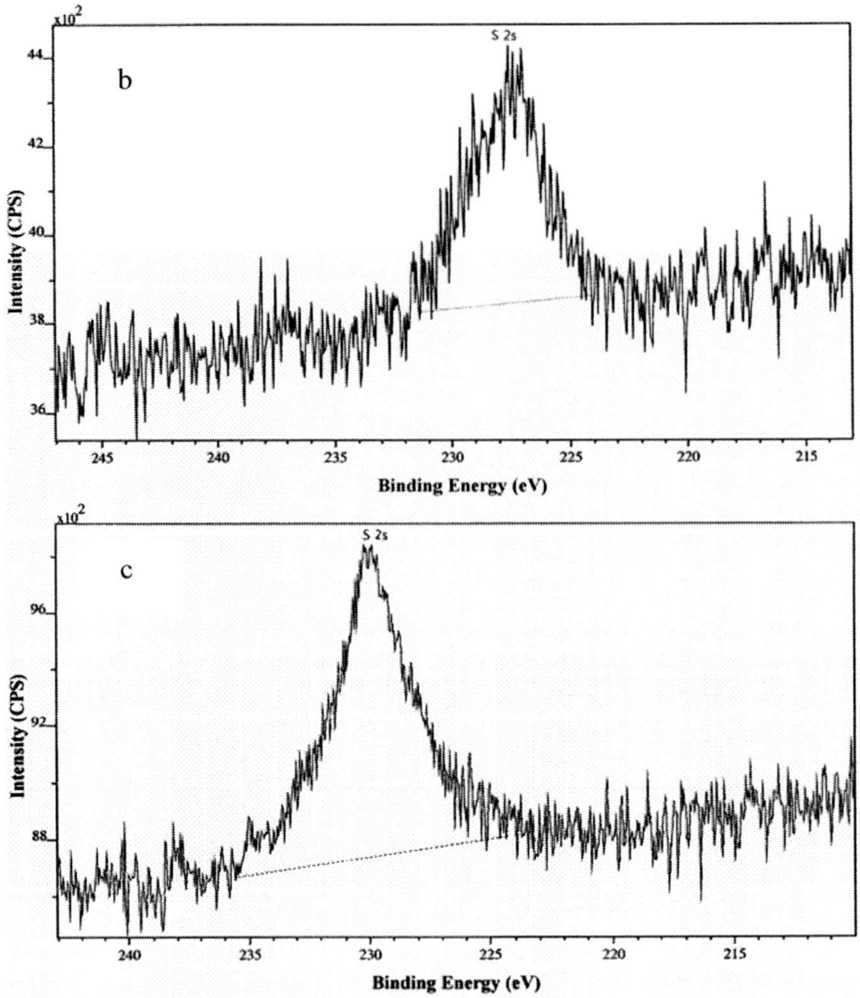

Figure 12. The S 2s XPS spectra of the P(VFA-*co*-VAm)/silica particles before the coupling of Boc-S-benzyl-L-cysteine (a), after the first (b) and fifth (c) coupling of Boc-S-benzyl-L-cysteine.

The second deprotection gives hybrid particles with S 2s peak raw area of 1790 CPS, showing that the hybrid material was stable in TFA after the second deprotection. The S 2s peak after the fifth coupling have a raw area of 4785 CPS. The increase of the aminoacid amount bound to silica hybrid particles was demonstrated by the increase of the intensity of the S 2s peak. It was thus demonstrated that the P(VFA-*co*-VAm)/silica hybrid particles can be used in the peptide synthesis.

5. Conclusion

In the first part of this chapter, the main reviews and a selection of the recently published original articles which have brought new contributions in the field of the applications of the membranes modified by sequential adsorption of polyelectrolytes, the applications of some multicomponent stimuli-responsive systems, and polyelectrolyte-colloids reactive systems have been briefly reviewed. In the second part, the adsorption of P(VFA-*co*-VAm) onto silica particles, the crosslinking reaction of the adsorbed P(VFA-*co*-VAm) with ECH, and the coupling reaction of Boc-S-benzyl-L-cysteine with the free amino groups from the surface of the P(VFA-*co*-VAm)/silica hybrid material have been investigated. The adsorption of P(VFA-*co*-VAm) onto silica particles from salt-free aqueous solution followed the Freundlich adsorption isotherm equation, and the adsorption of P(VFA-*co*-VAm) from KCl 10^{-1} mol·L^{-1} followed the Langmuir adsorption isotherm equation. The enthalpic factor favors the adsorption process ($\Delta H^0 < 0$), but the entropic factor did not ($\Delta S^0 < 0$). Subsequent derivatization reaction of the accessible primary amino groups of P(VFA-*co*-VAm)/silica hybrid particles with ECH offered a hybrid material with possibilities for subsequent derivatization reactions such as: coupling aminoacids, peptides, obtaining ion exchange resins, etc. Using the Boc-S-benzyl-L-cysteine, a oligopeptide was formed onto the surface of P(VFA-*co*-VAm)/silica hybrid particles.

ACKNOWLEDGEMENTS

The authors thank Prof. Dr. Stefan Spange for providing the P(VFA-*co*-VAm) and Dr. Cornelia Bellmann for her kind help in the getting of the electrokinetic results on silica particles.

REFERENCES

[1] Warner, M.G.; Hutchison, J. E. In: *Synthesis, Functionalization and Surface Treatment of Nanoparticles*; Baraton, M.-I.; Ed.; American Scientific Publishers: San Francisco, CL, 2003, pp 67-89.

[2] Castelvetro, V.; De Vita, C. *Adv. Colloid Interface Sci.* 2004, *108-109*, 167-185.

[3] Dutta, J.; Hofmann, H. In: *Encyclopedia of Nanoscience and Nanotechnology*; Nalwa H. S.; Ed.; American Scientific Publishers: North Lewis Way Stevenson Ranch, CL, 2004, Vol. 9, pp 617-640.

[4] Berwald, S.; Meier-Haack, J. In: *Handbook of Polyelectrolytes and their Applications*; Tripathy, S. K.; Kumar, J.; Nalwa H. S.; Eds.; American Scientific Publishers: Los Angeles, CL, 2002, Vol. 3, pp 99-114.

[5] Decher, G.; Hong, J. D. *Ber. Bunsenges. Phys. Chem.* 1991, *95*, 1430-1434.

[6] Decher, G.; Hong, J. D. *Makromol. Chem. Macromol. Symp.* 1991, *46*, 321–327.

[7] Decher, G. *Science* 1997, *277*, 1232–1237.

[8] Bertrand, P.; Jonas, A.; Laschewsky, A.; Legras, R. *Macromol. Rapid Commun.* 2000, *21*, 319-348.

[9] Hammond, P. T. *Curr. Opin. Colloid Interface Sci.* 2000, *4*, 430-442.

[10] Van Ackern, F.; Krasemann, L.; Tieke, B. *Thin Solid Films*, 1998, *327-329*, 762-766.

[11] Krasemann, L.; Tieke, B. *J. Membr.Sci.* 1998, *150*, 23-30.

[12] Krasemann, L.; Tieke, B. *Mat. Sci. Eng. C* 1999, *8/9*, 513-518.

[13] Krasemann, L.; Tieke, B. *Chem. Eng. Techn.*. 2000, *23*, 211-213.

[14] Krasemann, L.; Toutianoush, A.; Tieke, B. *J. Membrane Sci.* 2001, *181*, 221-228.

[15] Tieke, B.; Krasemann, L.; Toutianoush, A. *Macromol. Symp.* 2001, *163*, 97-111.
[16] Toutianoush, A.; Krasemann, L.; Tieke, B. *Colloids Surf. A* 2002, *198-200*, 881-889.
[17] Toutianoush, A.; Tieke, B. *Mat. Sci. Eng. C* 2002, *22*, 459-463.
[18] Meier-Haack, J.; Rieser, T.; Lenk, W.; Lehmann, D.; Berwald, S.; Schwarz, S. *Chem. Ing. Techn.* 1999, *71*, 839-844.
[19] Meier-Haack, J.; Lenk, W.; Lehmann, D.; Lunkwitz, K. *J. Membr. Sci.* 2001, *184*, 233-243.
[20] Lenk, W.; Meier-Haack, J. *Desalination* 2002, *148*, 11-16.
[21] Budd, M.;. Ricardo, N. M. P. S.; Jafar, J. J.; Stephenson, B.; Hughes, R. *Ind. Eng. Chem. Res.* 2004, *43*, 1863-1867.
[22] Sullivan, D. M.; Bruening, M. L. *J. Membr. Sci.* 2005, *1-2*, 161-170.
[23] Stroeve, P.;Vasquez, V.; Coelho, M. A. N.; Rabolt, F. J. *Thin Solid Films* 1996, *284/285*, 708-712.
[24] Leväsalmi, J. M.; McCarthy, T. J. *Macromolecules* 1997, *30*, 1752-1757.
[25] Van Ackern, F.; Krasemann, L.; Tieke, B. *Thin Solid Films*, 1998, *327-329*, 762-766.
[26] Zhou, P.; Samuelson, L.; Alva, K. S.; Chen, C. C.; Blumstein, R. B.; Blumstein, A. *Macromolecules* 1997, *30*, 1577-1581.
[27] Sullivan, D. M.; Bruening, M. L. *Chem. Mater.* 2003, *15*, 281-287.
[28] Urairi, M.; Tsuru, T.; Nakao, S.; Kimura, S. *J. Membr. Sci.* 1992, *70*, 153-162.
[29] Tsuru, T.; Nakao, S.; Kimura, S. *J. Membr. Sci.* 1995, *108*, 269-278.
[30] Krasemann, L.; Tieke, B. *Langmuir* 2000, *16*, 287-290.
[31] Harris, J. J.; Stair, J. L.; Bruening, M. L. *Chem. Mater.* 2000, *12*, 1941-1946.
[32] Stanton, B. W.; Harris, J. J.; Miller, M. D.; Bruening, M. L. *Langmuir* 2003, *19*, 7038-7042.
[33] Stair, J. L.; Harris, J. J.; Bruening, M. L. *Chem. Mater.* 2001, *13*, 2641-2648.
[34] Bruening M. L.; Harris, J. J.; Sullivan, D. M. *Polym. Mater. Sci. Eng.* 2001, *85*, 525-526.
[35] Sullivan, D.M.; Bruening, M. L. *J. Am. Chem. Soc.* 2001, *123*, 11805-11806.
[36] Toutianoush, A.; Tieke, B. In: *Novel Methods to Study Interfacial Layers*; Moebius, D.; Miller, R.; Eds.; Elsevier, Amsterdam, 2001, pp 416-425.
[37] Tieke, B. *Mat. Sci. Eng.* C 2002, *22*, 135-139.
[38] Toutianoush, A.; Jin, W.; Deligöz, H.; Tieke, B. *Appl. Surf. Sci.* 2005, *246*, 437-443.
[39] Jin, W.; Toutianoush, A.; Tieke, B. *Langmuir* 2003, *19*, 2550-2553.
[40] Farhat, T. R.; Schlenoff, J. B. *Langmuir* 2001, *17*, 1184-1192.
[41] Miller, M. D.; Bruening M. L. *Langmuir* 2004, *20,* 11545-11551.
[42] Miller, M. D.; Bruening, M. L. *Chem. Mater.* 2005, *17*, 5375-5381.
[43] Balachandra, A. M.; Dai, J. H.; Bruening, M. L. *Macromolecules* 2002, *35*, 3171-3178.
[44] Dai, J. H.; Balachandra, A. M.; Lee, J. I.; Bruening, M. L. *Macromolecules* 2002, *35*, 3164-3170.
[45] Hollman, A. M.; Bhattacharyya, D. *Langmuir* 2004, *20*, 5418-5424.
[46] Lajimi, R. H.; Ben Abdallah, A.; Ferjani, E.; Roudesli, M. S.; Deratani, A. *Desalination* 2004, *163*, 193-202.
[47] Toutianoush, A.; Schnepf, J.; El Hashani, A.; Tieke B. *Adv. Funct. Mat.* 2005, *15*, 700-708.
[48] Tieke, B.; Toutianoush, A.; Jin, W. *Adv. Colloid Interface Sci.* 2005, *30*, 121-131.

[49] Rieser, T.; Lunkwitz, T.; Berwald, S.; Meier-Haack, J.; Kessler, B.; Simon, F. *Polym. Mater. Sci. Eng.* 1997, *77*, 351-352.

[50] Müller, M.; Rieser, T.; Lunkwitz, K.; Berwald, S.; Meier-Haack, J.; Jehnichen, D. *Macromol. Rapid Commun.* 1998, *19*, 333-336.

[51] Müller, M.; Rieser, T.; Lunkwitz, K.; Meier-Haack, J. *Macromol. Rapid Commun.* 1999, *20*, 607-611.

[52] Dragan, E. S.; Mihai, M.; Schauer, J.; Ghimici, L. *J. Polym. Sci. Part. A Polym. Chem.* 2005, *43*, 4161-4171.

[53] Zhao, B.; Brittain, W. J. *Prog. Polym. Sci.* 2000, *25*, 677-710.

[54] Sevick, E. M.; Williams, D. R. M. *Macromolecules* 1994, *27*, 5285-5290.

[55] Israels, R.; Gersappe, D.; Fasolka, M.; Roberts, V. A.; Balazs, A. *Macromolecules* 1994, *27*, 6679-6682.

[56] Kontturi, K.; Mafe, S.; Manzanares, J. A.; Svarfvar, B. L.; Viinikka, P. *Macromolecules* 1996, *29*, 5740-5746.

[57] Ito, Y.; Park, Y. S. *J. Am. Chem. Soc.* 1997, *119*, 2739-2740.

[58] Park, Y. S.; Imanishi, Y. *Chem. Mater.* 1997, *9*, 2755-2758.

[59] Zhang, H.; Ito, Y. *Langmuir* 2001, *17*, 8336-8340.

[60] Ito, Y.; Ochiai, Y.; Park, Y. S.; Imanishi, Y. *J. Am. Chem. Soc.* 1997, *119*, 1619-1623.

[61] Ito, Y. *Nano-technology* 1998, *9*, 205-207.

[62] Ito, Y.; Park, Y. S.; Imanishi, Y. *Langmuir* 2000, *16*, 5376-5381.

[63] Sui, Z.; Schlenoff, J. B. *Langmuir* 2003, *19*, 7829-7831.

[64] Hollman, A. M.; Scherrer, N. T.; Cammers-Goodwin, A.; Bhattacharyya, D. *J. Membr. Sci.* 2004, *239*, 65-79.

[65] Iwata, H.; Oodate, M.; Uyama, Y.; Amemiya, H.; Ikada, Y. *J. Membr. Sci.* 1991, *55*, 119-130.

[66] Ying, L.; Kang, E. T.; Neoh, K. G.; Kato, K.; Iwata, H. *J. Membr. Sci.* 2004, *243*, 253-262.

[67] Yamaguchi, T.; Ito, T.; Sato, T.; Shinbo, T; Nakao, S. *J. Am. Chem. Soc.* 1999, *121*, 4078-4079.

[68] Baldi, A.; Lei, M.; Gu, Y.; Siegel, A. R.; Ziaie, B. *Sensors Actuat.* 2006, *B114*, 9-18.

[69] Park, Y. S.; Ito, Y.; Imanishi, Y. *Macromolecules* 1998, *31*, 2606-2610.

[70] Wu, L. F.; Tuo, X. L.; Chen, H.; Chen, Zh.; Wang, X. G. *Macromolecules* 2001, *34*, 8005-8013.

[71] Pyrasch, M.; Tieke, B. *Langmuir* 2001, *17*, 7706-7709.

[72] Zucolotto, V.; He, J. A.; Constantino, C. J. L.; Barbosa Neto, N. M.; Rodrigues Jr., J. J.; Mendonça, C. R.; Zílio, S. C.; Li, L.; Aroca, R.F.; Oliveira Jr., O. N.; Kumar, J. *Polymer* 2003, *44*, 6129-6133.

[73] Zucolotto, V.; Barbosa Neto, N. M.; Rodrigues Jr., J. J.; Constantino, C. J. L.; Zilio, S. C.; Mendonca, C. R.; Aroca, R. F.; Oliveira Jr., O. N. *J. Nano-sci. Nano-tech.*, 2004, *4*, 855-860.

[74] Zucolotto, V.; Strack, P. J.; Santos, F. R.; Balogh, D. T.; Constantino, C. J. L.; Mendonca, C. R.; Oliveira, O. N. *Thin Solid Films* 2004, *453-454*, 110-113.

[75] Dragan, S.; Schwarz, S. *Macromol. Symp.* 2002, *181*, 155-166.

[76] Dragan, S.; Schwarz, S. *Prog. Colloid Polym. Sci.* 2003, *122*, 8-15.

[77] Saremi, F.; Tieke, B. *Adv. Mater.* 1998, *10*, 388-391.

[78] Toutianoush, A.; Saremi, F.; Tieke, B. *Mater. Sci. Eng.C* 1999, *8/9*, 343-352.
[79] Ziegler, A.; Stumpe, J.; Toutianoush, A.; Tieke, B. *Coloids Surf. A* 2002, *198-200*, 777-784.
[80] Dante, S.; Advincula, R.; Frank, C. W.; Stroeve, P. *Langmuir* 1999, *15*, 193-201.
[81] Soga, K.G.; Zuckermann, M. J.; Guo, H. *Macromolecules* 1996, *29*, 1998-2005.
[82] Zhulina, E.; Balazs, A. C. *Macromolecules* 1996, *29*, 2667-2673.
[83] Sidorenko, A.; Minko, S.; Schenk-Meuser, K.; Duschner, H.; Stamm, M. *Langmuir* 1999, *15*, 8349-8355.
[84] Rühe, J.; Ballauff, M.; Biesalski, M.; Dziezok, P.; Gröhn, F.; Johannsmann, D.; Houbenov, N.; Hugenberg, N.; Konradi, R.; Minko, S. Motornov, M.; Netz, R. R.; Schmidt, M.; Seidel, C.; Stamm, M.; Stephan, T.; Usov, D.; Zhang, H. *Adv. Polym. Sci.* 2004, *165*, 79-150.
[85] Husemann, M.; Morrison, M.; Benoit, D.; Frommer, G., Mate, C. M.; Hinsberg, W. D.; Hedrick, J. L.; Hawker, C. J. *J. Am. Chem. Soc.* 2000, *122*,1844-1845.
[86] Minko, S.; Usov, D.; Goreshnik, E.; Stamm, M. *Macromol. Rapid Commun.* 2001, *22*, 206-211.
[87] Minko, S.; Patil, S.; Datsyuk, V.; Simon, F.; Eichhorn, K.-J.; Motornov, M.; Usov, D., Tokarev, I.; Stamm, M. *Langmuir* 2002, *18*, 289-296.
[88] Minko, S.; Müller, M.; Motornov, M.; Nitschke, M.; Grundke, K.; Stamm, M. *J. Am. Chem. Soc.* 2003, *125*, 3896-3900.
[89] Lemieux, M.; Minko, S.; Usov, D.; Stamm, M.; Tsukruk, V. V. *Langmuir* 2003, *19*, 6126-6134.
[90] Lemieux, M.; Usov, D.; Minko, S.; Stamm, M.; Shulha, H.; Tsukruk, V. V. *Macromolecules* 2003, *36*, 7244-7255.
[91] Tokareva, I.; Minko, S.; Fendler, J. H.; Hutter E. *J. Am. Chem. Soc.* 2004, *126*, 15950-15951.
[92] Retsos, H.; Gorodyska, G.; Kiriy, A.; Stamm, M.; Creton, C. *Langmuir* 2005, *21,* 7722-7725.
[93] Xu, C.; Fu, X.; Fryd, M.; Xu, S.; Wayland, B. B.; Winey, K. I.; Composto, J. R. *Nano Lett.* 2006, *6*, 282-287.
[94] Schüler, C.; Caruso, F. *Macromol. Rapid Commun.* 2000, *21*, 750-753.
[95] Caruso, F.; Schüler, C. *Langmuir* 2000, *16*, 9595-9603.
[96] Caruso, F.; Fiedler, H.; Haage, K. *Colloids Surf.* A 2000, *169*, 287-293.
[97] Lvov, Y.; Caruso, F. *Anal. Chem.* 2001, *73*, 4212-4217.
[98] Liang, Z.; Wang, C.; Tong, Z.; Ye, W.; Ye, S. *React. Funct. Polym.* 2005, *63*, 85-94.
[99] Yang, W.J.; Trau, D.; Renneberg, R.; Yu, N. T.; Caruso, F. *J. Colloid Interface Sci.* 2001, *234*, 356-362.
[100] Caruso, F.; Spasova, M.; Salgueiriño-Maceira, V.; Liz-Marzán, L. M. *Adv. Mater.* 2001, *13*, 1090-1094.
[101] Gittins, D. I.; Susha, A. S.; Schoeler, B.; Caruso, F. *Adv. Mater.* 2002, *14*, 508-512.
[102] Rogach, A.; Susha, A.; Caruso, F.; Sukhorukov, G.; Kornowski, A.; Kershaw, S.; Möhwald, H.; Eychmüller, A.; Weller, H. *Adv. Mater.* 2000, *12*, 333-337.
[103] Kidambi, S.; Dai, J.; Li, J.; Bruening M. L. *J. Am. Chem. Soc.* 2004, *126*, 2658-2659.
[104] Shi, X.; Han, S.; Sanedrin, R. J.; Zhou, F.; Selke M. *Chem. Mater.* 2002, *14*, 1897-1902.

[105] Dautzenberg, H.; Jaeger, W.; Kötz, J.; Philipp, B.; Seidel, C.; Stscherbina, D. *Polyelectrolytes: formation, characterization and applications*; Carl Hanser, Munich, 1994.

[106] Bauer, D.; Killmann, E.; Jaeger, W. *Prog.Colloid Polym. Sci.* 1998, *109*, 161-169.

[107] Huguenard, C.; Widmaier, J.; Elaissari, A.; Pefferkorn, E. *Macromolecules* 1997, *30*, 1434-1441.

[108] Gailliez-Degremont, E.; Bacquet, M.; Laureyns, J.; Morcellet, M. *J. Appl. Polym. Sci.* 1997, *65*, 871-882.

[109] Sukhorukov, G. B.; Donath, E.; Lichtenfels, H.; Knippel, H.; Knippel, M.; Budde, A.; Möhwald, H., *Colloids Surf. A* 1998, *137*, 253-266.

[110] Donath, E.; Sukhorukov, G. B.; Caruso, F.; Davis, S.A.; Möhwald, H. *Angew. Chem.* 1998, *110*, 2324-2327.

[111] Spange, S.; Meyer, T.; Voigt, I.; Eschner, M.; Estel, K.; Pleul, D.; Simon, F. *Adv. Polym. Sci.* 2004, *165*, 43-78.

[112] Meyer, T.; Rehak, P.; Jäger, C.; Voigt, I.; Simon, F.; Spange, S. *Macromol. Symp.* 2001, *163*, 87-96.

[113] Voigt, I.; Simon, F.; Esthel, K.; Spange, S. *Langmuir* 2001, *17*, 3080-3086.

[114] Voigt, I.; Simon, F.; Esthel, K.; Spange, S.; Friedrich, M. *Langmuir* 2001, *17*, 8355-8361.

[115] Voigt, I.; Simon, F.; Komber, H.; Jacobasch, H. J.; Spange, S. *Colloid Polym. Sci.*, 2000, *278*, 48-56.

[116] Poptoshev, E.; Rutland, M. W.; Claesson, P. M. *Langmuir* 1999, *15*, 7789-7794.

[117] Serizawa, T.; Yamamoto, K.; Akashi, M. *Langmuir* 1999, *15*, 4682-4684.

[118] Jacobasch, H. J.; Simon, F.; Werner, C.; Bellmann, C. *Technisches Messen: Sensoren, Geräte, Systeme* 1996, *63*, 447-452.

[119] Bajpai, A. K.; Sachdeva, R. *J. Appl. Polym. Sci.* 2002, *85*, 1607-1618.

[120] Bucatariu, F.; Simon, F.; Spange, S.; Schwarz S., Dragan, S. *Macromol. Symp.* 2004, *210*, 219-228.

In: New Trends in Ionic (Co) Polymers and Hybrids
Editor: Ecaterina Stela Dragan, pp. 145-164

ISBN 978-1-60021-611-4

Chapter 6

Ionic Hybrid Hydrogels

Ecaterina Stela Dragan** and Maria Cazacu
"Petru Poni" Institute of Macromolecular Chemistry,
Aleea Grigore Ghica Voda 41 A, RO-700487 Iasi, Romania

Abstract

The swelling capacity of the classical hydrogels, generally consisting in hydrophilic networks, can be controlled by co-networking with a hydrophobic phase resulting in amphiphilic co-networks. Such phase can be formed by a polysiloxane well known for its hydrophobic character. Polysiloxanes, on the other hand, can be chemically modified by introduction of ionic groups to induce hydrophilicity and other interesting properties depending on the cationic centers content and the substituent structure. Recent developments in this direction were briefly reviewed in the paper. Own results on the synthesis and characterization of some novel ionic organic/siloxane hydrogels were also presented. Ionic organic/siloxane crosslinked copolymers containing quaternary ammonium salt (QAS) groups in the backbone were obtained by a polyaddition reaction between 1,3-bis(3-aminopropyl)tetramethyldisiloxane and siloxane oligomers, on the one side, and some reactive cationic oligomers containing preformed QAS groups, on the other side. Ionic hybrid hydrogels containing QAS sequences in the side chain were obtained by the quaternization reaction of either polyethylene glycol chains bearing tertiary amine groups or oligo(N,N-dimethylaminoethylmethacrylate) with a polyfunctional chloroalkylated siloxane. Thermooxidative stability and swelling properties of the ionic hybrid hydrogels in different solvents were evaluated.

Keywords: ionic hybrid hydrogel, siloxane-organic network, polyaddition reaction, Menshutkin reaction, QAS, solvent uptake, TGA.

* E-mail: sdragan@icmpp.ro

List of Abbreviations

AP	α,ω-Bis(3-aminopropyl)oligodimethylsiloxane
AP_0	1,3-Bis(3-aminopropyl)tetramethyldisiloxane
APCNs	amphiphilic co-networks
CM_0	1,3-Bis-(chloromethyl)-1,1,3,3-tetramethyldisiloxane
CP_0	1,3-bis-(chloropropyl)-1,1,3,3-tetramethyldisiloxane
CM	α,ω-bis(chloromethyl)oligodimethylsiloxane
CMCl	α,ω-trimethylsiloxy-oligodimethylmethylchloromethylsiloxane
DMDAP	N,N-dimethyl-1,3-diaminopropane
IPN	interpenetrated network
QAS	quaternary ammonium salt groups
PDMS	poly(dimethyl siloxane)
PEG	polyethylene glycol
PEGA	PEG chains with tertiary amine groups in the backbone and in the side chain
PEGDGE	PEG diglycidyl ether
PEO	poly(ethylene oxide)
PHEMA	poly(2-hydroxyethyl methacrylate)
PNIPAM	poly(N-isopropylacrylamide)
PTMO	poly(tetramethylene oxide)
RAFT	reversible-addition-fragmentation chain transfer
RCO	reactive cationic oligomer
TEOS	tetraethoxysilane

1. Introduction

A hydrogel is defined as a polymer network capable of imbibing a large volume of water but retaining its three-dimensional structure after swelling [1,2]. The swollen equilibrium state results from a balance between the osmotic driving force that causes the water to enter into the hydrophilic polymer, and the forces exerted by the polymer chains in resisting to expansion [3]. The three-dimensional structure of a swollen hydrogel can be maintained either by chemical (covalent or ionic-covalent bonds) or physical (hydrogen bonding, van der Waals interaction, hydrophobic interaction, chain entanglements, crystallinity, or ionic complexation) crosslinking [1-4]. The applicability of the chemically crosslinked, thermosetting hydrogels is often limited by the lack of processability due to their physical brittleness. In contrast, thermoplastic hydrogels, formed by physical crosslinking, allow solvent casting or thermal processing. Furthermore, the introduction of noncovalent intermolecular interactions allows a supramolecular engineering of the material properties. Secondary interactions between different polymer chains can be also exploited to enhance the miscibility of components that are incompatible in the absence of these interactions.

Due to their high water content and the corresponding rubbery nature, hydrogels are similar to a variety of natural living tissues, having widespread applications as biomaterials (contact lenses, sutures, dental materials, and controlled drug delivery systems). By selecting the functional groups along the polymer chains, hydrogels can be designed as stimuli-responsive systems, or "smart hydrogels", which undergo large changes in volume in response to physical or chemical changes in the environment, such as: pH, ionic strength, solvents, temperature, electric and magnetic field and light [2]. Furthermore, applications of stimuli-responsive hydrogels include mechanochemical transducers that can be used as switchers, microactuators, and as a new type of artificial muscle. Their soft and tissue-like physical properties are known to minimize frictional irritation upon implantation. In addition, their surface hydrophilicity is responsible for less protein adsorption and less cellular adhesion, resulting in improved biocompatibility. Hydrogels demonstrated also a relative low trombogeneity. Therefore, they have found large applicability in reconstructive surgery for artificial organs, tissue engineering, cartilage, muscles, immunoinsulation membranes, drug delivery, etc. By their high hydrophilicity, hydrogels are also recommended as drying agents for foods, sanitary goods, to retain water in soil in drought affected area, and for controlled delivery of pesticides and fertilizers. Superwater-absorbent resins can absorb several hundred or thousand times as much water as their own weight to an equilibrium value, which make them very useful in medicine, sanitary goods, food industry, chemical processings, and so on.

Among the hydrogels, those bearing ionic groups occupy a special place, numerous contributions being dedicated to the preparation and characterization of these materials, the presence of ionic or ionizable centers increasing the number of external stimuli which could support their use as “smart materials” [5-30]. The synthesis of ionic hydrogels is typically performed via free-radical copolymerization of the ionic or ionizable monomer with a monomer containing two or more double bonds, the most utilized being N,N’-methylenebisacrylamide, with or without a nonionic monovinylic monomer. Ionic hydrogels usually possess either weak ionic sites, which ionization is pH dependent [5,6,10,11,12,14-22,24-27,29,30], or strong and pH independent ionic centers [6-9,13,23].

A special category of hydrogels recently developed by Okay et al. [4,13,16], either nonionic [4] or ionic [13,16], is that of superfast responsive hydrogels, which main characteristic is their swelling equilibrium is established in few seconds comparative with the usual hydrogels which need hours or days to accommodate the good solvents.

Although the number and diversity of existing hydrogels are impressive, they have been developed using a relatively small number of polymers and conventional methods of chemical or physical crosslinking. Their functions and biocompatibility are often compromised since their structures are not well defined. It is highly desirable to design hydrogels with precisely controlled characteristics such as gelation temperature, swelling, elasticity, rigidity, porosity, biodegradability, bioerosion, and fast response to chemical and/or physical external stimuli such as pH, temperature, light, and ionic strength. A less explored route to design new hydrogels with a high added value and tailored properties is the preparation of ionic organic/inorganic hybrids based on chemical/physical interactions, which offer a large variety of unexplored structures and properties.

1.1. Organic/Inorganic Hydrogels

Generally, the organic-inorganic hybrids concept is employed to combine the organic and inorganic properties in more sophisticated materials with a higher added value, their potential for various applications being rapidly evaluated. According to their morphological combinations, the organic-inorganic hybrids can be classified in the following categories:

- with an inorganic matrix, where organic materials are embedded;
- with an organic matrix, where inorganic materials are embedded;
- the interpenetrating networks, where inorganic and organic polymeric networks are independently formed without mutual chemical bonds;
- true hybrids, where inorganic and organic composites with mutual chemical bonds are formed [31, and references therein].

True organic-inorganic hybrids have been developed by either chemical modification of the polysiloxane backbone with various organic functional groups, the term polysiloxane reflecting a polymeric structure that contains repeating silicon-oxygen groups as backbone [1] or by utilizing a clay as a crosslinker instead of the organic crosslinkers [31]. The family of materials containing co-continuous polysiloxane and hydrophilic phases (amphiphilic co-networks, APCNs) are called siloxane hydrogels. By their hydrophilic/hydrophobic nature, APCNs swell both in water and hydrocarbons, and respond to changes in the medium by morphological isomerization ('smart' networks) [1].

The hydrophilic organic polymers usually used as constituents for hydrogels are poly(methacrylic acid), poly(acrylamide), poly(N-isopropylacrylamide) (PNIPAM), poly(2-hydroxyethylmethacrylate) (PHEMA), poly(vinyl alcohol), and poly(ethylene oxide) (PEO). Among these polymers, PEO attracted a particular interest for creating protein- and cellular-resistant surfaces [1]. Hydrogels based on PEO, as a hydrophilic block, and poly(tetramethylene oxide) (PTMO) or poly(dimethyl siloxane) (PDMS), as hydrophobic blocks, were synthesized by a polycondensation reaction as candidates for biomaterials (Scheme 1). Short blocks of hydrophilic/hydrophobic segments were coupled with aliphatic diisocianate without using a chain extender, producing urethane linkages. By varying PEO length (M_W = 2000 or 3400 Da), hydrophobic components (PTMO or PDMS) or polymer topology (block or graft copolymers), various physical hydrogels were prepared [1].

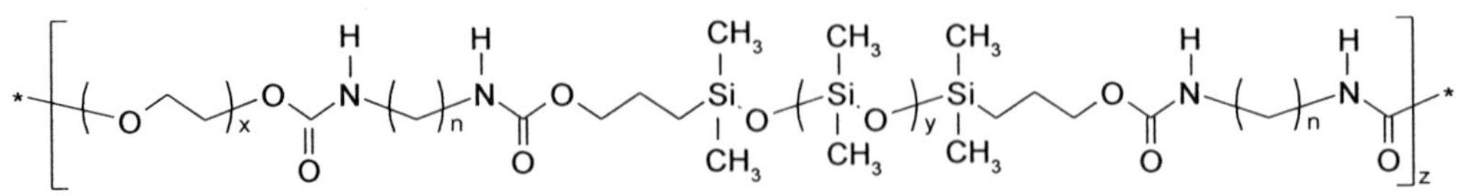

Scheme 1.

Laser-induced surface grafting of PHEMA on PDMS and a sequential method for preparation of interpenetrating polymer networks of PDMS/PHEMA have been used for surface and bulk modifications. The structures of the host PDMS and the guest PHEMA networks are shown in Scheme 2 [32].

Swelling measurements performed on networks containing between 15 and 60% PHEMA showed that swelling increased linearly in the range 7 - 20%, with the fraction of PHEMA. This linear behavior can be related to the connected hydrogel domains. For the PHEMA fraction in this range, the hydrogel domains are connected together so that water can easily diffuse to these domains [32].

One of the most important destinations of hydrogels is that of contact lens materials. To design successful contact lens materials, a number of characteristics must be optimized:

optical transparency, chemical and thermal stability, wetability to tears, mechanical properties, biological compatibility, and high oxygen permeability.

Two basic methods for the preparation of hydrogels with high oxygen permeability are usually employed. The first approach consists in the development of high water content hydrogels.

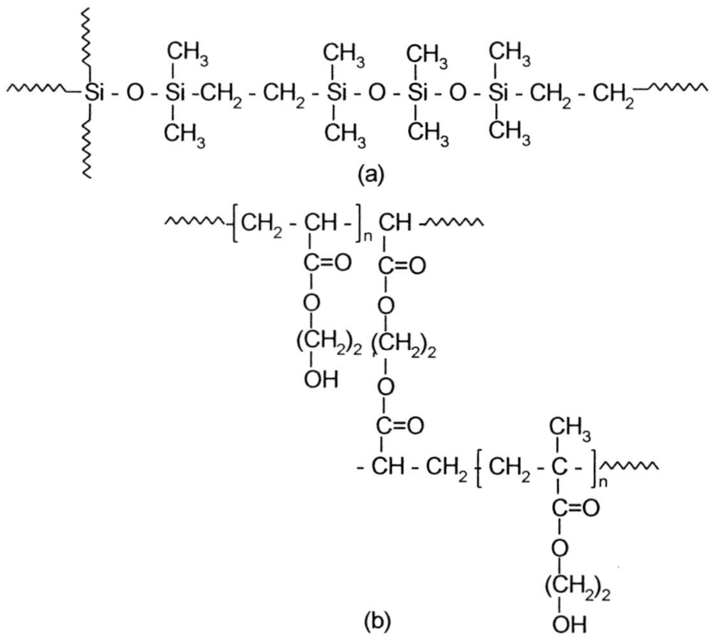

Scheme 2.

However, high water content materials that have in excess up to 80% water typically exhibit poor mechanical properties and, in dry environments, induce epithelia dehydration that may cause adverse physiological responses. The second approach follows the development of high oxygen permeable hydrogels by the copolymerization of methacrylate- or vinyl-functionalized PDMS with hydrophilic monomers. PDMS, due to its low modulus of elasticity, optical transparency, and high oxygen permeability, has been extensively studied for use in contact lenses. PDMS possesses an oxygen permeability about 70 times higher than the oxygen permeability of the PHEMA hydrogel. There are, however, several limitations to overcome before designing hydrogels based on PDMS. The biggest one is that silicone is hydrophobic and insoluble in hydrophilic monomers. The copolymerization of methacrylate-functionalized silicones with hydrophilic monomers generally results in opaque, phase-separated materials. Künzler and Ozark described the synthesis of silicone hydrogels based on methacrylate end-capped siloxanes containing diethylene glycol propyl methyl ether, 5-hexyl-1,2-diol, and 3-propyloxy-1,2-propane diol side chains and made a comparative evaluation of their application potential as hydrogels for contact lenses [3].

It was recently shown that hydrogels made from a hydrophilic polymer backbone substituted either with fluoro- or siloxanyl- groups can display very high oxygen permeability [33]. Optimizing oxygen permeability without losing transparency, mechanical integrity and surface hydrophilicity requires controlled phase separation via careful macromolecular design. Copolymerizing hydrophobic macromonomers with hydrophilic monomers can form self-reinforcing hydrogels. Hydrogels with high moduli were thus formed, and their mechanical integrity was postulated to originate from hydrophobic interactions of the macromonomer combs [33]. In this context, Pai et al. have recently reported the application of the reversible-addition-fragmentation chain transfer (RAFT) polymerization to prepare controlled block structures containing siloxanyl and fluoro functionality. The success of using N,N-dimethylacrylamide and 2-(butyl perfluorooctanefluorosulfonoamido) ethyl acrylate to design hydrogels for contact lenses led to the development of this system even further using RAFT polymerization. Additionally, the PDMS was incorporated to enhance the oxygen permeability. Such amphiphilic block copolymers might be suitable in the preparation of physically crosslinked hydrogels [33].

1.2. Ionic Organic/Inorganic Hydrogels

It is of a great interest to introduce ionic units into the polysiloxanes [34]. For example, the hydrophobicity of the siloxanes can be reduced by introduction of the polar or ionic groups such as quaternary ammonium salts or pyridinium salts at the chain end [35]. The pendant quaternary ammonium chloride units improve the compatibility of the polysiloxane with organic polymers and bring antimicrobial and anti-electrostatic properties to the materials to make them applicable as fabric-conditioning agents, bactericides, textile finishing and plastic processing aids [34].

It has been recently shown that an interpenetrated network (IPN) consisting in linear quaternary ammonium salt (QAS) groups-substituted polysiloxane of high molecular weight dispersed in a PDMS network had fungicidal properties, by both direct contact and diffusion, and a much better durability than conventional fungicidal silicone mastics used for sanitary applications [36]. Polysiloxanes bearing pendant QAS groups are particularly attractive as biocidal polymers with long-lasting or permanent activity (i.e., bactericidal, fungicidal, and algicidal). The literature reported results on the bactericidal activities of some soluble or insoluble silicone funtionalized with QAS salt groups [36-39]. The amphiphilic character of the poly(dialkylsiloxane) chain may be another advantage of this polymer as a carrier of biocidal groups. At an interface, polysiloxanes show a tendency to form a layer cleaning to the hydrophilic face, with their hydrophilic inorganic part, and to the hydrophobic face with their hydrophobic organic groups [37].

Ionic hydrogels can be used as controlled release systems that respond to an external substrate or trigger by swelling or de-swelling. One example is a glucose-sensitive system for insulin-controlled release based on pH-sensitive hydrogel [40]. For certain application, such as humidity sensors, the hydrophylicity of the gel must be limited and controlled. The measurement and control of humidity are important in many areas, including industry (paper, food, electronic), domestic environment (air conditioner), medical (respiratory equipment), etc. [41]. So far, organic polymers [42-44], polymer composites [45-47], and modified polymers [48] with hydrophilic properties have been used as components for humidity sensor devices. Polyelectrolytes containing hydrophilic groups such as –COOH, SO_3H, $-N^+R_3Cl^-$, etc. are excellent materials for sensing low humidity but these cannot operate at high humidity because of their solubility in water [41]. Such shortcomings have been overcome by blending with a hydrophobic polymer, or by chemical modification of the hydrophobic polymers to generate ionic groups resulting in humidity sensitive materials with improved properties.

The hydrophobic character of polysiloxanes is well known and commonly used in water repellency [49]. Polysiloxanes can be chemically modified by introduction of various hydrophilic functions, which considerably affect their properties. Thus, Sauvet et al. have recently reported the introduction of quaternary ammonium salt groups (QAS) and hydroxyalkyl groups in polysiloxanes of various topologies to induce hydrophilic properties. The polysiloxanes containing a low proportion of highly hydrophilic groups proved to be very hygroscopic and strongly swelled in water [49]. Crosslinked organo-polysiloxane having hydrophilic groups such as NH_2, $N^+(CH_3)_3Cl$, SO_3H, OH were grafted on a pressed silica gel or a sintered alumina plate to make humidity sensors [41]. The presence of the hydrophilic functions led to particularly strong water affinity due to the electrostatic and hydrogen bond forces [49].

Inorganic-organic hybrid charged membranes have attracted particular attention due to their excellent properties, such as structural flexibility, thermal and mechanical stability. As extremely important classes of hybrid materials or membranes, the zwitterionic hybrids have attracted considerable attention owing to their double-functional group characters and it is expected to be applied in piezodialysis for separation of inorganic salts from water-soluble organic substances [50]. A series of novel organic-inorganic zwitterionic hybrid membranes were synthesized by Liu et al. via a sol-gel process of hybrid precursor. The hybrid precursor was obtained via the reaction of 3-glycidoxypropyltrimethoxysilane with N-[3-(trimethoxysilyl)propyl] ethylene diamine, followed by the reaction with g-butyrolactone to create ion pairs in the membrane. The most probable structure of the resulted membranes is illustrated in Scheme 3 [50].

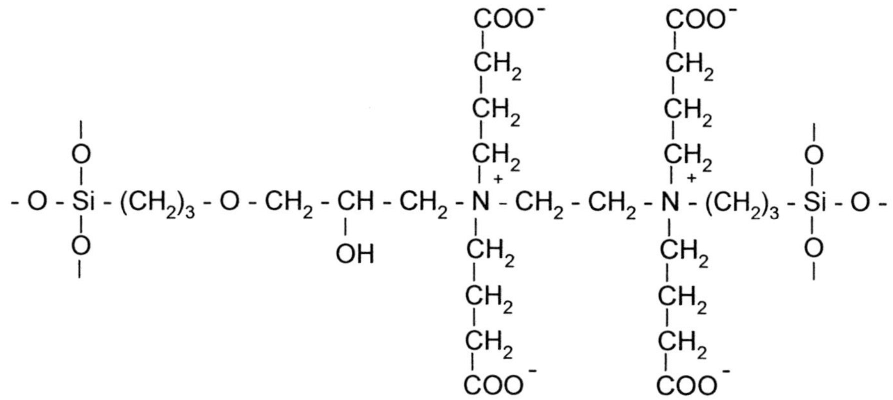

Scheme 3. Hybrid charged zwitterionic polymer.

Amination of oligo-chloro(methyl)siloxane either with an excess or insufficient amount of 4,4'-dipyridyl led to water-soluble or crosslinked polymers, respectively. Radicals obtained for the first time on monoquaternized dipyridyl oligosiloxane were studied with the EPR method. Redox properties were observed for these siloxanes at their interaction with divalent metal chlorides [51].

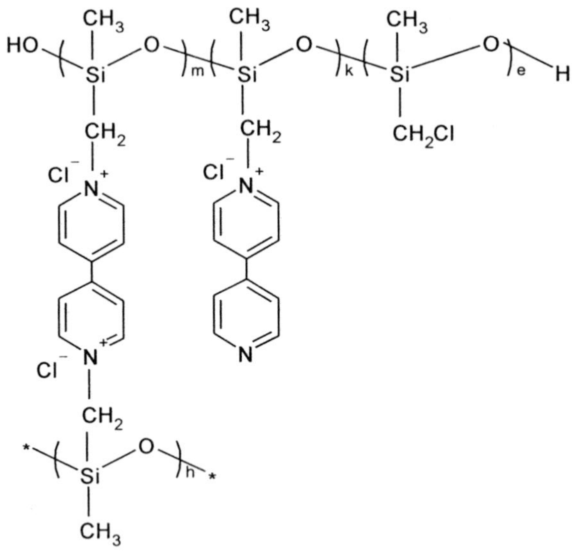

Scheme 4.

In the synthesis of organosilicon ion exchangers used for the sorption of metal ions, or as catalysts for the purification of hormones and antibiotics, the alkylsilanation reaction of amines has been successfully applied by Kolesnik et al. namely, amination of oligo-chloromethyl-methyl-siloxane (I) with an excess or insufficient amount of 4,4`-dipyridyl (II) (Scheme 4) [51]. The preparation and characterization of organosilicon dipyridylium complexes is not only of theoretical, but also of practical interest, such compounds exhibiting redox and electrochromic properties. Such compounds containing piperazine as a crosslinker have been synthesized in order to be used as humidity sensors [52].

2. Novel Ionic Hybrid Hydrogels Based on Functionalized Siloxanes

In the recent investigations developed in our group, we were interested in the synthesis and characterization of some new ionic organic/inorganic hybrid materials (hydrogels) [53,54] as promising candidates for gas-separation membranes, materials with sanitary applications drug delivery systems, or as novel polymer electrolytes when they are complexed with inorganic salts. Thus, some new QAS-containing crosslinked polymeric structures based on organic and siloxanic moieties, having ionic groups either in the backbone [53] or as side groups, were prepared [54].

For this purpose, two approaches were employed:

- A reactive cationic oligomer, RCO, containing preformed QAS groups in the backbone and reactive groups like 2,3-epoxypropyl and 2-hydroxy-3-chloropropyl on the end of the chain was synthesized according to the previously described procedure [55]. 1,3-Bis(3-aminopropyl)tetramethyldisiloxane (AP_0), and α,ω-bis(3-aminopropyl)oligodimethylsiloxane (AP), having a certain molecular weight, and synthesized according to a well-known procedure [53] were used as bis-amino siloxanes in these reactions.

- Cationic polymers with polyethylene glycol (PEG) sequences and tertiary amine groups in the backbone and in the side chain (PEGA) have been prepared by the polyaddition reaction of PEG diglycidyl ether (PEGDGE) with N,N-dimethyl-1,3-diaminopropane (DMDAP), according to the procedure presented in detail elsewhere [56]. 1,3-Bis-(chloromethyl)-1,1,3,3-tetramethyldisiloxane (CM_0), 1,3-bis-(chloropropyl)-1,1,3,3-tetramethyldisiloxane (CP_0), α,ω-bis(chloromethyl)oligodimethylsiloxane (CM) having an average numerical molecular weight (calculated on the basis of the ^{1}H-NMR spectra) M_n = 675, n being about 6, was synthesized according to the own procedure [57], and α,ω-trimethylsiloxy-oligodimethylmethylchloromethylsiloxane, CMCl, having 7% methylchloromethylsiloxane units on the chain were used as reactive chloromethylsiloxanes in the Menschutkin reaction with PEGA [54] or with oligo(N,N-dimethylaminoethylmethacrylate) (ODMAEM).

2.1. Ionic Hybrid Hydrogels Containing QAS in the Backbone Synthesized by Polyaddition Reactions

To obtain QAS organic/inorganic crosslinked materials by polyaddition reactions, RCO and some 3-aminopropyl-end functionalized siloxanes were mixed at different molar ratios. HCl (1 mL of 36 wt.-% HCl solution/100 g reaction mixture), added as a catalyst, was mixed carefully, until homogeneous and clear mixtures were achieved. The mixture was again degassed under vacuum for few minutes and cast onto a thin teflon sheet. The obtained membranes were kept for some weeks at room temperature to finalize the reaction. The crosslinked polymers were extracted first with water to remove the possible unreacted cationic precursor and then, with acetone to remove the unreacted siloxane.

The addition of a siloxane containing end diamine groups to the epoxy groups, of the RCO, containing preformed QAS groups in the backbone, led to ionic organic/inorganic polymeric materials of integral type with the general structure presented in Scheme 5. For this purpose, an amine terminated siloxane, AP_0 or AP, was reacted in bulk with the cationic oligomer RCO, at different molar ratio (Table 1).

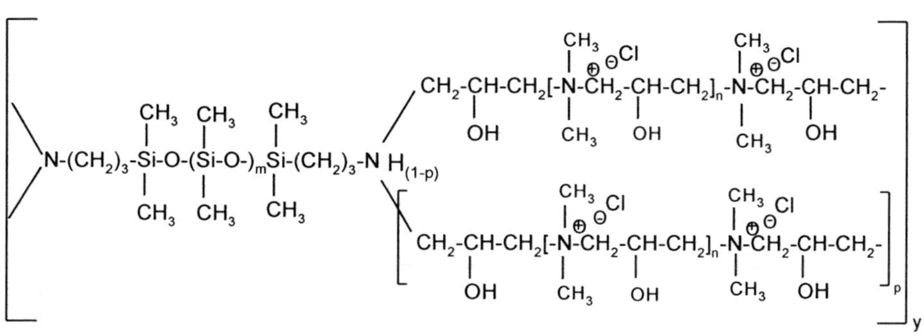

Scheme 5. General structures of the ionic hybrids containing QAS in the backbone.

Reaction conditions and some characteristics of the ionic hybrids of integral type (QAS groups in the backbone) were summarized in Table 1. With AP_0 as siloxane comonomer, the RCO:siloxane molar ratio in the hybrid copolymer varied from 1.5/1 (sample $RCOS_1$) up to 2:1 (sample $RCOS_2$), depending on the feed molar ratio between the epoxy and amine groups, and this showed that the crosslinking degree also increased with the increase of the content of the bifunctional component.

The quaternization reaction has been supported also by IR spectroscopy, the main characteristic bands and their changes being summarized as follows: the characteristic bands of the epoxy groups (860, 920, and 1260 cm^{-1}) disappeared after curing and the bands at 810 and 1270 cm^{-1}, characteristic for the Si-CH_3 bond, were visible for all ionic hybrids [53].

The thermooxidative stability of the ionic hybrid crosslinked materials was evaluated by thermogravimetry. Some representative data of the registered curves are summarized in Table 2. As can be observed, by comparing the TGA data, the thermal stability of the ionic organic/inorganic hybrids depended mainly on their content in siloxane component, a higher termostability being found for the sample $RCOS_1$ which had a RCO/siloxane ratio of 1.5/1. The char yield also increased with the increase of the siloxane content.

Table 1. Ionic Organic/Inorganic Hybrid Materials Having QAS Ionic Groups in the Backbone Synthesized by a Polyaddition Reaction

Sample Code[a]	Feed molar ratio epoxy:amine groups	Cl, % calc.[b] found		RCO: siloxane molar ratio in copolymer[c]	Copolymer aspect
$RCOS_1$	1:1	14.9	11.8	1.5:1	yellowish membrane
$RCOS_2$	1.5:1	17.4	11.7	2:1	yellowish membrane
$RCOS_3$	2:1	19.4	15.2	1.9:1	colourless, transparent membrane
$RCOS_4$	1:1	5.9	4.5	1:1	colourless, transparent membrane

[a] Samples $RCOS_1$, $RCOS_2$ and $RCOS_3$ were synthesized with AP_0 as bisamino siloxane; sample $RCOS_4$ was synthesized with AP as bisamino siloxane.

[b] Calculated on the basis of the feed molar ratio of the precursors.

[c] Determined on the basis of silicon content.

Table 2. The Main Thermogravimetric Data for the Crosslinked Hybrids Containing QAS in the Backbone

Sample[a]	wt.% QAS groups/copolymer	T_{on} [b]	T_{10} [c]	T_{of} [d]	% Residue
$RCOS_1$	0.15	50	200	568	15
$RCOS_2$	0.18	55	220	600	6
$RCOS_3$	0.19	42	140	595	5

[a] The same significance like in Table 1;

[b] The onset temperature;

[c] The temperature at which 10 wt.% loss was observed;

[d] The offset temperature.

However, a high content of the organic component is necessary for a high crosslinking degree, and therefore a compromise between the thermal stability and the crosslinking degree is normally necessary.

The hybrids presented above have been also characterized by XPS measurements, a preferential distribution of the siloxane segments on the film surface being found [53]. A dramatic increase of the silicon atom concentration on the surface compared with the values determined in bulk by elemental analysis was observed.

2.2. Ionic Hybrid Hydrogels Containing QAS in the Side Chain and PEG Sequences In The Backbone

The Menschutkin reactions between the haloalkyl-functionalized siloxanes (CM_0, CP_0, CM and CMCl) and PEGA were performed at a molar ratio between the target functional groups of 1:1. Reactions were carried out in refluxing n-propanol. NaI, in a molar ratio of 1:2, related to the chlorine content, was added as a catalyst. For alkyl halides, it is known that the reactivity increases with the increase of the halogen atom volume. Therefore, the chloroalkyl groups in the siloxanes have been converted into iodopropyl groups by halogen substitution with NaI, the iodoalkyl groups reacting with tertiary amines much more easily than chloroalkyl groups [58]. The reaction was monitored by the ionic chlorine content [54]. When the maximum of Cl_i (related to the calculated value) was obtained, the reaction mixture was cooled and filtered off in order to remove the unsolved inorganic salt. The filtrate was then slowly poured on a Teflon sheet and exposed to IR beam for ca. 48 h in order to allow complete solvent evaporation and reaction finalization. The resulted network was removed from the sheet and maintained in vacuum ca. 48 h before characterization.

The Menschutkin reaction between the chloroalkyl groups belonging to a siloxane component and tertiary amine groups pending or within the chain of amino-functionalized PEGA was performed in order to obtain new ionic organic/inorganic hybrid materials described by the general structure presented in Scheme 6.

Some characteristics of the ionic organic/inorganic hybrids with QAS in the side chain generated from PEGA are collected in Table 3.

Table 3. Some Characteristics of the Ionic Hybrid Hydrogels with Ionic Groups Located in the Side Chain

Sample	Siloxane reactant	Reaction time[a] (h)	Cl, wt % Calc.[b]	Found[c]	Aspect
PEGA-CM_0	CM_0	3.5	4.9	4.3	yellowish film
PEGA-CP_0	CP_0	3.5	4.7	4.4	yellowish film
PEGA-CM	CM_4	3.5	3.6	3.7	white film
PEGA-CMCl	CMCl	4.0	1.7	1.7	white-yellowish film

[a] The time at which the ionic chlorine content remained constant;

[b] Determined by considering that all covalent chlorine was converted in ionic chlorine.

[c] Reported to the product weight;

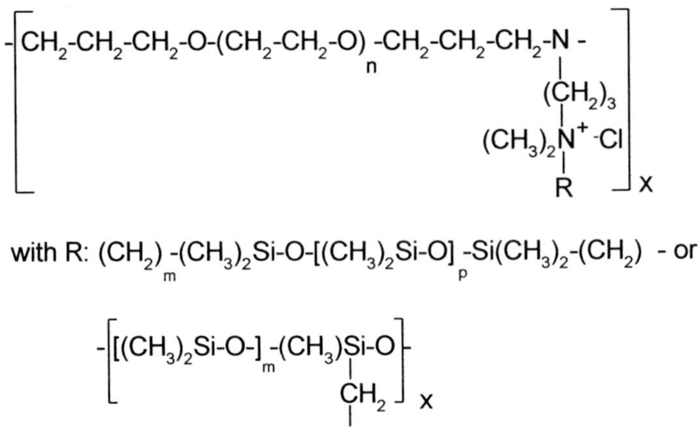

Scheme 6. General structure of the hybrids containing pendent QAS groups.

The molar ratio of 1:2 between the chlorine and tertiary amine groups has been selected to quaternize mainly the pendent tertiary amine groups of PEGA, these groups being not sterically hindered. Even if all the hybrids collected in Table 3 were crosslinked materials, a very good agreement between the calculated and found values of chlorine were found. Furthermore, the use of NaI as catalyst led not only to a high acceleration of the reactions but also to a very good reproducibility of the hybrid characteristics.

The solvent-uptake capacities of the hybrid networks were evaluated by measuring the weight increase due to the solvent uptake, relative to the dry extracted network. The solvent in excess was removed by centrifugation at 500 g for 1 h and weighed, this value being W_w. The gels were dried until a constant weight as it was described above, this value being W_d. The gravimetric swelling ratio (W) is defined as: W (%) = (W_w - W_d)/W_d x 100 [18]. Table 4 shows swelling capacities between 800 and 2500%, in chloroform, and 2000% to unlimited in water [54]. The swelling in water up to about 100 times and in chloroform up to 25 times show that these gels behave as both hydrogels and organogels.

The thermooxidative stability of the ionic hybrid crosslinked materials was evaluated by thermogravimetry. Some representative data of the registered curves are summarized in Table 5. As can be observed, by comparing the TGA data, the thermal stability of the ionic hybrid materials of this type depended mainly on their content in QAS groups and on the siloxane structure.

Table 4. Equilibrium Swelling of Crosslinked Structures in Different Solvents

Sample	Swelling, %				
	n-Propanol	Water	DMF	Acetone	Chloroform
PEGA-CM_0	2156	4560	1988	146	1310
PEGA-CM_4	unlimited	2093	unlimited	221	2521
PEGA-CP_0	1614	unlimited	1553	138	977
PEGA-CMCl	1548	9281	221	228	834

[a]After 1 day immersion in a solvent at room temperature.

Table 5. The Main Thermogravimetric Data for the Crosslinked Structures

Sample[a]	wt.% QAS groups/copolymer	T_{on} [b]	T_{10} [c]	T_{of} [d]	% Residue
PEGA-CM_0	0.13	190	240	385	7
PEGA-CP_0	0.12	180	255	575	8
PEGA-CMCl	0.057	190	285	385	50

[a] The same significance like in Tables 3 and 4;
[b] The onset temperature;
[c] The temperature at which 10 wt.% loss was observed;
[d] The offset temperature.

As Table 5 shows, the decomposition of all ionic hybrids started at relatively low temperatures owing to the presence of QAS groups and aliphatic oxyethylene sequences. On the other hand, the sample PEGA-CMCl synthesized with a long siloxane chain bearing pendent chloromethyl groups was the most thermostable hybrid hydrogel.

2.3. Ionic Hybrid Hydrogels Containing QAS in the Side Chain Derived from ODMAEM

CM_0, CP_0 and CM, already presented in previous section, were used as chloroalkyl functionalized siloxanes in the reaction with tertiary amine groups of ODMAEM, with a molar mass of around 600 Da. The siloxane and ODMAEM, in various molar ratios (Table 6) were solved in n-propanol to prepare a solution of 0.37 mole Cl/L. In the most cases, NaI in a molar ratio of 0.5/1, related to the chlorine content, was added. The mixture was refluxed on a thermosetting bath in a flask equipped with magnetic stirrer, reflux condenser and a stopcock through which samples were withdrawn at different intervals. The progress of reaction was followed by the examination of the product consistence remained after solvent evaporation under IR beam. The reaction times were between 35-40 h. Finally, the reaction mixture was slowly poured on a Teflon sheet to assure an easy release of the volatile products. The samples were exposed to IR beam for ca. 48 h in order to allow complete evaporation and complete reaction. The resulted networks as transparent to opaque and flexible to rigid, depending on the reactant type and their molar ratio, were removed from the sheet as free-standing films and maintained in vacuum ca. 48 h. Purification was made by extraction, first with acetone by immersion for 24 h, then with water for 72 h at least, and finally again with acetone, followed by drying in vacuum at 50 °C ca. 48 h. Chloro and silicon contents were determined as a rough measure of the siloxane and ionic groups, compared with those corresponding to the idealized structures (Schemes 7 and 8), their values being collected in Table 6.

The crosslinking degree as well as the type of the free functional groups were designed by the feed molar ratio between the reactive functional groups (Schemes 7 and 8).

Table 6. Ionic Hydrogels Obtained by the Quaternization of ODMAEM with Chloroalkyl-Functionalized Siloxanes

Sample	Reaction partners			Characteristics of the products	
	Chloroalkyl-Siloxane (CAS)[a)]	Feed molar ratio, ODMAEM/CAS	Conc., mole Cl/L	% Si found (calc)	% Cl found(calc)
AT-$CM_0$1:1	CM_0	1:1	0.37	1.02 (10.3)	12.14 (13.0)
AT-CM1:1	CM	1:1	0.37	16.25 (22.9)	-
AT-$CP_0$1:1	CP_0	1:1	0.37	5.09 (9.5)	7.52 (12.1)
AT-$CM_0$1:2	CM_0	1:2	0.37	1.75 (14.4)	7.21 (18.3)
AT-$CP_0$1:2*	CP_0	1:2	0.37	8.21 (12.8)	16.3 (16.2)
AT-$CM_0$1:2, x3[b]	CM_0	1:2	3x0.37	1.87 (14.4)	8.71 (18.3)
AT-CM1:2*	CM	1:2	0.37	22.02 (26.9)	6.95 (8.5)
AT-$CM_0$1:1*	CM_0	1:1	0.37	3.6 (10.3)	-
AT-$CM_0$1:1*, x3[b]	CM_0	1:1	3x0.37	3.8 (10.3)	(13.0)
AT-CM1.6:1*	CM	1.6:1	0.37	3.6 (19.0)	3.8 (6.0)

a) CM_0 – 1,3-bis-(chloromethyl)-1,1,3,3-tetramethyldisiloxane; CP_0 – 1,3-bis-(chloropropyl)-1,1,3,3-tetramethyldisiloxane; CM - α,ω-Bis(chloromethyl)oligodimethylsiloxane;

b) the concentration of siloxane component was three times higher;

*) the syntheses were performed in the presence of NaI added at a molar ratio 0.5:1 towards chlorine content.

The swelling behavior of the hybrid networks was evaluated by measuring the solvent uptake, relative to the mass of the dry extracted network (Table 7). The crosslinked samples previously dried in vacuum at 50 °C, and maintained on P_2O_5 for one week, were used in order to determine the swelling capacity in various media. Distilled water, aqueous solution with pH = 2, aqueous solutions of electrolyte with concentrations of 0.2 M NaCl, and 0.2 M Na_2SO_4 have been used for this purpose. Each sample was soaked for 72 h in selected solvent, this time being almost enough as the swelling equilibrium to be established. The solvent uptake has been determined by gravimetric method as it has been described in the previous section.

As it is known [59,60], there are a multitude of parameters that can compete to the swelling behavior of a polymer network, like hydrophilic/hydrophobic balance in the network, the presence of ionic or ionizable groups in the polymeric segments and crosslinking extent of the network. The strongly swollen ionic polymer networks may undergo conformational transitions as a function of the changes of the medium parameters such as temperature, solvent polarity, ionic strength, etc. A great attention has been addressed to the influence of the characteristics of the external medium like solvent quality (dielectric constant) [10], and of the counterion nature [11,15] on the swelling behavior of ionic hydrogels, a very strong influence of the charge density on the gel collapse being reported by Philippova et al. [6] and by Kawaguchi and Satoh [9].

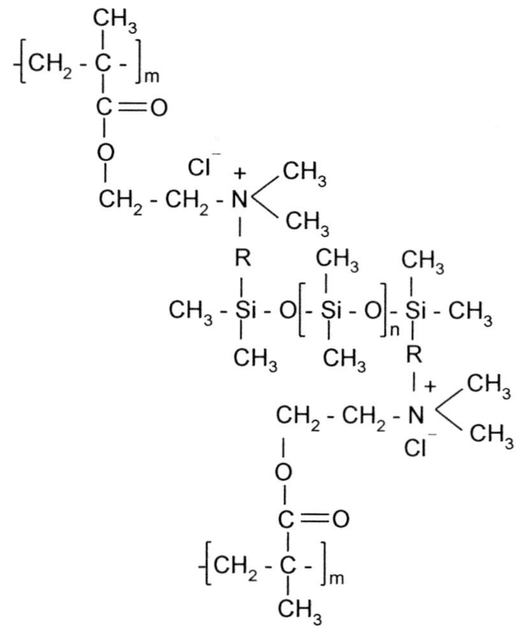

Scheme 7. The idealized structure of a completely crosslinked structure (AT-$CM_0$1:1, AT-CM1:1, AT-$CP_0$1:1, AT-$CM_0$1:1*, AT-$CM_0$1:1*), n = 0 or 6, R: CH_2 or $(CH_2)_3$.

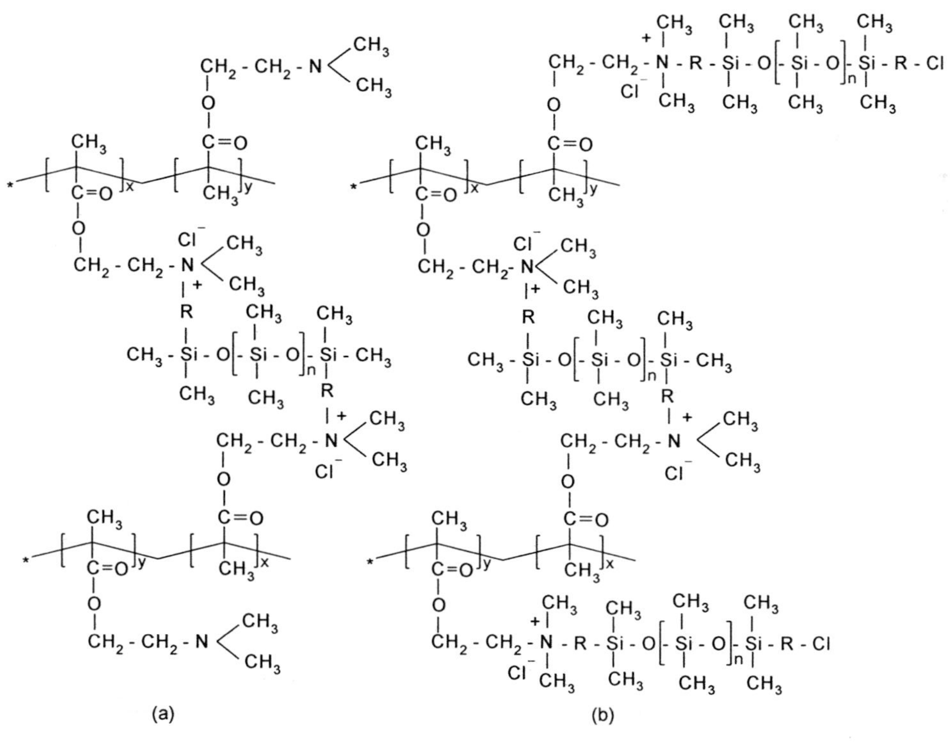

Scheme 8. The idealized structure of the product obtained by using: a - amine groups in excess (AT-CM1.6:1*), n = 6, R: CH_2; b - chloro-alkyl groups in excess (AT-$CM_0$1:2, AT-$CP_0$1:2*, AT-$CM_0$1:2,x3, AT-CM1:2*), n = 0 or 6, R: CH_2 or $(CH_2)_3$.

Table 7. Solvents Regains of the Ionic Hydrogels Obtained by the Quaternization of Oligo-DMAEM with Chloroalkyl – Siloxanes at a Temperature of 20 °C

Sample	Swelling capacity (%wt/wt) in:			
	water	aqueous solution, pH = 2	NaCl, 0.2 M	Na_2SO_4, 0.2 M
AT-$CM_0$1:1	2952	1653	885	1073
AT-CM1:1	2292	-	1950	820
AT-$CP_0$1:1	11526	1153	1072	852
AT-$CM_0$1:2	2528	1152	2060	2016
AT-$CP_0$1:2*	73	-	95	80
AT-$CM_0$1:2, x3	-	7449	1921	1601
AT-CM1:2*	4600	5340	-	967
AT-$CM_0$1:1*	234	1443	1623	363
AT-$CM_0$1:1*, x3	130	3381	42	35
AT-CM1.6:1*	2380	2896	-	881

In our case, the increase of the crosslinking degree would lead to the increase of both the hydrophobic content (due to crosslinking agent based on the siloxane) and the ionic groups formed by the Menshutkin reaction. In addition, the presence of high flexible siloxane between crosslinking points permits mobility or relaxation of macromolecular chains in the matrix, providing enough space for accommodation of water molecules in the network [59]. The degree of swelling is mainly influenced by the charge density of the chains, and by the crosslinking degree, these two factors having opposite effects for the ionic hybrids under study, because any new crosslink means a higher rigidity of the gel and a higher hydrophilicity by the increase of the QAS sites density.

A clear influence of the hydrogel structure and solvent can be seen in Table 7. Thus, the most hydrophilic (11526% in water) proved to be AT-$CP_0$1:1, obtained by crosslinking of ODMAEM with CP_0 in a molar ratio of 1:1. The most hydrophobic appears to be AT-$CP_0$1:2*, with a swelling degree of 73% in water, obtained by using chloromethyl groups in excess towards amine groups (2:1), and in the presence of NaI. The influence of the counterion nature on the swelling degree of the hybrid hydrogels has been also investigated in the paper by comparing the values of the swelling degrees in 0.2 M NaCl aqueous solution with those found in 0.2 M Na_2SO_4 aqueous solution. As Table 7 shows, for almost all samples investigated the swelling degree in aqueous solution of 0.2 M Na_2SO_4 is lower than that in aqueous solution of 0.2 M NaCl and much more lower than in water without any electrolyte. The divalent counterion (SO_4^{2-}) is more effectively bound to the ionic centers than the monovalent (Cl^-), a corresponding collapse of the gel taking place in this case. Sometimes the swelling degrees were higher at pH 2 than in water and this shows that a higher content of free tertiary amino groups are present in the gel, and therefore the swelling of the gel is pH-dependent (samples AT-CM1:2*, AT-$CM_0$1:1*, AT-$CM_0$1:1*, x3, and AT-CM1.6:1*).

The type of the crosslinking agent possessing longer or shorter alkyl spacer between siloxane sequence and functional groups also would influence the swelling behavior of the network. Therefore, it is difficult to correlate the swelling results with the reaction parameters or idealized structures. Detailed studies on the synthesis and networks characterization are the scope of our future paper.

Most of the synthesized crosslinked structures were investigated by TGA, the main data of the registered curves being presented in Table 8. The networks have comparable thermal stabilities, more significant differences being visible in T_{of} and residue values. The last characteristic appears to be in concordance with the low values for the found silicon content.

Table 8. The Main Thermogravimetric Data for Crosslinked Structures

Sample[a]	T_{on} [b]	T_{10} [c]	T_{40} [d]	T_{of} [e]	% Residuu
AT-$CM_0$1:1	50	215	278	600	2
AT-$CP_0$1:2*	55	230	285	670	5
AT-$CM_0$1:2, x3	72	217	270	535	1.5
AT-$CM_0$1:1*	58	205	260	635	8
AT-$CM_0$1:1*, x3	60	241	300	750	11
AT-CM1.6:1*	50	215	300	750	5

[a] The same significance like in Tables 6 and 7;
[b] The onset temperature;
[c] The temperature at which 10% wt loss is observed.
[d] The temperature at which 40% wt loss is observed.
[e] The offset temperature.

However, the most important characteristic of these ionic hydrogels, which constitutes the main reason for their synthesis and investigation, is the very high swelling degree in water, which makes them potential candidates for novel ionic hybrid hydrogels with applications in sanitary goods, controlled delivery systems, etc.

CONCLUSION

A short literature survey on the hydrogels and their applications has been first presented with a special concern on the ionic organic/inorganic hydrogels.

Some original results on the synthesis by less investigated approaches of novel siloxane-organic networks having preformed ionic groups or which are formed as crosslinking occurs, have been presented. (1) A polyaddition reaction between a reactive cationic oligomer, RCO, containing preformed QAS groups in the backbone and reactive groups like 2,3-epoxypropyl and 2-hydroxy-3-chloropropyl on the end of the chain and two bis-amino siloxanes (AP_0 and AP). (2) The Menschutkin reaction performed between either cationic polymers with PEG sequences and tertiary amine groups in the backbone and in the side chain (PEGA), or ODMAEM, with 1,3-Bis-(chloromethyl)-1,1,3,3-tetramethyldisiloxane (CM_0), 1,3-bis-(chloropropyl)-1,1,3,3-tetramethyldisiloxane (CP_0), α,ω-bis(chloromethyl)oligodimethylsiloxane (CM) and α,ω-trimethylsiloxy-oligodimethylmethylchloromethylsiloxane, CMCl. In all cases, siloxane derivatives were used as crosslinking agents for an organic oligomer or polymer.

Complex systems having siloxane moieties as a hydrophobic component and organic sequences with cationic groups as hydrophilic part have been obtained. Due to the presence of the latter, most of the prepared structures are able to swell in water behaving like hydrogels. The swelling ability can be designed by proper choice of the hydrophilic/hydrophobic balance. The ionic hybrids have been characterized by elemental analysis, IR spectroscopy, swelling degree in different solvents and thermogravimetric analysis.

ACKNOWLEDGEMENT

The authors gratefully acknowledge the financial support of this research from the Projects CEEX-MATNANTECH 40/2005 and 880/2006.

REFERENCES

[1] Park, J. H.; Bae, Y. H. *Biomaterials* 2002, *23*, 1797-1808.
[2] Hilt, J. Z.; Gupta, A. K.; Bashir, R.; Peppas, N. A. *Biomed. Microdevices* 2003, *53*, 177-184.
[3] Künzler, J.; Ozark, R. *J. Appl. Polym. Sci.* 1995, *55*, 611-619.
[4] Dinu, M. V.; Ozmen, M. M.; Dragan, E. S.; Okay, O. *Polymer* 2007, *48*, 195-204.
[5] Savina, I. N.; Mattiasson, B.; Galaev, I. Y. *Polymer* 2005, *46*, 9596-9603.
[6] Philippova, O. E.; Pieper, T. G.; Sitnikova, N. L.; Starodoubtev, S. G.; Khokhlov, A. R.; Kilian, H. G. *Macromolecules* 1995, *28*, 3925-3929.
[7] Starodoubtev, S. G.; Khokhlov, A. R.; Sokolov, E. L.; Chu, B. *Macromolecules* 1995, *28*, 3930-3936.
[8] Ozmen, M. M. ; Okay, O. *Eur. Polym. J.* 2003, *39*, 877-886.
[9] Kawaguchi, D.; Satoh, M. *Macromolecules* 1999, *32*, 7828-7835.
[10] Nishiyama, Y.; Satoh, M. *J. Polym. Sci. Part A Polym. Phys.* 2000, *38*, 2791-2800.
[11] Nishiyama, Y.; Satoh, M. *Macromol. Rapid Commun.* 2000, *21*, 174-177.
[12] Pinkrah, V. T.; Beezer, A. E.; Chowdhry, B. Z.; Gracia, L. H.; Cornelius, V. J.; Mitchell, J. C.; Castro-Lopez, V.; Snowden, M. J. *Colloids Surf. A* 2005, *262*, 76-80.
[13] Ozmen, M. M. ; Okay, O. *J. Appl. Polym. Sci.* 2006, *99*, 319-325.
[14] Kim, D.; Park, K. *Polymer* 2004, *45*, 189-196.
[15] Minato, T.; Satoh, M. *J. Polym. Sci. Part A Polym. Phys.* 2004, *42*, 4412-4420.
[16] Ozmen, M. M. ; Okay, O. *Polymer* 2005, *46*, 8119-8127.
[17] Yamamoto, K.; Serizawa, T.; Muraoka, Y.; Akashi, M. *Macromolecules* 2001, *34*, 8014-8020.
[18] Kim, S. J.; Yoon, S. G.; Kim, S. I. *J. Appl. Polym. Sci.* 2004, *91*, 3705-3709.
[19] Kim, S. J.; Shin, S. R.; Lee, S. M.; Kim, I. Y.; Kim, S. I. *Smart Mater. Struct.* 2004, *13*, 1036-1039.
[20] Kim, S. J.; Lee, K. J.; Kim, S. I. *J. Appl. Polym. Sci.* 2004, *91*, 1097-1101.
[21] Elliot, J. E.; MacDonald, M.; Nie, J.; Bowman, C. N. *Polymer* 2004, *45*, 1503-1510.
[22] Yazici, I.; Okay, O. *Polymer* 2005, *46*, 2595-2602.
[23] Komarova, G. A.; Starodubtsev, S. G.; Khokhlov, A. R. *Macromol. Chem. Phys.* 2005, *206*, 1752-1756.
[24] Marsano, E.; Bianchi, E.; Vicini, S.; Compagnino, L.; Sionkowska, A.; Skopinska, J.; Wisniewski, M. *Polymer* 2005, *46*, 1595-1600.
[25] Zhang, Yu. X.; Wu, F. P.; Li, M. Z.; Wang, E. J. *Polymer* 2005, *46*, 7695-7700.

[26] Nam, K.; Watanabe, J.; Ishihara, K. *Polymer* 2005, *46*, 4704-4713.
[27] Karadağ, E.; Üzüm, O. B. *Polym. Bull.* 2005, *53*, 387-392.
[28] Noh, J.-G.; Sung, Y.-J.; Geckeler, K. E.; Kudaibergenov, S. E. *Polymer* 2005, *46*, 2183-2190.
[29] Martel, B.; Ruffin, D.; Weltrowski, M.; Lekchiri, Y.; Morcellet, M. *J. Appl. Polym. Sci.* 2005, *97*, 433-442.
[30] Tang, C.; Yin, C.; Pei, Y.; Zhang, M.; Wu, L. *Eur. Polym. J.* 2005, *41*, 557-562.
[31] Xiang, Y.; Peng, Z.; Chen, D. *Eur. Polym. J.* 2006, *42*, 2125-2132.
[32] Abbasi, F.; Mirzadeh, H. *J. Polym. Sci. Part B Polym. Phys.* 2003, *41*, 2145-2156.
[33] Pai, T. S. C.; Barner-Kowollik, C; Davis, T. P., Stenzel, M. H. *Polymer* 2004, *45*, 4383-4389.
[34] Huang, Z.; Yu, Y.; Huang, Y. *J. Appl. Polym. Sci.* 2002, *83*, 3099-3104.
[35] Aoyagi, T.; Nakamura, T.; Yabuchi, Y.; Nagase, Y. *Polym. J.* 1992, *24*, 545-553.
[36] Sauvet, G.; Dupond, S.; Kazmierski, K.; Chojnowski, J. *J. Appl. Polym. Sci.,* 2000, *75*, 1005-1012.
[37] Sauvet, G.; Fortuniak, W.; Kazmierski, K.; Chojnowski, J. *J. Polym. Sci. Part A Polym. Chem.* 2003, *41*, 2939-2948.
[38] Hazizza-Laskar, J.; Helary, G.; Sauvet, G. *J. Appl. Polym. Sci.* 1995, *58*, 77-84.
[39] Gottenbos, B.; van der Mei, H. C.; Klatter, F.; Nieuwenhuis, P.; Busscher, H. J. *Biomaterials* 2002, *23*, 1417-1423.
[40] Traitel, T.; Kost, J.; Lapidot, S. A. *Biotechnol. Bioeng.* 2003, *84*, 20-28.
[41] Adhikari, B.; Majumdar, S. *Prog. Polym. Sci.* 2004, *29*, 699-766.
[42] Sakai, Y.; Matsuguchi, M.; Hurukawa, T. *Sensors Actuators* B 2000, *66*, 135-138.
[43] Sakai, Y.; Sadaoka, Y.; Matsuguchi, M.; Sakai, H. *Sensors Actuators B.* 1995, *24-25*, 689-691.
[44] Penza, M.; Cassano, G. *Sensors Actuators B* 2000, *68*, 300-306.
[45] Wang, J.; Lin, Q.; Zhou, R.; Xu, B. *Sensors Actuators B* 2002, *81*, 248–253.
[46] Sormani, P. R.; Viswanath, A. K.; Aiyer, R. C.; Radhakrishnan, S. *Sensors Actuators B.* 2001, *80*, 141–148.
[47] Suri, K.; Annapurni, S.; Sarkar, A. K.; Tandon, R. P. *Sensors Actuators B* 2002, *81*, 277-282.
[48] Pi-Guey, S.; Cherng, I. C.; Ren-Jang, W. *Anal. Chim. Acta* 2001, 449, 103-109.
[49] Fortuniak, W.; Rozga-Wija, K.; Chojnowski, J.; Labadens, F.; Sauvet, G. *React. Funct. Polym.* 2005, *61*, 315-323.
[50] Liu, J.; Xu, T.; Fu, Y. *J. Membr. Sci.* 2005, *252*, 165-173.
[51] Kolesnik, Yu. R.; Goloshchapov, A. V.; Koval, I. V.; Teslenko, V. V.; Grekov, A. P.; *Polymer Sci. USSR* 1990, *32*, 1922-1928.
[52] Cazacu, M.; Ioanid, A.; Ioanid, G.; Racles, C.; Vlad, A. *Appl. Organometal. Chem.* 2006, *20*, 494-498.
[53] Dragan, S.; Cazacu, M.; Tóth, A. *J. Polym. Sci. Part A Polym. Chem.* 2002, *40*, 3570-3578.
[54] Cazacu, M.; Ghimici, L.; Dragan, E. S. *J. Polym. Sci. Part A Polym. Chem.* 2004, *42*, 3720-3728.
[55] Dragan, S.; Barboiu, V.; Ghimici, L.; Stoica, Gh. *J. Macromol. Sci.Chem.* 1991, *A 28*, 283-296.
[56] Dragan, S.; Ghimici, L. *Angew. Makromol. Chem.* 1994, *215*, 107-119.

[57] Cazacu, M.; Marcu, M.; Caraiman, D.; Lazarescu, S.; Vlad, A. *Rev. Roum. Chim.* 1998, *43*, 757-763.
[58] Kolesnik, Yu. R.; Golosceapov, A. V.; Kovali, I. V. *Ukr. Khim. J.* 1990, *56*, 655-659.
[59] Bajpai, S. K.; Johnson, S. *React. Funct. Polym.* 2005, *62*, 271-283.
[60] Gan, L. H.; Deen, G. R.; Gan, Y. Y.; Tam, K. C. *Eur. Polym. J.* 2001, *37*, 1413-1478.
[61] Dragan, E. S.; Cazacu, M., Manuscript in preparation.

In: New Trends in Ionic (Co) Polymers and Hybrids
Editor: Ecaterina Stela Dragan, pp. 165-234
ISBN 978-1-60021-611-4

Chapter 7

SELF-ASSEMBLED MULTILAYERS: CONSTRUCTION, PROPERTIES, AND APPLICATIONS

Ecaterina Stela Dragan** and Florin Bucatariu
"Petru Poni" Institute of Macromolecular Chemistry,
Aleea Grigore Ghica Voda 41 A, RO-700487 Iasi, Romania

ABSTRACT

The self-assembled multilayers (SAMs) built with a large variety of building blocks such as synthetic polyelectrolytes, biopolymers, organic compounds, organic and inorganic colloidal particles and their applications, were briefly reviewed in this Chapter, on the basis of the the newest information, being well known that the number of publications in this field increases continuously with every year. The mechanism of the multilayer growth and factors which influence the stability of these nanoarchitectures are still being investigated with interest for both fundamental and practical reasons. The main tools which allow to control the properties of these nanostructured materials and to tailor nanoobjects for specific applications have been discussed in detail, with a special concern on the influence of ionic strength, pH, solvent quality, and temperature, the influence of the last parameter being less investigated so far.

Keywords: building blocks, ionic strength, counterion, self-assembled multilayers, ESAM, HSAM, polyelectrolytes, biopolymers, azo dyes, methylene blue, colloidal particles, microcapsules.

LIST OF ABBREVIATIONS

AGA	alginic acid
AFM	atomic force microscopy

* E-mail: sdragan@icmpp.ro

AMPS	2-acrylamido-2-methylpropanesulfonic acid
CHI	chitosan
CMC	carboxymethylcellulose
CMP	carboxymethylpullulan
CNT	carbon nanotube
SWCNT	single walled carbon nanotube
CV	cyclic voltametry
Cyt	cytochrome c
DMA	dynamic mechanical analysis
DMAA	N,N-dimethylacrylamide
DMPG	dimyristoyl phosphatidyl glycerol
DS	dextran sulfate
ESAM	electrostatically self-assembled multilayers
FTIR-ATR	Fourier transform infrared spectroscopy with attenuated total reflection
GA	glucoamylase
GOD	glucose oxidase
HA	hyaluronic acid
Hb	hemoglobin
HSA	human serum albumine
HSAM	hydrogen bonded self-assembled multilayers
IgG	immunoglobulin G
LbL	layer-by-layer
LDHs	layered double hydroxides
Lys	lysozyme
Mb	myoglobin
MB	methylene blue
MF	melamine formaldehide
MWCNTs	multiwalled carbon nanotubes
NaPSS	poly(sodium styrenesulfonate)
NIPAM	N-isopropylacrylamide
NMVA	N-methyl-N-vinylacetamide
NMVF	N-methylvinylformamide
PAA	poly(acrylic acid)
PABA	poly(aniline boronic acid)
Pasp	poly(L-aspartic acid)
PAZO	poly{1-4[4-(3-carboxy-4-hydroxyphenylazo)benzensulfonamido]-1,2-ethanediyl sodium salt}
PCA_5 -	polycation of ionene type containing about 95 mole % of N, N-dimethyl-2-hydroxypropyleneammonium chloride units
PDMAEM -	poly(dimethylaminoethylmethacrylate)
PDMS -	poly(dimethylsiloxane)
PEI -	poly(ethyleneimine)
PMAA -	poly(metacrylic acid)
PM4VP -	poly(N-methyl-4-vinyl pyridinium iodide)
PAH -	poly(allylamine hydrochloride)

PANI	polyaniline
PDADMAC	poly(diallyldimethylammonium chloride)
PEL	polyelectrolyte
PEO-	poly(ethylene oxide)
PET	poly(ethyleneterephtalate)
PGA	poly(glutamic acid)
PLB	poly(L-lysine-co-ε-biotinyl-L-lysine)
PLL	poly(L-lysine)
POD	peroxidase
Poly(VAm-co-NVF)	poly(vinylamine-co-N-vinylformamide)
PRM	protamine
PS	polystyrene
PTFE	poly(tetrafluoroethylene)
PVPy	poly(4-vinylpyridine)
PVPON	poly(vinylpyrrolidone)
PVS	poly(potasium vinyl sulfonate)
QCM	quartz crystal microbalance
QDs	quantum dots
SANP-	sodium 9-antracenpropionate
SDS	sodium dodecyl sulfate
SFM	scanning force microscopy
SMCs	smooth muscle cells
SPRS	surface plasmon resonance spectroscopy
SWCNT	single walled carbon nanotube
XPS	X-ray photoelectron spectroscopy

1. INTRODUCTION

Self-assembled multilayer (SAMs) nanoarchitectures constructed with a large variety of building blocks and stabilized mainly by electrostatic and hydrogen bonding interactions have been selected for discussion in this chapter. Electrostatic self-assembly has been developed first by Iler who alternately deposited oppositely charged layers of colloidal particles, such as silica and alumina, onto a glass surface [1]. The real progress of this technique started in 90's when Decher et al. assembled oppositely charged bipolar amphiphiles [2] and polyelectrolytes [3-5]. This technique proved to be very attractive by its versatility, numerous components being used as building blocks to generate new advanced materials whose properties can be controlled at a nanoscale level by some characteristics of the deposition medium. Any surface, irrespective of geometry, can be used as substrate for the SAM construction. The layer-by-layer (LbL) deposition is easy to perform manually or automatically in any laboratory and allow a large freedom to the imagination of researchers to combine different building blocks to get new performances at a nano scale. The main publications in this field are time to time reviewed by renowned scientists [6-13]. Therefore, from the beginning the authors of this review ask to be forgiven for their daring to prepare a new review on this topic concerning the supramolecular architectures. It was difficult to select the most representative contributions in the field considering the high level of journals hosting the authors who belong to very powerful research groups involved in this type of nanoconstruction. On the other side, the main references which constitute the basis of one of the fastest developed field will be mentioned with priority.

To guide the reader, a rough classification of the SAMs materials will be performed from the beginning, considering the structure of the main building block as a criterion. Thus, the systems which will be discussed in this review and the references where they can be easily found have been summarized in Table 1.

Table 1

Main building block	References
Strong polyelectrolytes	[6-10], [14-59], [61-127]
Weak polyelectrolytes	[135-223]
Proteins and nucleic acids	[224-245]
Low molar mass organic compoundsdyes other low molar mass organic compounds	[255-270] [271-274]
Colloidal particles organic inorganic	 [291-302] [125,303-333]

2. SAMs Based on Strong Polyelectrolytes

Strong polyelectrolytes (PEL) are maybe the most often used as building blocks in the construction of multilayers for fundamental reasons (mechanism of the LbL thin films formation and factors controlling their construction) as well as for generation of novel knowledge-based nanocomposite structures for special applications [6-10,14-59,61-127]. However, only a few of references have been selected for this chapter, mainly the latest ones. The ability to tailor such materials with a predictable behavior under certain conditions depends on the well understanding of the factors which influence their growth and structure.

2.1. Factors Controlling the Multilayer Buildup

The main factors which have been investigated to monitor the multilayer thin film formation and properties are as follows [4-6,14-28]:

- Preparation mode
- Polyion structure and molar mass
- Ionic strength
- Temperature

The influence of other parameters such as: polyion concentration and dipping time have been established from the beginning [6-10] and will not be discussed here in detail.

2.1.1. Preparation Mode

The most simply and very well known mode to buildup electrostatically SAMs (ESAMs) thin films is alternately dipping of a substrate in two beakers each containing the aqueous solutions of the oppositely charged polyions, intercalating 2-3 rinse steps in pure water after each polyion adsorption, with drying after each adsorption step [2-5,15,17,24,25,29,37-40,51-55,59,63,65,87] or without drying [4,5,20-23,25-28,31-36,50,54,57,58,70-74,81-87,95,101-103] in between. To determine if there is an influence of drying on the film formation, different investigations have been performed comparing the data for films obtained by drying after each layer with those dried only after deposition of a total number of layers [22,25,52]. For most of the samples, the adsorbed amounts and the surface roughness were significantly higher when films have been dried after each deposition step. The atomic force microscopy (AFM) images of the films prepared with and without drying after each adsorption step showed a higher surface roughness for the films dried after each adsorption step. On the other hand, when the films were built by the deposition of PELs from 1 M NaCl, no influence of drying has been observed [22].

This preparation mode could be manually [2-5,14-27,48,51,53,54-57,59,65,74,76] or automatically performed [22,23,25,52,55,59,64,68,71,76-78,82,95,98]. Usually, no influence of the preparation mode was observed. The kinetic studies showed that the equilibrium of adsorption of a PEL layer is reached within 10-20 min [4,14,25]. A background of four PEL layers is usually necessary prior the regular, stepwise growth to be established [4-7,19]. It was found that the substrate surface charge has a little influence on the SAMs growth [16,21]. An influence of the substrate restricted to the first adsorbed layers has been reported by Hoogeveen et al. [16], the structure and interaction between oppositely charged PELs governed the adsorbed amount after that. Significant differences of PEL adsorption as a function of the substrate surface charge have been recently reported by Schwarz et al. in the system poly(dimethylaminoethylmethacrylate hydrochloride) (PDMAEM-HCl)/poly(sodium styrenesulfonate) (NaPSS), by employing a large variety of surface analytical methods to follow the film growth both in dry state (X-ray photoelectron spectroscopy, XPS, AFM, UV spectroscopy) and wet state (zeta potential, ellipsometry and surface plasmon resonance spectroscopy, SPRS) [57].

The solvent usually used in the ESAMs buildup is water, but mixed solvents or organic solvents have been also utilized. The influence of solvent quality on the growth of PEL multilayers has been studied by Dubas and Schlenoff [22] and Caruso et al. [29], the solvent quality being varied by the addition of ethanol to the aqueous solutions of PELs used for deposition of layers. The multilayer growth has been followed by SPRS, quartz crystal microbalance (QCM) and AFM. It was expected that the presence of ethanol to reduce the solubility of the oppositely charged PELs, poly(allylamine hydrochloride) (PAH) and NaPSS and to influence the layer thickness because changing the dielectric constant of the mixed solvent, conformational changes of the PEL chains occur. The chains adopt a more compact conformation, demonstrated also for other polycations in mixed solvents [128-130]. Therefore, increasing the amount of ethanol in the deposition solution resulted in a strong increase of the multilayer film thickness and mass loading [29]. Close to the precipitation point, the film thickness sharply increased and the film exhibit an exponential growth pattern resulting from the adsorption of PELs in highly coiled conformations. By this study it was demonstrated that the solvent quality could be used as a parameter to tune the multilayer properties like thickness, permeability, porosity.

Other preparation modes have been meantime suggested. Thus, Schlenoff et al. [30], using poly(diallyldimethylammonium chloride) (PDADMAC) and NaPSS as oppositely charged polyions, compared the dip-coating mode with a spray-coating mode, the thickness of the whole PEL thin film being almost the same for the same contact time of the substrate with the PEL solution. Such method could be advantageous for fast modification of large area membranes and for external coatings as corrosion protection or adhesion modification. Ferry et al. recently suggested a novel pending drop technique which allow for the synthesis of free-standing nanocomposites [31]. According to this method, a lipid (dimyristoyl phosphatidyl glycerol, DMPG) monolayer deposited on a pendant drop was compressed to present a defined surface charge density and then has been used as a charged molecular template for the LbL adsorption of PAH/NaPSS polyion pair. The significance of this technique is that the geometry of the drop interface permits precise measurements of the dynamic surface tension providing insight into the dynamics of multilayer buildup, inaccessible by other methods. Thus, free-standing nanocomposites of the structure DMPG-(PAH/NaPSS)$_{n=1-3}$ were prepared on the surface of the pendant drop by precise subphase exchange (polycation, polyanion, and intermittent flushing) and the dynamics of the synthesis were followed via in situ surface tension measurements.

A dynamic and fast LbL deposition method to construct region-selective multilayer thin films from PDADMAC and poly{1-4[4-(3-carboxy-4-hydroxyphenylazo) benzensulfonamido]-1,2-ethanediyl sodium salt}(PAZO), combining the advantages of LbL deposition fluidic device, was reported by Kim et al. [32]. Multilayer films fabricated according to this method had a similar quality of the thickness and roughness compared with those fabricated by the dipping method, processing time being only of 90 s. Free-standing films have been also fabricated by this method.

2.1.2. Polyion Structure and Molar Mass

Numerous polycations and polyanions have been used to buildup multilayer thin films with special properties such as: conjugated polyions [14,33-35] as new approaches to photo voltaic devices, redox polyions [18,21,36], photosensitive polycationic diazoresins able to transform ionic stabilized multilayers in covalently-attached thin films [37-40], homopolymers [32,41-46,49] and amphiphilic copolymers with photoactive properties [47], etc. Complete charge reversal of the surface, i.e., overcompensation after each adsorption stage, is the key to subsequent deposition of the oppositely charged PELs. A linear increase of the adsorbed amount of PELs is usually observed with an increasing number of layers, mainly when fully charged homopolymers are used as building blocks [4,6,7].

The structure of the ionic repeat unit and the charge density are the most important characteristics which strongly influence the properties of the ESAMs thin film [20,26,48-59]. Strong PELs bearing fully ionized groups are preferred for the study of this parameter because their charge is independent of the solution pH. When a weak PEL was used, the studies were performed under pH conditions which ensure that the PEL remains in its most highly charged form (for example, protonated poly(allylamine), PAH). In the followings, our attention will be focused on the role of charge density on the construction/desorption balance in multilayer films based on such building blocks. The charge density has been varied either by the quaternization of the tertiary amines contained in weak polybases [16] or by the synthesis of ionic/nonionic copolymers with a strict control of the ratio between the two types of comonomers [7,48-52,54]. Thus, Hoogeveen et al. performed the first systematic study of the influence of polyion charge density on the multilayer buildup employing quaternized poly(vinylimidazole), the quaternization degree being varied in the range 0 up to 0.52, and fully deprotonated poly(acrylic acid) (PAA) [16]. The multilayer growth and stability have been followed by the measurements of electrophoretic mobilities and hydrodynamic layer thickness. At low charge density, no complexation of the oppositely charged PELs occured on the surface; at a medium charge density, the charge complexation occured but has been followed by a partial desorption of the complex; with highly charged polyions very stable multilayers could be formed [16]. A critical charge density of 0.12 at an ionic strength, I, of 0.001 mol.L^{-1}, and of 0.18 for $I = 0.01$ mol.L^{-1} have been evidenced. Weaker binding of PELs to a surface resulting in desorption has been attributed by Laschewsky et al. to the charge mismatching of the anionic and cationic PELs used to buildup the ESAMs thin films [7,49].

The existence of a critical charge density is strongly correlated with the location of charges on the main chain or on pendant side chains (their accessibility), and also with the structure of the nonionic comonomer, which could induce further nonelectrostatic interactions and thus may influence the global behavior of the copolymer. Kolarik et al. examined the buildup of ESAMs thin films onto gold surfaces from water-soluble cationic and anionic PELs different by the chemical structures, molar mass (5 x 10^4 to 1.5 x 10^7 Da), and charge densities (20-100%) using QCM and AFM to follow the multilayer growth [48]. They observed that when the charge density of the cationic PELs decreased, the number of layers that could be deposited also decreased. At the same time, the film became much more open and heterogeneous. High molar mass cationic PELs, such as Zetag 32, 64, and 89, could be used to build highly heterogeneous architectures with the fully charged NaPSS of a low molar mass (7 x 10^4 Da). The charge density decreased in the order Zetag 89 > Zetag 64 > Zetag 32, and the adsorbed amount accordingly increased.

In their study on the competition between surface/PEL and PEL/PEL interactions, Klitzing et al. [50] used a statistic copolymer consisting in a positively charged monomer, DADMAC, and a neutral monomer, N-methyl-N-vinylacetamide (NMVA) [P(DADMAC-*co*-NMVA)] with a variable linear charge density of f = 89%, 75%, 50%, 24%, and 14%. The molar mass was almost the same, 100 000 Da. X-ray reflectivity spectra gave information on the total thickness of the films. The authors identified three regions for the multilayer growth:

- $f \geq 75\%$, thick films were built, which became slightly thinner with increasing degree of charge. This was assigned to the transition from a more coiled to a flat chain conformation of the adsorbed chains because of the increasing electrostatic repulsion between the charges along the chain, accompanied by the decrease of film roughness.
- $f \leq 50\%$, in this region the change in surface charge after adsorption of the polycation seems not to be high enough for the formation of a multilayer system, and therefore only thin films were built. This fact was almost independent of the charge density (between 14% and 50%). No neutral polymer (polyNMVA) was adsorbed on the surface and this led to the conclusion that at least a low polymer

charge (between 0 and 14%) is necessary for the adsorption of a polymer chain onto a charged surface.

- the charge overcompensation threshold was located between 50% and 75%.

Almost the same charge density threshold, 0.574-0.711, below which the multilayer did not growth, has been found by Sun et al. [56] in the preparation of ESAMs thin films using PDADMAC and pyrene labeled copolymers of 2-acrylamido-2-methylpropanesulfonic acid (AMPS) and N,N-dimethylacrylamide (DMAA), with charge densities of 0.20-0.999 AMPS mole fraction. A monotonous increase of the multilayer thickness has been reported for copolymers with a mole fraction of AMPS higher than 0.574, an abrupt decrease of the adsorbed amount after 13th layer for polyanion with a mole fraction of AMPS of exactly 0.574, and no adsorption for polyanion with a mole fraction of AMPS of 0.449.

To bring more light on the role of the charge density as driving force in the multilayer growth, Caruso et al. investigated the influence of polycation charge density on the growth of multilayer thin films with statistic copolymers of DADMAC with either NMVA [51] or acrylamide [54]. In the first study, copolymer models with nearly the same molar mass (ca. 10^5 g.mol^{-1}) of DADMAC with NMVA, varying the percentage of DADMAC from 24 up to 100 %, and different techniques have been employed to probe the amount of PELs adsorbed on each layer and to gain insight into the factors that govern the ESAMs buildup. Such PEL models allowed exclusion of possible chain length effects and also the nonelectrostatic effects are diminished, the polymers being water soluble even in the absence of ionic comonomer. A critical charge density between 53 and 75 mol % of DADMAC has been identified in 0.5 M NaCl aqueous solution, below which no significant layer growth was possible [51]. By using multi-techniques measurements, the authors assumed another mechanism for the inhibition of multilayer growth with copolymers under the critical charge density than the absence of overcompensation of the surface. In agreement with the model of Hoogeveen et al. [16] discussed above, Caruso et al. consider that when PEL copolymers below the critical charge density are adsorbed on oppositely charged surfaces, the individual copolymer chains are very weakly bound. Exposure to the oppositely fully charged PEL, NaPSS, in the next adsorption step, results in removal of the weakly bound copolymer that probably prefer to form complexes in solution. Concerning the differences in the film characteristics with highly charged copolymers, the authors observed higher adsorbed amounts when the copolymer with 75 mol % DADMAC was used compared with fully charged PDADMAC. At a lower charge density, the PEL must form loops and tails leading to rougher and thicker layers comparative with fully charged PELs which are adsorbed in a flat conformation, mainly in the absence of neutral salts [60]. Therefore, the adsorbed amount increased as the charge density decreased from PDADMAC to the copolymer containing 75 mol % of positively charged monomer. A higher film thickness with a decreasing charge density of the adsorbing PELs can also be explained by the condition of surface charge overcompensation [3-6]. With PELs of decreasing charge density, more PEL chains have to be adsorbed on the surface to achieve charge reversal. On the other hand, when the charge density of PEL is too low, the attraction to the surface is weakened to such an extent that the adsorbed amounts decreased.

In a more recent study, Caruso et al. demonstrated that multilayer formation can be achieved with PELs of very low charge density (as low as 8%), with no critical minimum charge density, in contrast to previous investigations [54]. Random copolymers composed of DADMAC and acrylamide (AM) in different ratios, in alternation with fully charged NaPSS, bave been used to construct multilayers. No critical charge density limit has been observed in this system. Synergistic nonelectrostatic interactions cooperate to the growth of multilayers with PELs of a very low charge density, because the amide groups of the AAm can form hydrogen bonds with the other copolymer chains and therefore are likely to contribute to multilayer growth. Partial desorption of the PELs for film built with copolymers of low charge density upon the adsorption of the next layer has also been reported [54].

Laschewsky et al. have investigated the multilayer buildup with NaPSS and poly($DADMAC_x$-*co*-$NMVF_{1-x}$) with DADMAC content varying in the range 0-1 as polycations, the PEL adsorption being performed only from salt-free aqueous solutions [52]. The monomer N-methylvinylformamide (NMVF) is less hydrophobic comparative with NMVA, the possible additional hydrophobic interactions being thus minimized. Below 9% DADMAC, the reliable layer deposition was zero. With the increase of the linear charge density, the system passed first by a charge dependent "Debye-Huckel" regime, that is the content of the ionic comonomer varied between 10 and 40 mol %, and the thickness increment continuosly decreased down to a constant value of 3-4 Å. At higher charge densities, the multilayer grew in a charge screening regime, where counterion condensation controlled the multilayer growth.

Other combinations of PELs have been employed by Serizawa et al. [53], the charge density of polycation being controlled by the hydrolysis degree of poly(vinylamine-co-N-vinylformamide) (poly(VAm-co-NVF)), with NaPSS used as fully charged polyanion, the deposition being performed from salt-free aqueous solutions. The threshold VAm content for zigzag and stepwise growth has been identified at 30%, the highest amount of polycation being adsorbed at this charge density. After this value, the adsorbed amount continuosly decreased and stabilized around 50% VAm, no variation being observed after that. Three distinct assembly patterns have been revealed by plotting the adsorbed amount against the assembly step: zigzag corresponding to hydrolysis degree lower than 30%; exponential, for VAm content between 30% and 50%; an almost linear increase when VAm content increased up to 100%. AFM images revealed molecularly smooth surfaces irrespective of the VAm content in polycation, ultrathin hydrogels resulting by swell in water.

Well ordered ESAM built with strong PELs as building blocks have been designed by Arys et al. by combining a special polycation of ionene type with different polyanions [84]. The level of order of the oppositely charged PEL layers in the multilayer was compared with the structure of the corresponding PEL complexes in solid state. Internally structured multilayers have been identified when poly(vinyl sulfate), poly(vinyl sulfonate) (PVS) and poly(bis(3-sulfopropyl)itaconate) have been used as polyanions, i.e., the appearance of order into the multilayers occurred when the oppositely charged PELs were capable to form structured complexes.

To summarize, the differences in the multilayer growth arising from the results discussed above reveal that not only the charge density is important but also the whole chemical structure of oppositely charged PELs, mainly of the nonionic comonomer, strongly influence the multilayer growth and morphology of ESAMs thin films. This explains why some polymers pairs did not show a critical charge density in the multilayer buildup [54], and others showed a smooth surface of the ESAMs thin films irrespective of the polyion charge density [53]. These findings led to the conclusion that both the surface/PELs interactions and PELs structure control the PEL adsorption and the growth of ESAMs thin films.

The influence of PEL molar mass on the multilayer growth has been investigated by some groups [7,21,48,54,58,59,75,112]. Different combinations of low and high oppositely charged PELs have been employed by Kolarik et al. in the study of the influence of PEL molar mass on the ESAMs growth [48]. No significant influence of the PEL molar mass on the thin films buildup has been observed when fully charged strong PELs have been adsorbed from solutions with a low ionic strength [7,48,58]. In contrast, when films have been prepared with PELs of lower charge density adsorbed from solutions containing higher amounts of salt, where the PEL chains are in a more coiled conformation, a significant influence of the PEL molar mass has been reported. "Stripping versus sticking" as a function of PELs molar mass has been investigated by Schlenoff et al. by using combinations of a polyanion (NaPSS) and a polycation, poly(methyl-4-vinylpyridine) (PM4VP), characterized by pH independent charge density, different molar masses and narrow molar mass distributions [59]. The use of polymers of ca. 10^4 and 10^5 Da has been considered representative for the two molar mass regimes (low and high). With a molar mass around 10^4 Da, the thickness increment realized on addition of the shorter polymer was partially lost on exposure to the solution of longer PEL. The multilayer growth was completely inhibited when both oppositely charged PELs of low molar mass have been employed. The balance stripping/sticking depended not only on the polyions molar masses, but also on their structure and efficiency in forming soluble or quasi-soluble interpolyelectrolyte complexes (Q-PECs). Thus, Schlenoff et al. observed a difference between the multilayer growth for the polyion pairs NaPSS low/PM4VP high and NaPSS high/PM4VP low. Even if both systems provided a net increase in film thickness, the first system led to significant lower thicknesses, the difference being explained by the hydrophilicity of PM4VP which probably makes it more efficient in forming Q-PECs and stripping it off the surface comparative with the second system.

2.1.3. Ionic Strength

The influence of low molar mass electrolytes on the properties of the ESA thin films has been remarked from the beginning, an increase of the thickness of one layer pair from 13 Å, in the absence of salt, up to 34 Å, at 0.9 mol of NaCl/L, being reported by Lvov et al. [4]. The influence of the ionic strength will be discussed not only as one of the main factors to control the ESAMs growth [16-18,20-25,27,28,48,49,59,61-72,80], but also as a tool to control the swelling/permeability and the stability (erosion) of the ESAM thin films [15,27,28,68,70,73,74,81].

The screening of the repulsion between charges along the same chain by the salt ions changes the polyion conformation, altering the Debye length of the electrostatic interactions. Therefore, in dilute aqueous solutions, polyion chains change their conformation from a rodlike shape to a random coil, AFM images showing that film roughness increases with the salt concentration. Globular surface subunits suggest the PEL coiling induced by salt. By their charge screening effect, salt ions decreased both intrachain repulsion and interchain attraction, more polyion being needed to reverse the surface charge, creating thicker films. The trends observed in the film growth in the presence of salt are similar with those observed when the charge density of PELs decreased, since in both cases the stiffness of the polymer chain is reduced.

2.1.3.1. Salt Concentration

A linear increase of the layer pair thickness with the NaCl concentration (c_{NaCl}) has been reported [7,19,22,58], though the film thickness growth proportional to the square root of the ionic strength of the adsorption solution [17] or even lower power of c_{NaCl} (0.05-0.15) have been also observed [27]. An interesting behavior of the ESAMs built with PELs (NaPSS and PAH) deposited from NaCl with increasing concentrations was evidenced by Ladam et al. [27], using in situ characterizations by means of streaming potential measurements and scanning angle reflectometry (SAR). Thus, a multilayer built in salt solutions expanded when was brought in contact with pure water, but the structural modifications of the film were fully reversible, i.e., the initial structure of the multilayer was recovered when water was replaced by the salt solution used for the multilayer buildup. Considering the structural characteristics in the multilayer thickness, the authors identified three zones: a zone of the first layers close to the substrate, zone I, slightly different that zone II (core zone), characterized by a constant increase of the thickness with the number of layers, the last zone, zone III, being constituted by the outermost region with different properties. Zone II, being a zwiterionic polyelectrolyte complex, should swell in salt solution, because salt ions will break ion pairs formed between oppositely charged PELs. The layer in zone III being more extrinsic compensated should swell in pure water, as it was observed. Zone I supports the necessity of a precursor film (background), which ensured a homogeneous surface.

To get further information on how the salt concentration influences the thickness, roughness and morphology of the multilayer, McAloney et al. [63] investigated by AFM the multilayer growth by adsorption of NaPSS and PDADMAC from NaCl with concentrations from 10^{-4} up to 1 M. The authors found the thickness of multilayer increased almost linearly, irrespective of the number of bilayers, up to 0.3 M NaCl. Above 0.3 M NaCl, the increase of the multilayer thickness was linear only after the third bilayer, a thickness/bilayer of ~ 46 nm being found comparative with ~ 6 nm/bilayer found for the first three bilayers. This result was consistent with the difference between the first few layers and the later ones. The film thickness remained small and approximately constant for salt concentrations lower than 0.1 M NaCl, and the film thickness increased linearly with the concentration of added salt, for 10 bilayer films. The morphology of these films examined by AFM was as follows: the first and second bilayers were flat and featureless, but small pores as well as some larger globular features were observed for the third bilayer. By the fifth bilayer, a vermiculate morphology was observed. A linear increase of the film thickness as a function of bilayer number has been also found by Harris and Bruening [28] for NaPSS/PAH system with both polyions adsorbed from 0.5 NaBr.

The influence of the ionic strength on the selectivity of LbL deposition of polyions on chemically patterned surfaces has been intensively investigated by Hammond et al. [12,17,23], in the range from 0.001 to 4 M NaCl. All PEL concentrations were thus selected as polyions to be under or close to their critical concentration, c*. The layer thickness increased linearly with bilayer number for these systems. The authors found the increase of the film thickness proportional to the square root of the ionic strength of the adsorption solution as it was previously showed [60]. The ionic strength has been used as a tool to alter polyion deposition and selectivity on the functionalized surfaces, for the patterning of complex 3D superstructures [17,23].

Izumrudov et al. performed an interesting study on the multilayer formation correlated with the interpolyelectrolyte complexes formation in NaCl of increasing concentrations [67].

Multilayer growth in the system poly(N-ethyl-4-vinylpyridinium) bromide (QPVP)/poly(methacrylic acid) sodium salt (PMAA) was followed in situ by Fourier transform infrared spectroscopy with attenuated total reflection (FTIR-ATR). It was found that: (1) the amount of polymer deposited goes through a maximum as a function of ionic strength; (2) the position of this maximum shifts to higher ionic strengths when molar mass of PMAA increased; (3) thicker multilayers were formed with longer PMAA chains. The inhibition of the multilayer growth for ionic strengths lower than 0.2 M NaCl was correlated with the stability of the PEL complex in solution. This study could be a model for the ionic strength dependence of the polyion deposition at the surface comparative with salt-induced phase separation in PELs mixture in solution.

Recent publications report on the presence of counterions in the LbL thin films prepared by the deposition of oppositely charged PELs from aqueous solution of NaCl with increasing concentrations [66,70]. Using XPS as a sensitive method for the characterization of the films prepared from NaPSS and PAH, Laurenço et al. [70] demonstrated that sodium and chloride counterions are present in the ESAM thin films.

Post-preparation swelling of NaPSS/PAH and PVS/PAH multilayers, in solutions of different ionic strengths, (0.1 – 3) mol/L NaCl, revealed reversible thickness changes, which leveled off at salt concentrations of 0.1 mol/L [15]. This behavior showed that the salt is able to penetrate into the film and to reversibly break some of the polyanion/polycation bonds. No changes were observed after that even up to a NaCl concentration of 3 mol/L, and this showed the high stability of these multilayer films. Smoothing effect by swelling (annealing) of ESAM thin films in aqueous solution of NaCl [62,71] or KBr [73] with increasing concentrations has been also observed for other polyion pairs, a clear influence of the polyions structure on the film swelling and stability being reported. Appreciable structural rearrangements in the (PDADMAC/NaPSS) multilayer have been detected by neutron reflectometry only for NaCl concentrations starting with 0.8 M [71]. The movement of PEL under annealing was modeled by Jomma and Schlenoff as a "limited source diffusion" [71], According to this model, PELs migrate much more slowly in the bulk of multilayer than at the surface due to the presence of additional extrinsic charge compensation at the surface.

2.1.3.2. Counterion Nature

Concerning the salt type, unlike Kolarik et al. who observed almost no influence in the LbL thin films built with PELs of different charge densities [48], a strong influence of the counterion nature on the layer thickness has been found by others [22,65,72,77,79], the anion effect being larger than the cation effect. The least hydrated counterions led to thicker films. To explain this influence, Hofmeister's series was invoked [65]. The least hydrated counterions bind stronger than more hydrated counterions [129,130], the hydration entropy of the anion being also consistent in describing the charge screening effect of the counterions and the correlation between the polycation deposition process and the thickness of the dry multilayer [65].

In a parallel with cation exchange on the slightly crosslinked sulphonated polystyrene gels, the ion that swell the cation exchanger the least is preferred, i.e., the least hydrated ion pair is the most stable [131]. NaPSS and a polycation of ionene type containing about 95 mol % of N, N-dimethyl-2-hydroxypropyleneammonium chloride units in the backbone (PCA_5) have been used to prepare PAN composite membranes with different solvent affinities controlled by the nature and concentration of the supporting electrolyte (NaBr and KBr) in

PEL deposition solutions [72]. The highest polyion amount adsorbed from 1 M KBr solution was explained by the stronger interaction between PSS macroion and K^+ comparative with Na^+. This specific interaction was also supported by the thickness of (PAH/NaPSS) films obtained when the polyions have been adsorbed from 1 M KCl comparative with 1 M NaCl, at the same number of bilayers and the same temperature [75]. The thickness of the films was almost the same when both polyions were adsorbed from 1 M CsCl, but much thicker in KCl when the concentration of salts was 2 M. Since the surface charge must be overcompensated, more material is necessary to be adsorbed when the PEL is weakly charged, this being the case of adsorption from KCl solution, suggesting a stronger binding of K^+ to PSS.

The influence of supporting electrolyte nature on the permeability of preformed NaPSS/PAH multilayers adsorbed on the surface of an electrode, has been recently investigated by Silva et al. [74] using cyclic voltametry (CV) and electrochemical impedance spectroscopy (EIS). The salt counterions reduce the interchain attraction (intrinsic compensation), contributing to the dissociation of PEL ion pairs, and thus induce a partial extrinsic compensation. The multivalent counterions are preferred in competition with monovalent ions, the result being decreasing the reduced viscosity of PEL in solution [129,130], and the lower critical concentration of salts in the formation of PEL complexes in the presence of salts [132]. At the same concentration of salt (0.1 M), Silva et al. observed that the influence of cations (Na^+, K^+, Ca^{2+}, Ba^{2+}, and La^{3+}) obeyed the Hofmeister series, as it was observed by others [65,72]. The thickness of the $(NaPSS/PAH)_3$ films after the immersion in different salts was measured by ellipsometry in the dry state, the differences being not significant. The authors concluded that the structural changes of the film were not revealed in the film thickness, but mainly in the entanglements of the PEL chains [74], in accord with previous results [62,63]. An interesting study on the specific influence of the counterion on the multilayer growth and ion permeability was reported by Ball et al. who compared the PAH/NaPSS multilayers built in the presence of either NaCl or NaSCN [79]. A strong interaction between PAH and thiocyanate counterion, a highly polarizable and kaotropic anion, was found, and therefore the maximum concentration of this salt was limited at 0.5 M. The condensation of the thiocyanate counterions on the PAH at pH 7.4 was evidenced by zeta potential measurements. Concerning the multilayer buildup, even if the thickness growth was linear with both salts, a slope of about three times higher was found in the presence of 0.5 M NaSCN compared with 0.5 M NaCl. As a consequence of the ion condensation, a higher amount of PAH had to be adsorbed on the PSS layer to ensure charge overcompensation and further multilayer growth [79]. Concerning the ion permeability, the multilayer buildup from 0.5 M NaSCN solutions allowed a fast and reversible permeation of ferrocyanide anions, whereas the same multilayer built from 0.5 M NaCl solution was completely impermeable to the electroactive anions. The difference has been made by the more hydrated and probably porous thin film prepared in the presence of thiocyanate counterions.

Therefore, for a given pair of strongly dissociated polycations and polyanions, the salt concentration in the deposition solution has a strong influence on the thickness of each polymer layer, which is approximately proportional to salt concentration, but the counterion/polyion interaction has also a strong influence on the multilayer growth and morphology.

2.1.4. Temperature

Usually, ESAMs thin films have been constructed at room temperature, less concern being devoted to the influence of temperature on the multilayer growth. Only lately some experimental investigations have been dedicated to the influence of temperature on the multilayer buildup [75-77,82], and on the properties of preformed multilayers [74,78], the oppositely charged PELs being either PAH and NaPSS [75,77,78] or PDADMAC and NaPSS [76,77]. Büscher et al. reported for the first time an almost linear increase of the multilayer thickness as a function of temperature, for (PAH/NaPSS) films adsorbed from KCl aqueous solutions, both 1 M and 2 M, at temperatures between 5 and 40 °C [75]. The surface roughness was almost constant up to 20 °C and dramatically increased after that. The authors assumed that the increase of polymer-solvent interaction, by reducing the range of electrostatic interactions along the polyion by small ions and by approaching the precipitation temperature (60 °C for NaPSS in 1 M KCl), caused an increase of the adsorbed amount of PELs [75,82]. Smoother surfaces were formed by decreasing the temperature of PEL adsorption. These results were confirmed by Tan et al. for (PDADMAC/NaPSS) multilayer films deposited at temperatures up to 70 °C, in the presence of 1 M NaCl, the influence of temperature for the films deposited from salt-free aqueous solutions being ruther weak [76]. Comparative with Büscher et al., further experimental investigations were performed by Tan et al. which led to the elaboration of a new mechanism to explain the dependence of multilayer thickness on temperature [76]. Thus, the authors observed a greater temperature dependence of the PDADMAC deposition than did the NaPSS. On the basis of the new results, the author assumed that "heat affects the PEL multilayer films in a manner similar to added salt", an increased tendency to send loops and tails into the solution above the film being the final result [76]. The temperature dependence of the $(PDADMAC/NaPSS)_n$ films growth, has been explained in correlation with the swelling (annealing) of preformed multilayers discussed above [76], a greater flexibility and fluidity leading to an increased interpenetration between PEL layers.

Salomäki et al. investigated the effects of temperature on the multilayer growth on two pairs of PELs: PDADMAC/NaPSS pair (in 0.1 M NaBr or NaF aq. solutions), which shows mostly a linear buildup, an initial exponential buildup being reported at high ionic strengths, and PAH/NaPSS pair (PAH in 1 M NaCl aq. solution) as a model for a linear buildup [77]. As Figure 1 shows, a predominantly linear buildup was observed when the $(PDADMAC/NaPSS)_n$ was deposited at 15 and 25 °C, while the multilayers deposited at 45 and 55 °C showed a clear exponential buildup regime, the PDADMAC being more temperature sensitive as the mass growth than NaPSS.

The effect of temperature on the $(PDADMAC/NaPSS)_n$ film growth has been stronger than on the $(PAH/NaPSS)_n$ system, the exponential growth of this assembly being evident only up to eight bilayers at a temperature of 80 °C [77]. To explain the influence of temperature on the multilayer growth, a different basis was looking for by Salomäki et al., who followed also the reduced viscosities of PDADMAC and NaPSS as a function of temperature. The decrease of the reduced viscosity was found for both PELs with the increase of temperature, and this showed PELs adopted more compact forms, this being an evidence for the similarity between the effect of temperature and high ionic strength. That is, both factors cause cleavage of the ionic bonds inside the ESAM film, allowing new rearrangements with increasing the entanglements and interpenetration between PELs chains [77]. A buildup model, based on the assumption that each deposition step led to a quasi-equilibrium between the PEL concentration in solution and the composition of ESAM film, was simulated to explain the influence of temperature. According to this theory, the rate of multilayer growth is determined by the growth exponent β, which dependence on temperature led to an Arrhenius-like function.

The effect of increasing temperature on the EIS response of the preformed multilayers on an electrode has been correlated with the decrease of the film resistance with the increase of the multilayer disorder favoring the mobility of small ions [74].

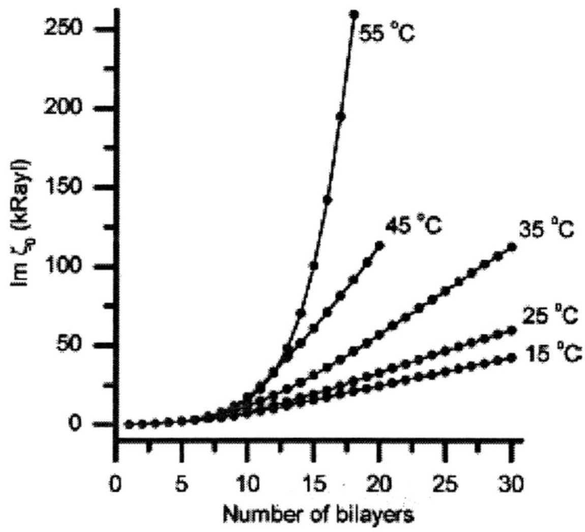

Figure 1. PSS/PDADMAC multilayers deposited in 0.1 M NaBr at different temperatures (reprinted from ref. [77] with permission from the American Chemical Society).

In conclusion, besides the well established experimental variables with a strong impact on the buildup of ESAMs films, temperature seems to be a significant factor of influence.

2.2. Mechanism of ESAM Thin Films Buildup

As a function of the PEL structure and the deposition conditions (ionic strength, solvent polarity, and temperature), two growing mechanisms of the ESAMs thin films formed with strong PELs as building blocks were identified [61,77,79,83,85-87]:

- a linear growth in which the thickness increment is constant during each deposition of PEL layer pair, the level of chain interpenetration being limited to 2-3 bilayers, this being the most found mechanism for these multilayers [6,7,10,61,81,83,87,112];
- a non-linear (exponential) growth of the film thickness as a function of the number of bilayers, this mechanism being normally correlated with low charge densities [50,51,53,54], high ionic strength [61,63], and high temperatures [77].

It was established from the beginning that the surface charge overcompensation is the key to buildup ESAMs [6]. Questions like how strong are PELs bound on the surface and how far loops and tails extend into the liquid phase arose very quickly [6,16,21,22,24]. The surface charge reversal is a critical step, i.e., the driving force of the multilayer growth. The level of overcompensation depends on the absence or the presence of neutral salts. Thus, in the absence of added salt, the adsorbed layers are thin and only slightly overcompensate the surface charge [22]. In the presence of sufficient salt, significantly higher amounts of PEL counterpart are adsorbed on the surface [21,22], much more than an equivalent monolayer of polymer repeat units. According to Schlenoff et al., salt ions permit localized rearrangements of polymer chains, the polymer in excess being accommodated in several layers, rather than in one layer of loops and tails, each adsorbed layer can penetrate into the existing SAM film to seek out the extrinsic charge [21]. The extensive interpenetration and limited mobility of the chains within PEL multilayers has been demonstrated almost at the same time by Schlenoff et al. using radioanalytical methods [21,22] and Lösche et al. using neutron reflectometry [58]. The ion and polymer content in PEL multilayers have been probed by Schlenoff et al. with PELs adsorbed from salt-containing (0.1 M NaCl) and salt-free aqueous solutions. ^{14}C-PM2VP and ^{35}S-PSS have been used as counterparts in the multilayers buildup. It was demonstrated that PELs charges quantitatively balance each other and no salt ions were detected in the deposited multilayers, surface charge being one-half of the charge within a single layer. The top layer has been kinetically stable and not stripped from the surface on exposure to the oppositely charged PEL [21,22]. Basically, the adsorption of PELs onto an oppositely charged surface is an ion exchange phenomenon, between the charged segments of PEL and the small ions compensating the surface charge [22,60]. According to this model, charged PEL segments compete with added salt ions to occupy the free charged surface sites. There are several experimental and theoretical studies concerning the ratio between intrinsic and extrinsic compensation of charges in the multilayer. According to the model of Schlenoff and Dubas, two main parameters characterize the ESAM film growth: the level of surface charge overcompensation, Φ, which is assumed to decrease exponentially from the film surface to the bulk, and the charge penetration length, l_{cp} [61]. At high salt concentrations, the excess charge is distributed over several layers, the thickness increment being controlled mainly by l_{cp}, which it was demonstrated to be 2.5 nominal layers for the system PDADMAC/NaPSS, by measuring the surface charge directly using radiolabeled sulfate as counterion. According to the model of Salomäki et al. [77], the LbL buildup is “inherently exponential, becoming linear whenever diffusion is not fast enough to carry the polymer within entire thickness of the film”.

The absence of counterions in the ESAMs films was demonstrated by Decher et al. [6,15,58] and Schlenoff et al. [18,21,30,61] by different techniques. However, there are experimental evidences which support the presence of counterions, i.e., an extrinsic compensation of polyion charges at a certain level, mainly for polyions with a charge density higher than the critical charge density where a strong counterion binding, Manning condensation [133,134], is expected [66,70]. With PAH as polycation, the charge stoichiometry of the PEL complexes in the bulk of multilayer was demonstrated by Riegler and Essler to be different than 1:1, i.e., the PAH charges overcompensate the NaPSS charges [66]. Therefore, the multilayer contains additional small counterions evidenced by XPS for films prepared in the presence of NaCl [70], their content depending on the charge density of polyions. At a distance between charges lower than the Bjerrum length, the Manning condensation determines the immobilization of counterions in the multilayer, an extrinsic compensation of charges being effective [66,70]. When the distance between charges is higher than the Bjerrum length, the multilayer films are free of counterions.

To design devices based on PEL multilayers with electronic properties, the understanding of the effects of the incorporation of small ions, the density of ionic crosslinks and the water content is strongly required. Dielectric properties of PAH/NaPSS multilayers, made both with and without NaCl in the dipping solutions used for sequential adsorption of PELs, have been investigated by Durstock and Rubner [78]. A bilayer thickness < 5 Å was found when the films were made without adding salt and about 16 Å/bilayer when a concentration of 0.1 M NaCl was used in both PAH and NaPSS solutions. Dielectric properties of these films have been not influenced by temperature up to 140 °C, but large changes were observed for higher temperatures, much larger value of the dielectric constant being found for the ESAMs built with added salt (~300 at ca. 350 °C) than without salt (~40 at ca. 350 °C). The large values of dielectric constant of the PAH/NaPSS multilayers imply the presence of small ions, the larger values found for multilayers built with added salt being considered an evidence that many ions become trapped in such films [78]. The authors discussed also the rigidity and related with this the mobility of ions in these multilayers compared with those made from two weak PELs (see in the next subchapter), a higher rigidity and hence a lower ionic mobility being found in the multilayers built from strong PELs. Changes in dielectric properties proved to be a very sensitive method in the detection of residual ions present in the multilayers, the level of their content being dependent on the nature of PELs and the conditions used in the multilayer construction. Although the overall concentration of small ions is low, it seems to be high enough to significantly influence the electrical properties of the devices based on such materials [78].

Two main questions have been often addressed concerning the ESAMs: haw many layers are necessary to screen the substrate influence and to enter in a linear regime of film growth and what is the hydrodynamic stability of the resulted nanostructured thin films. Electrokinetic investigations gave valuable information on the stability of ESAMs under shear stress [26,88,89]. The electrokinetic measurements, either electrophoretic mobility of colloid particles covered by multilayers or streaming potential of planar surfaces, showed to be an effective method for quantitative studies of PEL adsorption, desorption and reconformation. Thus, it was revealed that upon adsorption of PAH on mica approximately one half of the uncompensated charges of PAH were used to form bonds with mica, the adsorption of NaPSS leading to the change of surface charge [88]. A higher stability of the LbL multilayer was found when NaPSS was the last layer adsorbed. The hydrodynamic stability of the (PDADMAC/NaPSS) multilayers built on poly(ethyleneterephtalate) (PET) containing carboxylic groups on the surface, and poly(tetrafluoroethylene) (PTFE) has been investigated by Köstler et al., both polyions being adsorbed from 0.5 M KCl [89]. A strong influence of the last layer adsorbed on the electrokinetic stability has also been observed, a high stability being characteristic for the films covered with NaPSS comparative with those terminated with PDADMAC. The authors explain this difference by the higher hydrophilicity of underlying NaPSS layer, leading to slight separation inside the film, which would lead to more loops and tails of PDADMAC layer to the aqueous phase. The big difference between the glass transition temperatures of these polyions has been also taken under consideration as a possible cause.

2.3. Tuned Properties and Applications

It has been mentioned above that, normally, the ESAMs thin films are well organized nanoarchitectures with strong electrostatic interactions between oppositely charged groups. However, the irreversible association of the oppositely charged counterparts as PEL complexes in the ESAMs thin films may not be desirable for responsive systems. One of the most important properties of ESAMs is their permeability to ions and molecules, which strongly depend on the presence of pores or structural defects. Diffusion of small molecules through the multilayer films prepared by the LbL alternating deposition of PELs is very important for both fundamental and application purposes, the ESAMs thin films having high potential for numerous applications like anticorrosion coatings [94], control of the hydrophilic/hydrophobic properties of surfaces [80,81,90-93], separations by composite membranes [95-99,101-105], biomedical applications [109,110,112], sensors [7,100,114-118], etc.

2.3.1. Swelling Behavior of the ESAM Thin Films

For many applications such as: sensors, biomedical applications and surface patterning it is very important to control the swelling/wettability properties of various substrates [2-6,81,90-92]. Thus, Lingström et al. studied the influence of the $(PDADMAC/NaPSS)_n$ multilayers deposited on the chemical soft wood fibers and on SiO_2, at various electrolyte concentrations, in order to determine how the multilayer composition affects adhesion between the wood fibers and their wettability [80]. Paper testing showed that the tensile index increased by about 90% when 11 PEL layers were deposited. It was also shown that the last layer adsorbed onto the surface significantly influenced the tensile index, PDADMAC producing a higher value than the NaPSS as outermost layer. The strain and break increased as a function of the layer number. Treated fibers with NaPSS as the last layer adsorbed swelled much higher than those with PDADMAC. The amount of PEL adsorbed on wood fibers increased linearly except the first layer. Adsorption onto SiO_2 showed a more exponential increase comparative with wood fibers, and this support the influence of the substrate nature on the ESAMs growth.

The effect of the post-treatment with NaCl, NaOH and HCl solution with concentrations of 1 mM (duration treatment 48 h), of the (PAH/NaPSS) assemblies, after the deposition on natural mica, on the wetting characteristics has been investigated by Kolasińska and Warszyński [81]. The contact angle of the multilayer surface varied regularly with the PEL layer number, showing that the surface was more hydrophobic when the polycation was the last layer adsorbed compared with NaPSS as the last layer. It was shown that the post-treatment of the PEL multilayers with solutions of different pH influenced their wetting properties, the strongest influenced being observed at pH 11, where the PAH is weakly charged, the lower charge of PEL facilitating the structural changes in the multilayer film.

Investigating the swelling behavior of the (PAH/NaPSS) assemblies by ellipsometry and contact angle measurements, Wong et al. observed that the swelling/deswelling were completely reversible [90]. The relative degree of swelling decreased with increasing number of layers, showing that the last layers are loosely packed and consequently swelled better than the inner ones, in agreement with the model of Ladam et al. [27]. Multilayer films with NaPSS as the last layer adsorbed swelled much higher than those with PAH as the outermost layer. By contact angle measurements, Wang et al. studied the surface wettability of the

(PDADMAC/NaPSS) multilayers, irreversible compressed by a patterned poly(dimethylsiloxane) (PDMS) [91]. The contact angle of PDADMAC layer decreased with the increase of the layer number due to the interpenetration of PEL chains. However, when the multilayers were compressed, always PDADMAC as the last layer, the contact angle value were quite uniform and around 90°, the characteristic value for PDADMAC alone. In the process of compression, polymer chains became much closer, the hydrophilic groups being embedded in the multilayer, this being reflected in the decrease of the film wettability.

Köstler et al. studied the LbL deposition of $(PDADMAC/NaPSS)_n$ multilayers on PET and PTFE, selected due to their future biomedical applications [92]. A regular odd-even pattern even for the first bilayers on sufficiently smooth and hydrophilic substrates was observed by contact angle measurements. The effect of thicker layers and a more coiled conformation by deposition from salt containing solutions was also noticed. A special wetting behavior of PDADMAC terminated multilayers was found, i.e., the surface being less hydrophilic compared with NaPSS, in agreement with other systems mentioned above [80,81]. Superhydrophobic surfaces were obtained by a combination between LbL self-assembly of PDADMAC and NaPSS with the electrodeposition technique of Ag aggregates, on which a monolayer of n-decanthiol was chemosorbed [93]. This is considered a possible method to get self-cleaning surfaces with a contact angle as high as 154°.

Selective deposition of PELs on substrates offers the possibility of pattern formation and processing using ESAMs. Usually, the common means for preparation of patterned films, such as photolithography, ask for a highly efficient chemistry and involve many steps. Self-assembled monolayers functionalized with endgroups, which promote and prevent adsorption, can be selectively attached to a metal or metal oxide surface. The LbL deposition on these patterned surfaces will result in a patterned thin film [106-110], different from the ESAMs continuous films mainly concerned on the film growth and film roughness. Park and Hammond introduced the patterning PEL multilayers during adsorption, utilizing chemically patterned surfaces that guide polymer adsorption to specific regions based on selective deposition [12,106,109,110]. The authors presented a direct method of patterning PEL multilayers using an additive approach. In the process, a PEL multilayer was fabricated onto the surface of a PDMS stamp by the LbL deposition of polycation/polyanion.

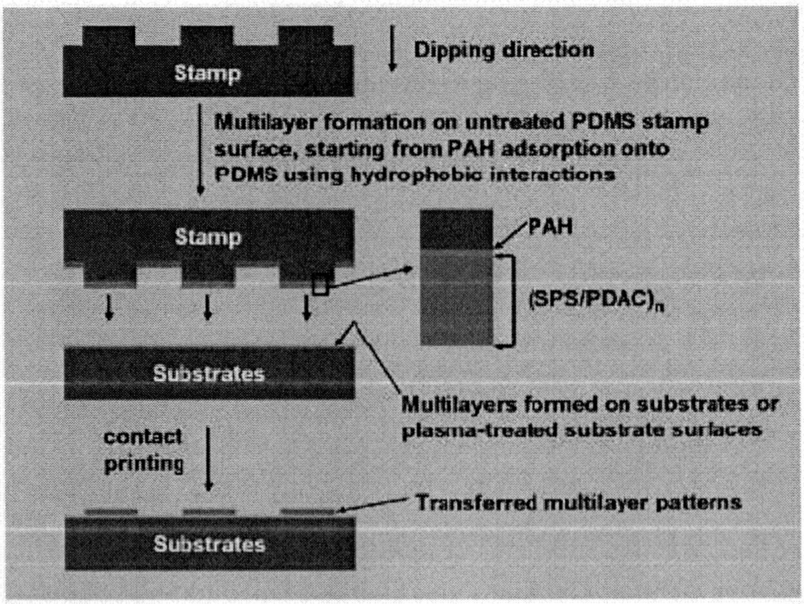

Figure 2. An illustration of the multilayer transfer printing process (reprinted from ref. [109] with permission from Wiley-VCH).

As Figure 2 shows, after the adsorption of PAH, as a background layer, the NaPSS/PDADMAC multilayers were formed onto stamp surface by alternating adsorption of PELs from 0.1 M NaCl. Then, the stamp was placed onto a substrate, which was oppositely charged from the multilayer top surface and the multilayer was transferred to the substrate resulting in a patterned multilayer thin film on the substrate, because the hydrophobic interactions between the PAH and PDMS surface were weaker than the electrostatic interactions between PDADMAC top surface and the negatively charged substrate. Complete transfer of the multilayer was illustrated by AFM imaging. The thickness of the transferred PAH/(NaPSS/PDADMAC)$_n$ multilayers ($5 < n < 11$) increased linearly with the number of layers. Because the films can be stamped onto a variety of substrates, this technique proved to be useful in electronic and biomedical systems. Films with electrochemical, optical, electronic, and sensor functionalities can be integrated directly into devices using the multilayer transfer printing.

2.3.2. Transport and Separations

Working with pyrene labeled copolymers of AMPS with DMAA with variable charge density as polyanions (ADPy), Sun et al. observed that nitromethane quencher diffused faster through the multilayer film made with a polyanion with lower charge density (ADPy-45, $F_{AMPS} = 0.711$) compared with the multilayer built with polyanion of a higher charge density (ADPy-100, $F_{AMPS} = 0.999$) [56]. The authors explained their results by the most streatched conformation of the highly charged polyanion ADPy-100, in the deposition solution, leading to a more compact layer on the positive PDADMAC layer, resulting in a lower permeability to small molecules of nitromethan. The stronger the electrostatic attraction, the more compact the multilayer structure and therefore a lower permeability to organic molecules was observed. The influence of film thickness and of the ionic strength on the diffusion characteristics can be summarized as follows: decreasing the PEL charge density, increasing the salt concentration in the deposition solution, and reducing the bilayer number accelerated the diffusion of small molecules in the ESAM films.

The composite membranes consist in a thin and highly selective surface layer, and a thick substrate layer that is porous and provides mechanical strength. The membrane surface, which plays an important role in the membrane performance, can be further modified by physical or/and chemical methods. Surface modification by physical methods is mainly based on ionic, H-bonds and hydrophobic interactions. One of the most convenient physical method for surface modification of the membranes is LbL self-assembly of oppositely charged PELs [20,72,95-103]. Film thickness depends on the PEL structure, number of PEL layers as well as the supporting electrolyte concentration in the deposition solutions.

The applications of the composite membranes require knowledge of the transportation and permeability of small molecules passing through these films. Thus, the LbL physically modified membranes have been used for separation of alcohol/water mixtures under pervaporation conditions [96,97,105], ion transport and transport of neutral organic substances under nanofiltration and reverse osmosis conditions [64,98,99,101-103], etc. By a careful choice of the PELs and by optimization of deposition conditions, composite membranes with excellent properties for alcohol/water and mono-/divalent inorganic anions and cations separation have been prepared by Tieke et al. [95-97]. The selective ion transport of these membranes was explained by their multibipolar structure consisting in alternation of positively and negatively charged PEL layers. The strength of the interaction between the PEL layers and oppositely charged small ions is proportional to the PEL charge density and to the permeating small ions, the permeation of di- and multivalent ions being thus much lower comparative with that of the monovalent ions. The selectivity is also influenced by the specific interaction between small ions and PELs [95,98]. Size-selective transport of uncharged organic compounds such as saccharides through LbL multilayer membranes, under nanofiltration conditions, has been recently reported by Bruening et al. [101-103]. A strong correlation between the outermost layer, the swelling, and the transport rate of glucose and sucrose through the LbL membranes has been observed, a higher swelling leading to minimal sugar rejection [103]. The nonlinear decrease in permeability of the (PAH/NaPSS) ESAM film, for $[Fe(CN)_6^{3-}]$ with the increase of bilayer number confirmed the different structure of the first two bilayers comparative with the rest of the film [28].

2.3.3. Biomedical Applications

Degradable and structurally related poly(β-amino ester)s as polycations and NaPSS were used to investigate the influence of polymer structure on the erosion profiles of multilayered PEL assemblies [55]. Both the charge density and hydrophobicity of polycations were found to play an important role in defining the rates at which multilayered assemblies eroded in physiological media. It was found that the erosion rate of these assemblies and the release of polyanion in solution decreased with the increase of the polycation hydrophobicity. The results reported by Lynn et al. may contribute to design multilayered assemblies that permit a strict control over the release of therapeutic agents from coated surfaces.

An interesting example of a synthetic macromolecule assisting as a catalyst in the separation and exchange of two fully charged and highly associated PELs was reported by Jomma and Schlenoff [69]. Starting from the interpolyelectrolyte exchanges reported for the PEL complexes containing polycarboxylic polyanions, the test of the reversible complexes formed in the (PDADMAC/d_5-PMAA)$_n$ was performed by self-exchange and exchange experiments with PMAA and NaPSS, respectively. In self-exchange experiments, a preference of PMAA over d_5-PMAA in the multilayer was found, and explained by the higher molar mass of the first polyanion. When NaPSS was used as displacer of d_5-PMAA, a surprissing behavior was observed: the exchange was much faster and complete when a mixture of PMAA and NaPSS (8:2) was used than with NaPSS alone, and no PMAA was detected in the multilayer. The "catalytic" effect of PMAA observed in this case has been compared with "assisted" macromolecular exchanges specific for nucleic acids.

As Serizawa et al. showed, ESAMs nanocoatings could be built on hydrogel substrates, such materials having potential for biomedical applications [111,112]. Cationic chitosan/dextran sulfate (CHI/DS) [111], and (PDADMAC/NaPSS) [112], have been alternately adsorbed from aqueous solutions containing 1 M NaCl on poly(vinyl alcohol) (PVA) hydrogel. When the second polyion pair was used in the LbL deposition, the films were stained with a cationic dye, methylene blue (MB). It was observed that the blue color of the hydrogel with LbL film adsorbed became denser, after the rinsing in water, as the number of deposited layers increased, indicating that the thickness of the ESAM films increased. Sequential and regular deposition of the oppositely charged PELs has been demonstrated by the contact angle measurements, the angle at even and odd steps being almost the same, irrespective of the number of adsorption cycles [112]. The LbL films peeled off from the hydrogel surface could be characterized by UV-vis, XPS and AFM. It was found that the film thickness almost linearly increased as the assembly step increased. The mean thickness/bilayer was estimated to be 11 nm, much smaller than that of a (PDADMAC/NaPSS) assembly built on a solid substrate (gold), indicating that the substrate nature could have a strong influence on the ESAMs growth [112].

2.3.4. Sensors

Many end-applications of the LbL modified surfaces concern those as sensors [7,39,114-118]. Some examples will be discussed in brief here.

Surface functionalization by LbL technique was used to deposit PAH/NaPSS, and biopolymers such as oligonucleotides on the electrospinned PS fibers containing sulfonic groups on the surface [114]. Nanocomposite PEL/gold fibers were prepared by immersing the PAH/NaPSS coated PS fibers into a dispersion of 4-(dimethylamino)pyridine-stabilized Au_{NP}. The structure of PAH/PSS multilayer was demonstrated by dissolving the inner PS fibers with THF, resulting in the formation of hollow PAH/PSS fibers. The authors consider these fibers have potential applications in filtration, sensing, catalysis and as release materials in medicine and agriculture. The construction and characteristics of a thermoresponsive thin film made by a LbL sequential deposition of two ionically modified random copolymers of N-isopropylacrylamide (NIPAM) (PAH-*co*-NIPAM and NaPSS-*co*-NIPAM) have been investigated by Jaber and Schlenoff [115]. Since many applications of NIPAM composites in temperature controlled drug release rely on the membrane permeability, the multilayer permeability was monitored by the transport of an electrochemically active probe molecule (potassium ferricyanide) through the film as a function of temperature. When the lower critical solution temperature (LCST) was reached, the film expelled water molecules and the transport of water soluble probe ion was substantially suppressed. By comparing the sensitivity to nitroaromatic quenchers of poly(phenylene ethylenes) (PPEs) in solution and in the LbL multilayers obtained by sequential deposition of PPEs and PDADMAC on silica microspheres, Wosnick et al. found that the coated microspheres showed similar fluorescence properties to polymers in solution, but with enhanced sensitivity (up 200-fold) in a variety of solvents [117].

Bucur and Schlenoff showed that an ultrathin multilayer film formed on an electrode enhanced the intensity of electrogenerated chemiluminescence (ECL) from the tris(2,2')bipyridylruthenium(II)/tripropylamine ($Ru(bpy)_3^{2+}$-TPrA) system [118]. CV revealed the highest ECL generated from $Ru(bpy)_3^{2+}$-TPrA system diffusing through the PEL multilayers on 16 layers of ESAM films built by alternately deposition of two strong polycations, either PDADMAC or poly(vinyl N-methylpyridinium iodide) (P4VMP) and NaPSS. The authors found that the preferential transport of species of lower charge within multilayers was responsible for enhancing the ECL output, and that the concentration of ions within the bulk of (P4VMP/NaPSS) ESAM was half of that within PDADMAC/NaPSS films. Greater luminescence enhancement with PDADMAC multilayers compared to P4VMP system was attributed to greater permeability and greater membrane ion concentration of the former.

Interesting results have been recently reported by Chen et al. who combined the precursor assembly and LbL deposition of a single charged functional species as sodium 9-antracenpropionate (SANP) and followed the (PDADMAC/NaPSS) multilayer as a nanocontainer for SANP and a nanoreactor for a photocycloaddition reaction (Figure 3) [119]. It was observed that positively charged binding sites and charge selectivity of the multilayer film were responsible for the linear deposition of the SANP/PDADMAC precursor assembly/NaPSS and for the fast loading of SANP. The electrostatic selectivity in the loading with single charged species has been followed by comparison of a positively charged species, 9-antracenetrimethylammonium bromide (ATAB) and another negatively charged species, sodium1pyrenesulfonate(SPS).

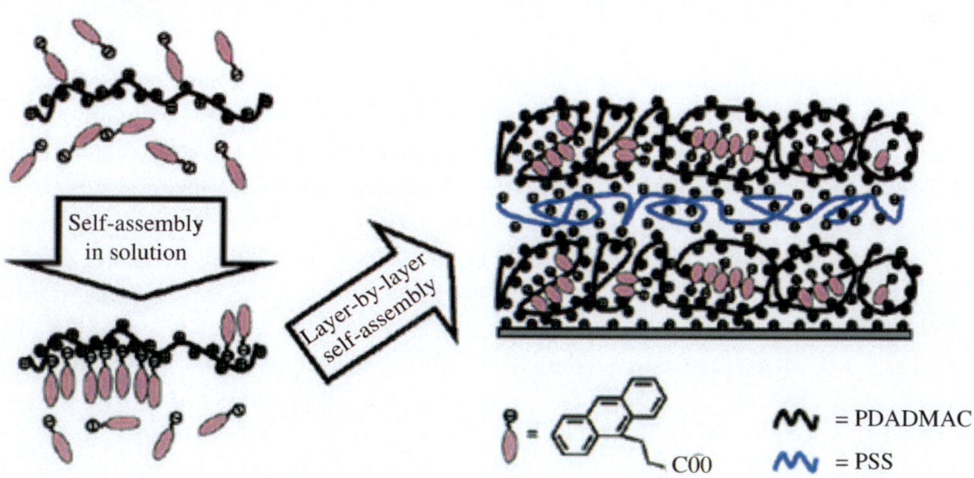

Figure 3. Illustration of hierarchical self-assembly: step 1, sodium 9-anthracenepropionate tethered to PDADMAC in solution; step 2, layer-by-layer selfassembly of PDADMAC-SANP/PSS (reprinted from ref. [119] with permission from the American Chemical Society)

It was observed that ATAB was barely adsorbed into the film but SPS could diffuse into the multilayer film as efficiently as SANP. The photocycloaddition reaction in the multilayer film took place with a higher quantum yield than that in solution, the reaction being favorised by the aggregation of the SANP molecules in the multilayer.

Carbon nanotubes (CNTs) by their electronic, mechanical and chemical properties constitute a very attractive material for nanodevices and nanotechnology. It was observed that the conductivity of CNTs is very sensitive to the surface modification, and therefore CNTs have been recently used as nanosubstrates for LbL deposition of oppositely charged PELs to alter their electronic properties to very sensitive nanoscale sensors [125-127]. Kim et al. [125] alternately deposited PDADMAC and NaPSS onto functionalized multiwall carbon nanotubes (fMWCNTs). PDADMAC/NaPSS polyion pair has been also deposited onto CNTs modified with cationic pyrene derivatives, which irreversible adsorb on the nanotube surface, to form a nanometer'thick amorphous polymer nanoshell around the CNTs [125]. The authors confirmed that the last PEL deposition step determined the composition of the outermost layer and that adsorption on one-dimensional objects as small as 2 nm in diameter is feasible. The surface of polymer layer provides a robust and flexible platform for attachment of various biological and chemical species. Kong et al. alternately deposited PDMAEM or hyperbranched poly(sulfone amine) and NaPSS onto the MWCNTs as nanosubstrate, previously functionalized with PAA or NaPSS grafted onto the surfaces by a "grafting from" method [127]. The covalent bonding of PELs to the CNTs surface improved their stability as aqueous dispersions, and opened new possibilities for their application as nanomaterials. It was demonstrated that the fMWCNTs have an extremely high efficiency for loading PELs by LbL self-assembly strategy, which will give the possibility to apply these nanomaterials in gene-delivery or other micro/nano transferring systems.

3. SAMs Based on Weak Polyelectrolytes

One of the most interesting categories of building blocks for SAMs construction is that of weak PELs, which degree of ionization can be controlled by pH of the surrounding solution. These nanostructured materials have been very intensively investigated last decade due to the complex behavior of weak PELs pairs over a wide range of pH which allows a fine control of the layer pair thickness, surface wettability, dielectric properties, surface roughness, film morphology, for specific applications [135-223]. Not only pH has been used to vary the SAMs properties, but also ionic strength of the deposition solutions [138,142,146,167,180,195]. Another particular aspect of weak PELs is the dramatic change of their pKa by the incorporation into the SAMs thin films comparative with solution-state values [100,139,158,169,180,181]. Thus, it was reported that the apparent ionization of a weak PEL incorporated into a SAM is influenced by both the external pH and the ionic strength as well as by the local microenvironment. The SAMs built with weak PELs, such as PAA and PAH as building blocks, allowed the fabrication of microporous [139,184] and nanoporous [142,183,184] thin films, by exposure them to acidic solutions (pH of 2.3-2.5), or by the post-exposure of the multilayers prepared from salt-containing deposition solutions to pure water. Their morphological changes and permeability could be thus precisely controlled making them attractive candidates for biomaterials and controlled release systems, as it will be shown latter.

The driving force in the multilayer construction with weak PELs as building blocks can be either electrostatic [135,138,146,167,180], at a pH where the PELs are fully charged, or hydrogen bondings [136,144,150-152,157,183,187]. The SAMs stabilized by hydrogen bonding interactions known that "hydrogen-bonded multilayers" will be called here HSAMs, to differentiate them from ESAMs resulted by electrostatic interactions. Normaly, there is a fine balance between electrostatics and hydrogen bondings, their role in the SAM film growth and stabilization being determined by the pH, ionic strength and local microenvironment.

3.1. Factors Controlling the Multilayer Construction

Some factors with an important role in the buildup of the SAM thin films from weak PELs were selected for discussion: pH value and the ionic strength of the deposition solutions, substrate nature and deposition method.

3.1.1. pH and Ionic Strength

Because pH-sensitive SAMs proved to have great potential for many applications, it is important to establish the boundaries between reversible and irreversible pH response, including morphological changes [161,169,181]. The role of electrolytes in the formation and deconstruction of SAMs with weak PELs as building blocks has been also deeply investigated [138,145,161]. Construction and deconstruction of multilayers made from a combination of a weak polyacid (PAA) and a strongly dissociated polycation (PDADMAC) has been studied by Dubas and Schlenoff [138] as a function of pH of the deposition solution, NaCl concentration, and PAA molar mass. Film thickness reached a maximum as a function of salt concentration, at around 0.3 M NaCl, when PAA was deposited at a pH of 11, and at around 0.15, when PAA was deposited at a pH of 5, with a high molar mass of PAA (84 500 Da), and then decreased quickly. Preformed multilayers decomposed rapidly and completely when exposed to NaCl of concentration > 0.6 M. The authors explained the apparent dissociation of multilayer PEL complexes by the competition between polymer/polymer ion pairs and polyion/ions of salt. Films built with PAA of low molar mass (5200 Da) were significantly thicker. Decomposition of multilayers has been also performed by protonating the PAA, thus decreasing polymer/polymer interactions, but an incomplete loss of polymer was observed in this case. To explain this behavior, the authors took into account two further effects of the pH decrease: PAA becomes more hydrophobic as it is protonated, and also fully protonated PAA could participate in hydrogen-bonding interactions. Therefore, an incomplete dissociation/salvation of the SAM components has been found. A model based on ion exchange/swelling of multilayers was elaborated to explain the stability and permeability as well as the dependence of film thickness on salt concentration. In the presence of salt ions, some extrinsic compensation is induced which level depends on the strength of the polymer/polymer interactions, yielding to bigger loops and at the multilayer decomposition at a certain salt concentration, as it is usually observed for the interpolyelectrolyte complexes at the critical salt concentration, a characteristic for a certain polyion pair [132,145].

Electrostatic interaction is the main driving force in the film growth when both weak PELs are deposited at a high degree of ionization [137,178]. To prevent the decomposition of

ESAMs films and to control their permeability, a heat-induced crosslinking with the transformation of ammonium carboxylate ion pairs into amide bonds was performed [137,178]. The decrease of the film permeability after the crosslinking has been observed [137].

A systematic investigation of the influence of the ionization degree of weak PELs on the growth of SAMs was recently done by Choi and Rubner by assembling several strong PELs with either PAA or PAH, to have all the time one pH independent PEL while varying the ionization of the counterpart [181]. A transition from thin flat layers to thicker deposited layers was observed when the charge density of the weak PELs decreased from their fully charged to values below 70-90% charged units. Rubner et al. also demonstrated that the pH of the deposition solution of PAH had a dramatic influence on the discontinuous swelling/deswelling transitions of PAH/NaPSS multilayers assembled at pH > 8.5, the hydrophobic interaction between the free amine groups being responsible for the shifting of pKa of the free amine groups of PAH to unusually low values. A pH-independent swelling was observed when the deposition of PAH was performed at pH < 8.5.

To vary under a strict control the total volume fraction of ionizable groups within a SAM film, pH dependent and pH nondependent PELs have been blended in the deposition solution [169,195]. To remain intact, a SAM thin film containing carboxylate groups must be able to accommodate a large quantity of negative charges without major rearrangements [169]. To overcome this problem, the pH-sensitive groups have been diluted by mixing a copolymer PDADMAC-*co*-PAA with PDADMAC in different proportions, this blend constituting the polycation, polyanion being NaPSS, the deposition being performed at pH 2 and 1 M NaCl. The stability of SAMs thus constructed, by increasing pH at 11, has been compared with that of $(PDADMAC/NaPSS)_{10}$ and $(\text{pure copolymer}/NaPSS)_{10}$, as extremes. The entire multilayer was lost, when the SAMs were made with pure copolymer, but they were stable with pure PDADMAC. Films terminated with NaPSS appeared to be more sensitive to pH jumps than those terminated with the polycation mixture, i.e., films with negative surface charges exhibited erosion of polymer, while positive-capped films with up to about a nominal copolymer mole fraction of 0.2 were completely stable. As Sui and Schlenoff showed, such films were able to accommodate the internal charge excess by charge extrusion to the surface and/or bulk rearrangements at a micro- or nano- scale, a certain polymer mobility being necessary. Ionization of carboxylic groups generates highly solvated cations into the bulk of multilayer film, dissociation of polyion/polyion interactions, and a corresponding increase in the chains mobility. The positive surface can compensate (soak) part of the negative charges appeared inside the multilayer film but, as this to occur, the PEL segments must be able to transfer the extrinsic charge to the surface by their mobility and local rearrangements [169].

A new approach to control the SAMs growth has been recently reported by Caruso et al. [180], PELs bearing weak and strong ionic groups hanged on the same chain being used as anionic components. The electrostatic linkages between strong ionic groups enhanced the film stability, while the weakly charged carboxylic groups made the SAM films pH responsive. Main parameters which have been varied were: the ratio between strong and weak ionic groups in polyanion, deposition pH and ionic strength. The ESAMs were thinner and smoother when no salt was present in deposition solutions, a less regular growth being also observed. The ionic strength influenced also the stability of ESAMs at different pH, due to the interplay of PEL conformation and ionization within the bulk of ESAM film [180].

3.1.2. Substrate Nature and Deposition Method

Plasma treated porous polyethylene (PPE) as a substrate was modified by deposition of a PEI/PAA bilayer. The effect of the solvent polarity and the surface wettability on the PEI and PAA adsorption was investigated [171], the density of charges within the PEL chains and the surface being maximum in highly polar solvents. The solvent polarity was adjusted by varying the ethanol/water ratio in deposition solutions. The mass of PEL adsorbed onto the PPE surface and the adsorption rate increased with the decrease of the solvent polarity. The presence of PEL bilayer on the surface increased the density and stability of the functional groups more than can be achieved through plasma treatment alone.

The alternate deposition of layers consisting in PAH at pH 7.5 and blends with different compositions of weak PEL (PAA) and strong PEL (NaPSS), at pH 3.5, onto colloidal particles of PS has been used to control the surface properties and hence the colloidal stability of the core-shell particles, as well as the thickness and morphology of the shell [195]. The stability of these coated particles was higher than that of the particles coated with PAA and PAH layers, even if the deposition of PAA occurred under conditions where PAA had a lower charge density, and this demonstrates the role of NaPSS in the stabilization of coated particles.

A very strong influence of the pH of the deposition solution on the multilayer growth and on the colloidal stability of the modified particles was observed by Caruso et al. with PAA and PAH alternately deposited onto PS and melamine formaldehyde (MF) particles [146]. A high stability was found when both PELs were in a high degree of dissociation, i.e., when the electrostatic interaction was the main driving force of the multilayer buildup. In this case, the net charge reversal was evidenced by ζ-potential measurements. The electrostatic repulsion between the colloid particles is necessary to protect them against aggregation. Even if the presence of NaCl in deposition solutions and in the washing solutions had a less important role in the colloidal stability of the particles, during and after the deposition process, the absence of salt resulted in a very low modification of the particles, due to the higher stiffness of the PEL chains. Generation of carboxylic acid groups in the (PAH/PAA) multilayers deposited on the colloidal particles has been also considered, since the presence of such groups is very important for post-building chemical reactions as it will be shown latter in the section dedicated to the applications of the ESAMs based on PAA.

It was reported that nonelectrostatic short range interactions have also a role in the SAMs growth with weak PELs as building blocks. Among these interactions, the hydrophobic interactions play an important role. The effect of the content and length of the alkyl chains grafted on the carboxymethylpullulan (CMP) on the multilayer built with strong and weak polycations as counterparts has been deeply investigated by Glinel et al. [187]. A nonlinear growth of layer thickness was found when hydrophobic derivatives of CMP have been used as polyanion compared with the linear growth characteristic for nonmodified CMP. A strong increase of the 10-bilayer thickness [PEI/CMP-xCn] with the increase of the alkyl chain length was evidenced and showed the longer the length of grafted hydrocarbon chains, the stronger the hydrophobic interactions, and a larger amount of material was deposited per bilayer [187].

The LbL technique and spin assembly have been used to construct multilayers with PEI as polycation and PAZO as polyanion in a mixture with sodium dodecyl sulfate (SDS) [172]. A higher amount of material was deposited by LbL than by spin assembly at the same number of bilayers. The competition between SDS and PAZO for the PEI binding sites have been observed, increasing concentration of SDS resulting in films with less adsorbed PAZO, until the critical micelle concentration of SDS was reached. SDS molecules interpenetrated throughout the PAZO and PEI layers increasing the surface hydrophobicity. Porcel et al. recently reported their results concerning the influence of the multilayer buildup method on the film growth process [175]. They have compared the well-known dipping method with the more recent spraying method in the construction of HA/PLL multilayers, the last method allowing a very fine control of the deposition parameters (mainly, the spraying time and the spraying rate). By both methods an exponential-to-linear transition was evident (more details in the next section).

3.2. Mechanism of SAMs Growth

The main factors controlling the level of interlayer interpenetration, the film composition and thickness of both the internal bilayers and the outermost layer of the SAMs thin films made with weak PELs as building blocks, and also the possibility to control the surface wettability have been systematically studied by Rubner et al. [135,161]. The most wettable surfaces were obtained when the outermost layer was an aliphatic polyacid such as PAA, whereas aromatic polyacids such as NaPSS and poly(thiophene acetic acid) produced surfaces with contact angles in the 50-60° range, no influence of the substrate nature being determined [135]. It was observed that the most wettable surfaces created with PAA/PAH ($\theta < 5^o$) exhibited also antifogging properties. The surface wettability has been followed also as a function of the bilayer number and it was observed that the surface wettability depended only on the nature of the outermost layer. The strong influence of the deposition solution pH of PAA on the interpenetration of PAA and PAH layers was evidenced by the oscillations of contact angle values as a function of layer number. The highest oscillations of the contact angle were evidenced when PAA was deposited at pH 2.5 and PAH at pH 2.5, 3.5 and 4.5. To estimate the extent of the chain interpenetration, the relative number of carboxylic groups available as binding sites was evidenced by a surface dying technique using MB as a testing probe (see later in section 5.2).

For SAMs formed with weak PELs as building blocks, molecular-level architecture is correlated with the pH of the PEL deposition solutions, relative to their pK_a values. A low incremental thickness was found when PAA and PAH were deposited at a pH around 6.5, when both PELs adsorb in their fully charged and extended state, the density of ion pairs being very high [28,135,161,169]. The multilayer growth was different when both PELs were deposited at a pH close to their pK_a, when they are only partially charged, the layer thickness being much larger, and highly interpenetrated [135,154,161,169].

Measurements of films thickness in dry state indicated that the deposited amount of PELs increased exponentially with the number of deposited layers, mainly with polypeptides and polysaccharides: poly(glutamic acid) (PGA)/poly(L-lysine) (PLL) [83,212,218,219], PAA/PLL [141], PLL/alginate [192,210], PGA/PAH [213,221], PLL/hyaluronic acid (HA) [140,201], CHI/sodium hyaluronan [164,185], but with synthetic PELs it was also observed [157].

For the exponential growth of the multilayer films formed with the polypetides and polysaccharides, a mechanism based on the PEL diffusion “in” and “out” of the core of the film, at the beginning only for polycation [83,140,164], and later for both polycation and polyanion has been demonstrated by using fluorescently labeled PELs [174,175]. An exponential increase of PAH/carboxymethylcellulose (CMC) and PAH/alginic acid (AGA) thickness was reported as the number of depositions increased. The thickness of $(PAH/CMC)_5$ and $(PAH/AGA)_5$ films in the dry state were approximately 150 and 90 nm, respectively [192].

To engineer films with desired secondary structures, SAMs were built from a weak polycation, PLL, and a blend of two weak polyanions, poly(L-aspartic acid) (Pasp) and PGA [162]. It was found that the proportions of the two polyanions incorporated into the SAMs thin films differed from those present in the deposition solutions, a preference for Pasp being observed. It was expected as the presence of a polypeptide in the SAM film to give secondary structures similar to those observed in proteins. Such secondary structures have been used as a source of information concerning the local interactions between PELs in the multilayer film. FTIR spectroscopy showed a large extent of β-structure for pure (PLL/PGA) multilayer film and no β-structure in (PLL/Pasp) films. On the other hand, in the films constructed with blends of polyanions, at all compositions, the content in β-sheets was higher than that expected for the content in PGA, assuming that PGA and Pasp worked independently. This result indicated a cooperative effect between PGA and Pasp in the formation of secondary structure [162].

Changes from the linear to exponential mechanism could be also controlled by adjusting the ratio between two polyanions in mixture, one known for the linear growth of the multilayer, NaPSS, and another one a weak PEL known for an exponential growth of the film, PGA, in combination with PAH. A hybrid growth low depending on both polyanions was found by Hübsch et al. for such system [85]. For films constructed with NaPSS/PGA mixtures, where PGA fraction was higher than 0.25, it was observed that at a certain point the exponential regime characteristic for the first deposition cycles abruptly changed to a linear regime, with a smaller slope than that observed at the end of the exponential regime. A mechanism which could explain this special behavior has been proposed, taking into account the differences between the diffusion coefficients of both polyanions that are assumed to diffuse "in" and "out" of the multilayer film. However, an exponential-to-linear transition in the buildup of the PAH/PGA multilayer and of the PLL/HA multilayers, after about 10-12 bilayers, has been reported by Michel et al. [221] and Porcel et al. [175], respectively. The transition from exponential to linear growth regim did not depend on the fabrication method, dipping or spraying [175]. This transition was explained by Salomaki et al. [77] and Porcel et al. [175] by the restructuring of the film which progressively forbids the diffusion of the PEL constituting the over part of the film. The forbidden zone regularly growth and keeps a constant increment of the thickness per bilayer.

In another interesting study, the PLL/HA was used as a system to build two compartments in the bulk of the multilayer, every containing five bilayers, separated by a compartment formed with the PAH/NaPSS system, also five bilayers, the last compartment working as a barrier between the first two [86]. It was demonstrated that PLL could freely move only in the compartments where it was used as building block and not in the compartment formed with PAH/NaPSS pair, new perspectives being thus opened to design nanoreservoirs in biomaterials coatings.

The stability of exponentially preassembled polypeptide multilayers, formed from (PLL/HA) [201] and (PLL/PGA) [219], as a function of various parameters such as pH [201,219], ionic strength [201], and temperature [219], was followed because their applications for biomedical purposes strongly depend on the film stability at physiological conditions.

3.3. Applications of SAMs Built with Weak PELs as Building Blocks

Because of their property to fast respond to the changes in the environment such as pH, ionic strength, solvent quality, SAMs films with different properties and applications have been constructed. Some of the numerous applications of such nanostructured materials are summarized as follows: separations and transport [28,143,147-149,154,156,161,176,177,194,217], responsive systems for controlled release [152,160,161,163,179,183,184,206,215], free-standing polymer membranes or sacrificial films in the preparation of SAMs membranes [178,193,197,200], corrosion resistant coatings [137], nanoreactors for nanoparticle synthesis [153,161,166,167,170,196], enzymatic nanoreactors [206,222], substrates for cells spreading and growth [140,163,164,189,198,214,220,223], surface patterning [189], sensors [100,141,156,192], catalysis [141,192], functionalization of hydrophobic surfaces [205], etc.

3.3.1. Separations and Transport

The charged top layer of ESAMs deposited on the ultrafiltration membranes can protect the membrane against fouling during the protein filtration [147] and control the performances of the membrane in pervaporation separation of water/alcohol mixtures [147,194], and in ion transport [148,149]. It is known that hydrophobic surfaces adsorb biological matters like proteins in the separation processes and a fouling of the membrane occurs. Meier-Haack and Müller [147] reported on the surface modification of some hydrophobic membranes by LbL technique, PAA and branched PEI being used as building blocks in the construction of the SAM film as skin layer. The fouling behavior of the composite membranes has been tested with human serum albumine (HSA) as aqueous solution. The lowest and the highest relative amount of adsorbed protein were found in the case of repulsive and attractive electrostatic interactions, respectively. Thus, when an electrostatic repulsion between the membrane surface and HSA took place (PAA as outermost layer), the adsorbed amount of the protein was very low, but when the surface was positive (PEI as outermost layer) a monomolecular protein layer was estimated [147].

Pervaporation experiments with water/2-propanol on composite membranes having $(bPEI/PAA)_n$ on the surface led to the following conclusions:

- six double layers of PAA and bPEI were sufficient to get a composite membrane with high water selectivity;
- the preparation conditions of the multilayer had a strong influence on the membrane performance, the raise of deposition temperature from 25 to 80 °C leading to an increase of the selectivity about 500 times comparative with the membrane prepared at 25 °C. The amide bonds could be formed at 80 °C, the covalent crosslinks leading to the decrease of the membrane swelling, the diffusion of the smaller water molecules being favorized.

A systematic study on the factors which could influence the performances of the composite membranes in the pervaporation process of water/2-propanol mixtures was reported by Lenk and Meier-Haack [194]. PEL concentration had a small influence on the selectivity of the composite membrane, but a strong influence was found for dipping time, molar mass of PAA and deposition temperature. A higher impact on the selectivity of the membrane was found when PAA with a high molar mass was used or when the temperature was raised from 25 °C up to 80 °C. The membrane modified with $(PEI/PAA)_4$ (M_w of PAA of 450000 Da), had a selectivity coefficient of 2500 comparative with that of the membrane built with PAA having M_w of 2000 Da, this strong difference being explained by the role of PAA to bridge the isolated spots of PELs formed after the first deposition on the membrane leading thus to a more complete coverage of the surface at a lower number of deposition steps. The highest impact on the membrane performance was found again for the deposition temperature, the formation of amide bonds leading to a tighter network stabilized by covalent crosslinks, and to the decrease of swelling.

Harris and Bruening observed a similar permeability for $[Fe(CN)_6]^{3-}$ of the $(PAH/PAA)_n$ and $(PAH/NaPSS)_n$ films made without added salt [28]. On the other hand, the permeability of PAH/PAA films strongly increased when the salt was used in the deposition solutions and showed the role of the constituent PELs in the design of SAMs films with controlled permeability. Bruening et al. enhanced the anion transport selectivity through porous alumina by deposition of $(PAA/PAH)_n$ multilayers, either by using as building blocks PAA and PAH partial derivatized with 2-nitrobenzyl groups [148], followed by postdeposition reactions (photolysis of photolabile groups) in the deposited multilayer, or by templating with Cu^{2+} [149]. According to the first procedure, net fixed charges, either negative or positive, have been generated after photolysis. The pH of the deposition solution was 5 for both PELs and therefore the electrostatic forces were the driving force in the multilayer growth. By diffusion dialysis experiments it was demonstrated that composite membranes prepared with 23%, 50%, and 65% 2-nitrobenzylesterified PAA exhibited Cl^-/SO_4^{2-} selectivities of 100, 150, and 170 respectively, after photolysis [148], i.e., the increase in ion-transport selectivity was correlated with the density of free charges (extrinsic compensated) into the SAM film. By the second strategy, Bruening et al. controlled the fixed charge density in the PAA/PAH multilayers by templating them with Cu^{2+} by alternate deposition of PAA partially complexed with Cu^{2+} and PAH, followed by removal of Cu^{2+} and deprotonation resulting in free COO^- groups, extrinsic compensated.

Cu^{2+} templated PAA/PAH membranes showed a 4-fold increase in Cl^-/SO_4^{2-} transport selectivity compared with the membrane modified with pure $(PAA/PAH)_n$ multilayer. Crosslinking of the templated multilayers through heat-induced amidation enhanced the Cl^-/SO_4^{2-} transport selectivity up to 610, without a diminuation of the flux compared to nontemplated membrane (Figure 4) [149].

Fused silica capillaries were coated with SAMs containing acrylic acid and imidazole units as pH sensitive functional groups. A precise control of the rate and direction of electroosmotic flow in microchanels has been accomplished by Sui and Schlenoff with such SAMs [154]. Ultrathin hydrogels with controlled ion permeability have been formed on a solid substrate by sequential amide formation between poly(vinylamine-*co*-N-vinylformamides) [P(VAm-*co*-NVF)s], with VAm contents of 20, 28, 42 mol %, and PAA using a water soluble carbodiimide [156]. The assembly amounts were controlled by the VAm content and by the copolymer concentration in the reaction solution. The authors found that the amount assembled from the copolymer with 28 mol % VAm was larger than that from the copolymer with 42 mol % VAm, because the high content in cationic groups led to a more extended polycation and therefore to a lower assembled amount.

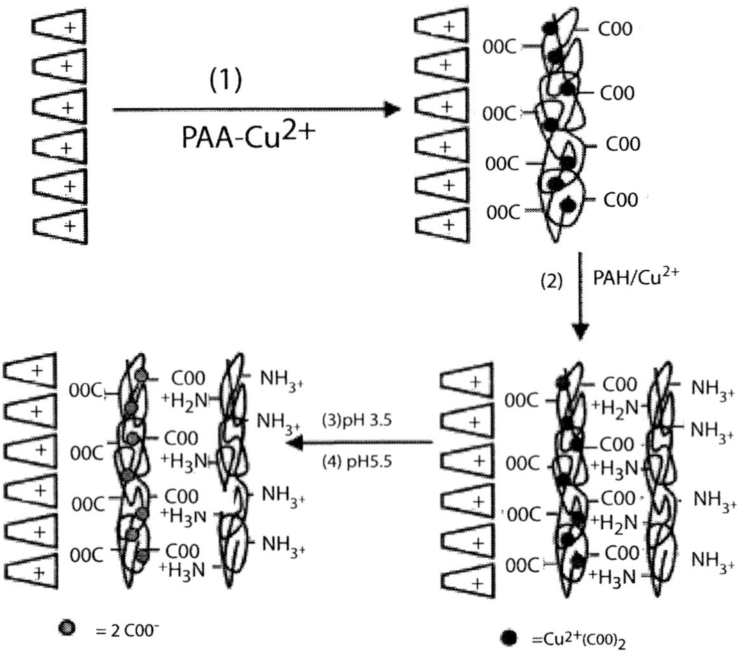

Figure 4. Preparation of Cu^{2+}-templated PEL films on porous alumina supports: Step 1, adsorption of partially Cu^{2+}-complexed PAA on porous alumina; Step 2, adsorption of PAH in the presence of uncomplexed Cu^{2+}; Step 3, removal of Cu^{2+}; Step 4, deprotonation of the free COOH groups of PAA. Repetition of steps 1 and 2 produces multilayer films (reprinted from ref. [149] with permission from the American Chemical Society).

These assemblies were stable in NaCl solution indicating the amide linkages were effectively formed. The ionic permeability was tested by CV with $Fe(CN)_6^{3-}$ and revealed that ions readily permeated the ultrathin hydrogels by diffusion [156].

Starting with Mendelsohn et al. [139], nanoporous and microporous SAMs thin films have been prepared from weak PELs as building blocks, PAH and PAA being the most widely employed, and their possible applications in drug release and biomaterials have been explored [142,183,184]. The microporosity in such thin films can be created simply immersing the SAMs briefly in a bath of low pH (pH = 2.4) followed by an exposure to neutral water [139] or by salt-induced structural changes [142]. Part of the interchain ionic bonds are cleaved by protonation of carboxylate groups followed by the reorganization and phase separation inside the bulk of multilayer, a large volume of pores being evidenced by AFM and SEM. Such microporous morphologies could be stabilized by the heat-induced formation of amide bonds [139]. Caruso et al. achieved nanoporous thin films by exposing to pure water the ESAMs prepared from salt-containing PEL solutions (0.2 M NaCl), washed with 0.2 M NaCl after each layer deposition. The dramatic changes in the film morphology with the formation of a nanoporous structure were demonstrated by scanning force microscopy (SFM) [142]. The multilayer film permeability was electrochemically (CV) evaluated using $[Fe(CN)_6]^{3-}$ as a probe, a better accessibility of the complex ion to the substrate demonstrating the presence of nanopores throughout the film. An interesting approach has been recently reported by Caruso et al. [183] who achieved nanoporous thin films by chemical crosslinking the PELs involved in electrostatic interactions (PAA and PAH) followed by removing the weak polybase [poly(4-vinylpyridine), P4VP], involved in hydrophobic interations with PAA, from a precursor composite thin film formed by assembly of PAA with a blend of PAH and P4VP. The growth of the precursor thin film and the nanoporous morphology of the final film have been controlled by the ratio between PAH and P4VP in the deposition blend (Figure 5).

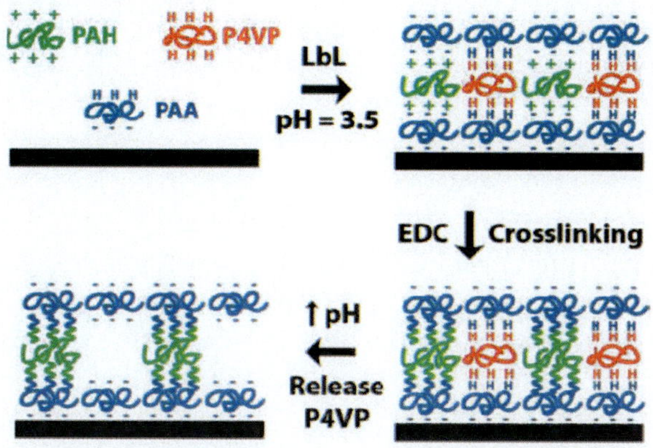

Figure 5. Procedure used for producing porous polymer films from PAA/(P4VP/PAH) multilayers (reprinted from ref. [183] with permission from Wiley-VCH).

3.3.2. Responsive Systems for Controlled Release

The LbL SAMs thin films with tunable uptake and realease properties attracted a great interest [152,160,161,163,179,183,202,206,215,216].

An interesting approach for the preparation of thermoresponsive thin films was recently reported by Serpe et al. [160]. According to their approach, a thermoresponsive component consitututed from poly(NIPAM-*co*-acrylic acid) microgels was used as polyanion and PAH as polycation in the construction of microgels thin films as confirmed by fluorescence microscopy. Microgel deposition temperature, microgel layer number and deposition pH influenced the film morphology and thin film thermoresponsivity. Burke and Barrett investigated by UV-vis spectroscopy the pH-dependent loading and release of small hydrophilic molecular probes (two water soluble dyes, Indoine Blue and Chromotrope 2R) from PAH and HA multilayer films [202]. The incorporation and release of dyes depended on the film swelling degree, the ability of the dye molecules to aggregate into the film, as well as the attractive and repulsive interactions between the dye and the functional groups of the SAM film. Polypeptide SAM films containing β-cyclodextrin-piroxicam inclusion complexes were developed by Benkirane-Jessel et al. as potential systems for biomaterials coatings with local anti-inflamatory drug release [215]. Embedding an antimicotic peptide (chromofungin) into a PLL/PGA multilayer film preserved the antifungal activity interacting with the fungal membrane and penetrating into the cell.

More information about the thermoresponsive systems formed by LbL technique with potential applications in controlled release can be found later in this subchapter, section "SAMs based on hydrogen bondings (HSAMs)" [152,163].

3.3.3. SAMs as Free-Standing Polymer Films/Sacrificial Layers

SAMs multicomposite thin films have potential applications such as separation membranes, sensors, micromechanical devices, wound dressing, or even artificial organs. Therefore, there is a great interest to develop a simple and fast method to release the thin film from a substrate as a self-standing membrane. Free-standing ultrathin films with thicknesses up to 1000 nm incorporating various nanomaterials have been recently reported [12,178,193,204].

Free-standing polymer membranes were prepared by LbL assembly of PAA and PAH adsorbed on electron microscopy grids as supports, without the need for sacrificial precursor coatings or substrate [178]. These films were sensitive to formic acid which decreased the pH to 2.5 and destabilized the $(PAA/PAH)_n$ film. Resistant membranes were obtained by intermolecular chemical crosslinking by heating at 120 ^{o}C for 10 h, with the formation of amide bonds. Both crosslinked and noncrosslinked membranes behave like hydrogels. Self-standing ESAM membranes at physiological conditions using pH-responsive films segments as sacrificial layers were reported by Ono and Decher [193]. A hybrid film consisting in two compartments was prepared first: a pH responsive film formed by H-bonds and an ESAM film on the top. The film assembly was performed by a spraying deposition, which can speed the deposition process by a factor of 50-150. The pH-sensitive film consisting in $(PAA/PEG)_n$ disintegrated at a neutral pH and released the self-standing ESAM film. The use of such mild conditions for removing a uniform polymer film could be very useful in the application of SAMs membranes.

It was demonstrated that CNT-containing membranes have interesting properties such as electrical conductivity, controlled molecular transport, and enhanced mechanical properties as it will be shown later in subchapter "Mechanical Properties of SAMs". Free-standing films with a composition of $[(PEI/MWCNT)_5(PEI/PAA)]_n$ with n = 20 were fabricated by Olek et al. [204]. Ellipsometry and UV-vis spectroscopy confirmed the sequential deposition of oppositely charged CNT and PEI indicating the high homogeneity of the composites with a linear growth as a function of the PEI/MWCNT bilayers. After every fifth deposition cycle, a MWCNT was replaced with a layer of PAA to introduce carboxylic groups involved in the stabilization of SAM by thermal and chemical (with glutaraldehyde) crosslinking. The high homogeneity of the free-standing hybrid films was demonstrated by SEM and AFM.

3.3.4. Nanoreactors

Perhaps the most studied SAM system is $(PAA/PAH)_n$, adsorbed on different substrates. One of the recently reported applications of such thin films is that of nanoreactor for the preparation of metal nanoparticles, a nanostructured composite resulting [153,161,166,167,170]. The key of the success in this type of synthesis is as the precursor to be a metal cation that can bind to the carboxylic groups of PAA [153,161,166]. Size controlled silver nanoparticles were synthesized within the PAA/PAH multilayers built at different deposition pH, the degree of ionization of PAA being varied accordingly, the content of carboxylic groups available to bind silver cations being thus controlled. Reduction of silver cations by a post–binding reaction produced silver nanoparticles, which diameters depended on the deposition pH, being higher for a lower pH, i.e., at a higher content in carboxylic groups. One implication of the strict control of the silver content in the nanocomposite multilayer is the ability to systematically change the optical properties of these films. A special type of dielectric mirror has been created by Nolte et al. using the in situ growth of silver nanoparticles with a strict control on the increase of the refractive index of certain regions, i.e., PAA/PAH bilayers, alternated with (PAH/NaPSS) bilayers, which remained unmodified and served as low index regions [170]. The authors demonstrated that the thickness and sequence of the silver-containing and non-silver-containing regions could be controlled and reproduced at the single nanometer length scale, allowing the creation of digitized rugate filters. Schuetz and Caruso formed metal (gold) or metal sulfide (PbS) nanoparticles in the PAH/PAA mulilayer films as nanoreactor deposited on PS spheres and protected against aggregation with three layers of strong PELs (NaPSS/PDADMAC/NaPSS) [167]. In this case, the $AuCl_4^-$ complex anion was used as a precursor for gold nanoparticles, the ammonium groups of PAH at a low pH, being involved in their binding, the gold nanoparticles resulting by a post-building chemistry, i.e., the reduction of the complex anion with $NaHB_4$.

The SAMs made from PLL as polycation and a mixture of PGA and Pasp, at different ratios, have been very recently reported as interesting candidates for biomimetic surface, since they display β-sheets, which content can be controlled by the PGA proportion in the polyanions mixture [196], as it was previously mentioned [162]. The mixed multilayers containing 0-100% PGA, ended with PLL, were able to induce nucleation and crystal growth of calcium phosphate. Phospholipide vesicles have been embedded into $(PGA/PAH)_n$ multilayers by Michel et al. [221,222]. Mineralization of calcium phosphate was induced by calcium ions as active ions in mineralization, spermine as an activator of crystal growth, and alkaline phosphatase as a catalyst included in embedded phospholipide vesicles which behave like submicronic reactors [222].

3.3.5. Substrates for Cell Growth and Adhesion

Biopolymers offer new possibilities to buildup biodegradable films or films that mimic the extracellular matrix by using polysaccharides such as HA, CHI, and proteins like collagen [140,164,165,220]. Richert et al. designed substrates for cell growth of numerous types of cells, such as chondrocytes and fibroblasts [164]. The growth of $(PLL/HA)_n$ films used for this purpose occured over two buildup regimes [140,164]. The first one was characterized by the formation of isolated islands that grow both by addition of new PELs on their top and by mutual coalescence of the islands. The second regime started after the eighth layer pair deposition in the working conditions, i.e., in the presence of 0.15 M NaCl, and was characterized by an exponential increase of the PEL adsorbed amount. Such films have potential to enhance the specific cell proliferation and adhesion. Proliferation of smooth muscle cells (SMCs) to SAMs films and the influence of the physical properties of the film on the cells adhesion have been investigated by Picart et al. [165,190]. Spreading of SMCs on PLL/HA multilayers with pre-attached collagen was highly dependent on film rigidity, stiffer films giving greater cell spreading, this result being consistent with the idea that cells are very tactile in their adhesive processes. The influence of CHI molar mass and the ionic strength of the PEL deposition solutions on the growth of $(CHI/HA)_n$ multilayers and cellular adhesion were investigated by Richert et al. [165]. It was also observed that the number of cells increased with the crosslinking degree of the SAM films. Selective adhesion of retinal cells on micropatterned SAMs lines adsorbed on PDMS surfaces using a microfluidic network was reported by Reyes et al. [198]. PEI and PAH were the polyions used as the top layer cellular adhesion material. Cells viability on the patterned SAMs was followed, PDMS being the cell-resistant surfaces.

Tissue surface modification by LbL technique was performed by Elbert et al. to evaluate the feasibility of this technique to create barrier materials on the surfaces of tissue-engineered implants [210]. The $(PLL/alginate)_n$ multilayers deposited on models of tissue surfaces were relatively bioinert, the interactions of cells with the underlying surface could be thus prevented, this property being correlated with the film thickness [210].

3.3.6. Surface Patterning

The forming of metal wires on a flexible substrate as ESAM films formed by deposition of PAH and PAA at pH 7.5 and 3.5, respectively, when both PELs where in a fully charged state has been reported. A pattern of metal circuits was realized by ink-jet printing of a catalyst on the multilayer film followed by an electroless plating technique to form the metal wire [189]. The film quality depended on the surface properties (plasma treated PET) because adhesion between the substrate, the ESAM, the catalyst layer, and the metal layer achieved by electroless plating are all depending on each other.

3.3.7. Sensors and Catalysts

Embedding a weak PEL (PMAA), whose charge density responds to changes of local environment, in an ESAM thin film may find application in surface sensors design. Thus, Xie and Granick investigated the local electrostatics within an ESAM film constructed with NaPSS and quaternized PVPy (QPVPy), PMAA being embedded as the second layer deposited on the substrate as a responsive layer to the local electrostatic environment [100]. Charge density changes in an ESAM thin film of $(QPVPy/NaPSS)_n$ deposited on the top of an embedded layer of PMAA have been monitored by FTIR-ATR spectroscopy, two bands being selected: 1701 cm^{-1}, assigned to the symmetric stretching vibrations of carbonyl group, $\nu(C{=}O)$, and the band at 1554 cm^{-1}, assigned to the asymmetric stretching vibrations of the carboxylate group, $\nu_{as}(COO^-)$. Ionization oscillation of embedded PMAA was modulated by changing surface electrical potential, i.e., the net charge of the outermost layer. The local electric field results in a local distribution of proton, which changes accordingly the local pH. Therefore, a negative surface (NaPSS as the outermost layer) led to the increase of the protons concentration near the surface and to the decrease of pH compared to the bulk; on the other hand, when QPVPy was the outermost layer adsorbed, the concentration of protons near the surface decreased leading to a higher pH. These changes of local pH led to oscillations in ionization degree of PMAA. A linear increase of the adsorbed amount of both strong PELs with the increase of layer number, at salt concentrations in the range 5 mM - 1 M was observed, the oscillations in ionization of PMAA being evident even at high ionic strength and at more than 10 layers. Hence, when charges have been added to the outermost layer by deposition of polyions on the top, electrostatic changes were induced far away within the multilayer. This approach may find application in the design of surface sensors for adsorbed ions and polyions [100].

LbL deposition of PLL and PAA on a gold electrode was conducted at pH 7, both polyions being fully charged at this pH [141]. A nonlinear increase of the multilayer thickness with the number of layers was found in the dry state by ellipsometry, the first polymer layers being thinner than the later deposited ones. A swollen and porous structure was found for these films. The negatively charged $Fe(CN)_6^{3-}$/ $Fe(CN)_6^{4-}$ as redox label was used to probe the electron transfer at the charged surface. The penetration of redox species into the multilayer and the subsequent electrochemical reactions have been controlled by the interaction between the charged species and the SAM film surface. The electrochemical behavior of the SAM thin film deposited on gold electrode was as follows: (1) the negatively charged redox label diffused freely into the polymer matrix when the outermost layer was positively charged (PLL), the concentration of the redox species being similar or higher than that of the bulk solution; (2) when the negatively charged PAA was the outermost layer, this worked like an electrostatic barrier and therefore the concentration of the redox label in the film matrix was lower than its concentration in the bulk solution. The interfacial electron-transfer resistance increased with the increase of the film thickness as it was also reported by Harris and Bruening [28].

It has been until very recently believed that the origin of the exponential growth of PLL/HA multilayer films is that PLL chains diffuse "in" and "out" of the films during preparation but HA does not. The free diffusion of only one of the used polyelectrolytes "in" and "out" of the film has been considered the origin of the exponential growth of the polypeptide and polysaccharide multilayer films [164,201]. In view of the fact that PEL chains can penetrate the interior of film, it is reasonable to assume that this type of SAM film is loosely packed and rather porous, as compared to the densely packed and ordered structure of linearly growing films. The interactions between $[Fe(CN)_6]^{3-/4-}$ ions and exponentially growing SAM films have recently been studied, and it was found that $[Fe(CN)_6]^{3-/4-}$ ions penetrate the films even when the films contain negative charges on the surface [192]. This is in contrast with the results discussed above for the gold electrode coated with a multilayer film of PLL and PAA, in which the response was suppressed when the surface was covered with PAA [142]. The binding properties of polysaccharide SAM films containing CMC and AGA for redox species were investigated by Noguchi and Anzai [192], interested in the electrochemical properties of $[Fe(CN)_6]^{3-/4-}$ ions confined in these films, as a function of the number of deposited layers. CVs of the $[Fe(CN)_6]^{3-}$ ion on the gold electrodes coated with the above mentioned multilayer films showed that the size and shape of the CVs were nearly identical to one another for $(PAH/CMC)_1$, $(PAH/CMC)_1PAH$, $(PAH/AGA)_1$, and $(PAH/AGA)_1PAH$ film-coated electrodes. However, for the electrodes coated with the $(PAH\text{-}PAA)_1$ and $(PAH/PAA)_1PAH$ films, the redox response to the $[Fe(CN)_6]^{3-}$ ion was severely suppressed when the negatively charged PAA was the outermost layer because of the electrostatic repulsion force. A reversible CV was observed only when the outermost surface was covered with positively charged PAH. The gold electrodes coated with the PAH/CMC and PAH/AGA multilayer films exhibited a redox response in the $[Fe(CN)_6]^{3-}$ solution irrespective of the sign of the electric charge on the surface of the film, assigned to the smooth diffusion of the $[Fe(CN)_6]^{3-}$ ions through the loose packed SAM films.

The confined $[Fe(CN)_6]^{3-}$ ion has been used as an electrocatalyst for oxidizing ascorbic acid in solution [192]. The authors consider that $[Fe(CN)_6]^{3-}$ ion-confined SAM film-coated electrodes may be useful for developing mediator type chemical sensors and biosensors, the use of polysaccharides as a material being beneficial because of its biocompatibility and low cost.

From the information presented above it is clear that the free diffusion of negatively charged redox label, irrespective of the surface charge, it is possible only in the SAMs films with both PELs able to freely move "in" and "out" the multilayer, the exponential growth of the film being not enough to find this behavior.

3.4. SAMs Based on Hydrogen Bondings (HSAMs)

The hydrogen bonding interactions as a driving force to buildup SAMs thin films has been communicated almost at the same time by Rubner et al. [208] and Zhang et al. [209], and has been quickly recognized as a means to enlarge the possibilities to construct and to control the properties of SAMs thin films. Important contributions in the field have Rubner et al. [150], Zhang et al. [136,144] and Sukhishvili et al. [151,152,211], but other valuable contributions must be mentioned [152,157,183,187,199,200,203].

One of the first explored systems for the multilayer built by hydrogen bondings was $(PVPy/PAA)_n$, both PELs being deposited from methanol solutions [136]. The nature of the interactions between PAA and PVPy has been established by IR spectroscopy; the absorption peaks present in the multilayer thin film showed that the carboxyl groups of PAA as well as pyridine groups of PVPy were not ionized and therefore the driving force was not electrostatic attraction but hydrogen bonding [136,144]. The main factors controlling the HSAM growth and characteristics were the concentration and molar mass of PVPy. The thickness of the thin film increased gradually from 14.9 to 22.7 nm with the increase of the PVPy concentration from 0.008 to 0.256 g/L, and this showed that the layer thickness could be controlled in the range of nanometer scale by changing the polymer concentration. The layer thickness increased also with the increase of the M_w of PVPy. The reconformation of the HSAMs yielding microporous films was investigated by Zhang et al. in NaOH aqueous solution and monitored by XPS, FTIR, UV-vis spectroscopy [144]. The film thickness increased linearly with the number of layer pairs. The dissolution of PAA by the destruction of hydrogen bonds into the basic solution did not destroy the whole film. The dissolution at a high pH (pH = 13) was very fast, but could be followed as a function of duration of contact by FTIR at a pH of 11. It was observed the disappearance of the characteristic band of carbonyl groups around 1718 cm^{-1} and the band at 1603 cm^{-1} shifts to 1595 cm^{-1} and could be explained by the disappearance of the association between pyridine ring and carboxylic groups. The reconformation of PVPy chains insoluble in basic aqueous solution produced a microporous film with potential applications in pharmaceutics and materials science.

PAA and PMAA as polyacids and poly(vinylpyrrolidone) (PVPON) and poly(ethylene oxide) (PEO) as polybases have been used by Sukhishvili and Granick [151] to construct HSAMs by deposition of counterparts at low pH, these thin films being erasable at high pH. Deposition pH and ionic strength were used as avenues to control the stability of these films. A strong influence on the fraction of hydrogen-bonded carboxylic groups in the multilayer, and hence on the critical pH at which these HSAMs thin films disintegrated, had the structure of the polyacid and polybase, i.e., the stronger the hydrogen bonding, the higher the pH at which a given ionization of polyacid was reached. A 2.1 times molar excess of PMAA than PVPON in the multilayer was found and was explained by the possibility as the carbonyl groups in the PVPON to form two hydrogen bonds with carboxylic groups of PMAA. A linear increase of the multilayer growth was characteristic for this polymer pair. An almost exponential growth of the total amount of polymer deposited was observed for the other system, the authors demonstrating that irregular deposition stemmed from weaker complexation between PEO and PMAA, a larger number of defects being possible in this case. The small peak at 1718 cm^{-1} assigned to the fraction of carboxylic groups bound to the ether groups of PEO has been also present in the IR spectrum of $(PVPON/PMAA)_n$ multilayer film. These HSAM films are destroyed at high pH, but an influence of PVPON molar mass on the critical pH was observed. Thus, critical pH decreased with 0.4 units when the PVPON molar mass decreased from ~1 300 000 to ~55 000 $g.mol^{-1}$. Incorporation and release of Rhodamine 6G in these HSAMs were followed to show that drugs can be incorporated and released as needed at preselected conditions.

Yang and Rubner followed new strategies for stabilizing HSAMs to environments of high pH, particularly to physiologic environments to extend their use as biomaterials [150]. Both thermal and photochemical approaches were used to transform hydrogen bonds in amide bonds thus making these multilayers insoluble at high pH. This made possible to use ink-jet printing and photolithography to create patterned multilayers thin films.

Kharlampieva et al. prepared HSAMs with promising properties for temperature-responsive containers by LbL deposition of poly(vinylcaprolactam) (PVCL) or poly(vinyl methyl ether) (PVME), as neutral thermoresponsive polymers, with PMAA at low pH [152]. The strength of the interlayer adhesion was correlated with the critical pH value for the multilayer growth and the critical ionization for the HSAM decomposition, these values being strongly dependent on the structure of the thermoresponsive polymer. HSAM thermoresponsive thin films were built by the interaction between poly(N-isopropylacrylamide) (PNIPAM) and PAA as a function of the PNIPAM deposition temperature [163]. Films prepared at a deposition temperature of 30 °C, i.e., close to the lower critical solubility temperature (LCST) of PNIPAM (32 °C) were smoother because the PNIPAM chains were adsorbed as compact coils, a higher density of the polymer on the surface being evidenced by SFM. The effect of preparation temperature upon the loading and release properties of these thin films was examined and it was observed that both processes were slower for the films prepared at higher temperatures, making these materials potential candidates for controlled uptake and release systems.

To monitor the local electrostatics within the SAMs thin films, new approaches have been employed, such as confining deconstructible HSAMs stacks between ESAMs layers [155]. $(PAA/PVPy)_n$ as HSAMs stacks have been alternated with ESAM formed layers of $(NaPSS/PAH)_n$, the characteristics of the film deconstruction (total film loss and rate of deconstruction) being dependent on the number of HSAMs layers confined between the ESAM layers. Such films have potential applications in drug delivery, patterning, free-standing multilayer films, and membrane science.

A new approach to buildup HSAMs films has been recently used by Ford et al. who grafted PVPy on singlewall CNT (SWCNT) (SWCNT-PVPy) [199]. LbL deposition of SWCNT-PVPy and PAA as building blocks led to a linear growth of the film, by hydrogen bonding between PVPy and PAA as driving force, demonstrated by FTIR spectroscopy.

Free-standing mechanically stable and interdigitated HSAMs assemblies were prepared by Hammond et al. by using PEO as hydrogen-bond-accepting polymer and PAA as a hydrogen-bond-donating polymer [200]. These films have been thermally and mechanically characterized to establish the conditions for fine-tuning free-standing films, for real applications. The ability to isolate large uniform sheets of the HSAM film allowed the first direct measurement of a glass transition by DSC and dynamic mechanical analysis (DMA), as well as the first direct tensile test of a HSAM assembly. The key was the choice of substrate, a hydrophobic surface such as Teflon and polypropylene with contact angles of 108° and 94°, respectively. The lack of the melting point of PEO at 65 °C showed that the hydrogen bonds within the film suppressed crystallization of PEO. It was also demonstrated that pH can be used as a control parameter of the film composition and morphology [200].

4. SAMs Based on Proteins and Nucleic Acids

The integration of biomacromolecules such as proteins and DNA in SAMs films is very interesting, but the release of these biological components from the LbL assembled films, where they have been embedded, should occur as close to the physiological conditions as possible owing to their sensitivity. Concerning the SAMs including proteins and nucleic acids, two situations will be considered for discussion:

- proteins and nucleic acids as building blocks, synthetic PELs being or not used as building blocks, embedded proteins are included here [6,7,224-245];
- adsorption or deposition of proteins (enzymes) onto the preassembled SAMs [246-254].

Even if this type of nanostructured materials attracted a great interest by their specific applications as biosensors, biocatalysts, membranes for biomedical aims, only a few references have been selected for discussion in this review, mainly the latest ones and those which were the first in this field, the space being not enough for all significant articles focused on this topic.

4.1. Proteins and Nucleic Acids as Building Blocks

The embedding of proteins into multicomponent ESAMs films by LbL deposition has been systematically investigated by Lvov et al. [225] who employed water soluble proteins as building blocks, such as: cytochrome c (Cyt), myoglobin (Mb), lysozyme (Lys), hemoglobin (Hb), glucoamylase (GA), and glucose oxidase (GOD), in combination with flexible synthetic PELs. The main concerns on the hybrid protein-containing ESAMs films are to preserve the protein biological recognition activity correlated with protein structure, and the influence of the assembly of protein in the LbL film on the protein mobility [224,226]. Multicomponent protein ESAMs films were designed by LbL deposition of Hb either at pH 4.5 (Hb being positively charged under its isoelectric point at pH = 6.8) with NaPSS, or at pH 9.2 (where Hb is negatively charged) with PEI as counterpart. A linear increase of the multilayer was found in both regions. Other hybrid multicomponet films where fabricated either by using two different proteins with intercalated oppositely charged PELs layers, or montmorillonite and GOD with PEI in between [225], thus opening new routes to sequential enzymatic reactions. To increase the sensitivity of the thin films with respect to detect immunoglobulin G (IgG), Caruso et al. immobilized anti-IgG in multilayer films by its alternate deposition with NaPSS, on planar substrates [226,227]. The film sensitivity was linearly dependent on the number of anti-IgG layers, i.e., on the amount of protein deposited, when the anti-IgG was separated by only one deposited layer of NaPSS, and was dependent only on the outermost deposited layer of anti-IgG when the proteins layers were separated by a dense NaPSS(PAH/NaPSS)$_2$ film. This showed the antibody permeation was restricted by the ESAM film intercalated and on the other hand the preservation of the biospecific activity after embedding in a multilayer film. A linear increase of the film as a function of the number of anti-IgG layers deposited on a precursor of (PAH/PSS)$_2$ was demonstrated by QCM and FTIR-reflection absorption spectroscopy (FTIR-RAS).

It was observed from the beginning that the employment of the flexible synthetic PELs as counterpart for proteins in the buildup of hybrid thin films has a positive role in the construction and stabilization of the multicomponent multilayer [224], but organized protein multilayers, with no synthetic PELs inside, has been also successfully performed via a special approach by Brynda and Houska [229]. Protein-containing hybrid multilayers deposited onto PS colloidal particles were fabricated by Caruso and Möhwald, by alternate deposition of fluorescein isothiocyanate-labeled BSA and anti-IgG with either positively (PDADMAC) or negatively (NaPSS) charged PELs [230] and by Schüller and Caruso from GOD alternated with PEI [233]. A linear growth of protein layer thickness with the number of depositions cycles and a similarity of the multilayer growth with the process developed on planar substrates were found if the aggregation of coated colloids was prevented [230]. The successful construction of hybrid ESAMs film on colloids opened perspectives for applications in immunosensing, biocatalysis, biosensors, etc. Colloidal silica was used as substrate for biocomposite thin film fabrication with protamine (PRM)/DS and PRM/BSA pairs as building blocks, by means of LbL technique [236]. Drug carrier and diagnosis devices are suggested as potential applications for this biocomposites.

Bio-catalitic materials have been fabricated by immobilization of enzymes either on planar substrates [237,242] or colloids [237,240]. Thus, Patel et al. [237] investigated the growth, characterization, and bio-catalytic activity of lignin peroxidase and manganese peroxidase LbL assembled with synthetic polycations (PDADMAC, PEI, and PAH) in the degradation of environmental pollutants. Immobilized enzymes were extremely stable and their catalytic activity increased with the increasing of PEL/enzyme layers. Bio-catalytic nanoparticles with urease immobilized in a multilayer with PDADMAC and NaPSS as flexible PELs as nanoreactors, and urease catalytic activity as a function of the multilayer buildup conditions were investigated by Liang et al. [240]. An interesting approach was recently used by Caruso et al. [242] who alternately deposited negatively charged nonstoichiometric complexes resulted by the interaction between peroxidase and NaPSS and oppositely charged PAH on porous polycarbonate (PC) membranes to prepare high surface area thin films for bio-catalysis. Removal of PC template generated PEL open or closed nanotubes. The film bioactivity was related on the pore size and porosity, and the number of enzyme-containing layers deposited. Bio-catalytic activity values up to 9-times higher were reported compared with the same film deposited onto nonporous quartz slides [242].

Not only electrostatic interactions were involved in the fabrication of LbL assembled protein/PEL hybrid films, but also specific recognition as Cassier et al. demonstrated for streptavidin/biotin nanostructured films [228]. Streptavidin has been either directly alternated with poly(L-lysine-*co*-ε-biotinyl-L-lysine) (PLB) or spacered by more PEL layers on a precursor film composed of a few PAH and NaPSS layers. The insertion of some spacer layers of flexible PELs has been used as a tool to control the surface roughness of each protein layer, and the addition of salt to the PLB solution showed a large improvement of the streptavidin binding, a larger number of biotin groups being exposed at the interface by the screening effects of the salt ions [228].

Starting with the first reported LbL assembled thin film containing nucleic acids as building blocks, fabricated by Lvov et al. [5], numerous researchers have been attracted to this interesting topic [6,7,87,238,239,241,244,245 and references therein]. Some recently reported results will be briefly presented. In their study on the growth of PAH/DNA multilayers, Gong et al. found a rough morphology, a highly heterogeneous structure, and an exponential growth of the film, the adhesion properties being strongly influenced by the structure of the last layer adsorbed [87]. Ren et al. [238] demonstrated that the polycation-based DNA nanoparticles can be successfully incorporated into multilayers. The authors used PEI to form a complex with DNA [molar ratio of PEI nitrogen to DNA phosphate (N/P ratio) was 5]. The $[(\text{PEI-DNA nanoparticle})/\text{PGA}]_n$ multilayers were deposited onto PEI-coated quartz substrates. These multilayers may provide a new approach to load and deliver non-viral DNA to targets in vivo. Johnston et al. [239] alternately deposited two-block homopolymer nucleotides ($\text{polyA}_{20}\text{G}_{20}/\text{polyT}_{20}\text{C}_{20}$). They also formed DNA multilayer films on silica particles. In this study, DNA layers were held together by hydrogen bonding of the base pairs. The multilayer buildup increased linearly with increasing layer number. The authors demonstrated that it is possible to build LbL DNA thin films using polymers with the same charge. The morphological changes in the film were achieved by treating the film with urea. The hollow capsules were obtained by assembling the DNA film on silica particles and dissolving the silica core using HF. Desigaux et al. [241] obtained supramolecular assemblies from DNA molecules and inorganic layered double hydroxides (LDHs). The LDH/DNA hybrids were characterized by a lamellar organization, with DNA molecules sandwiched between hydroxide layers. This study described a new self-assembly chemical approach to synthesize degradable inorganic materials at acid pH. The LDH/DNA-hybrid approach represents a new way to synthesize new labile, non-viral gene delivery systems. Recksiedler et al. [244] prepared poly(aniline boronic acid) (PABA)/ribonucleic acid (RNA) multilayer films. PABA interacted covalently with RNA through the formation of a boronate ester, a boron-nitrogen dative bond, as well as electrostatic interactions. It was observed a linear increase of multilayer with the number of PABA/RNA bilayers deposited. The PABA/RNA multilayer films were redox-active at neutral pH. Electrochemical control of PABA under these conditions allowed potential-induced controlled release of RNA from the multilayer at neutral pH.

4.2. Proteins Adsorption onto Preassembled ESAMs

The interaction of biological matters such as proteins is one of the main concern when surfaces modified by LbL deposition of PELs, such as membranes, come into contact with protein-containing fluids, a protein-resistant surface with anti-fouling properties being required, or when the whole hybrid film with one or more proteins as the last layer adsorbed must have specific biological activities [248,252,253].

The interaction between a series of positively and negatively charged proteins and (NaPSS/PAH) multilayers have been investigated by Ladam et al. [246,249]. It was observed that the protein adsorbed amount and the stability of the adsorbed protein layer depended on the outermost layer of PEL as well as on the ionic strength of the deposition solution and of the solution used for the multilayer treatment after the protein adsorption, the balance adsorption/desorption of protein being a very complex one. Thus, on the PSS-ending multilayers, a dense protein layer was formed, whereas on PAH-ending multilayers, films with thicknesses several times larger that the native protein dimension were evidenced. Several mechanisms of protein adsorption were proposed to explain the differences [246,249]. Adsorption of proteins on pH controlled multilayer films based on weak PELs has been also reported [250,254]. The effects of adsorption pH on the amount of protein (HSA) adsorbed on the multilayer films constructed from "protein-like" PELs like PLL and PGA, known for the exponential growth of the film, have been deeply investigated by Gergely et al. [250]. The adsorption pH varied in the range 3.0 - 10.5. The evolution of HSA adsorbed amount strongly depended on the PEL-ending the multilayer, and the adsorption pH. Thus, a limited amount of HSA has been adsorbed on PLL-terminated films, at low pH, because the protein has a positive charge below pH 4.6 and PLL is also positively charged in this region; a maximum of the adsorbed HSA amount was found at a pH of 6.5, the electrostatic interactions between the negatively charged protein and positively charged PLL being the motor. These results clearly showed that the pH of adsorption solution had a strong influence on the adsorbed protein amount on the SAMs films fabricated from "protein-like" PELs [250]. However, when the multilayer film was terminated with PGA, at high pH, HSA has been adsorbed even when PGA was the last layer deposited, PGA and HSA being both negatively charged here.

As many researchers observed, electrostatic interactions are the main driving force in the adsorption of proteins on the pre-assembled multilayer thin films, but hydrogen bonds and hydrophobic interactions play also a role [246,249,250,251,254].

5. SAMs from Low Molar Mass Organic Compounds

5.1. SAMs Thin Films Fabricated with Low Molar Mass Organic Compounds as Building Blocks

Starting with Cooper et al., who demonstrated that Congo Red (CR) and copper phtalocyanine tetrasulfonic acid could be assembled with a cationic polypeptide [255], multicharged azo dyes have been successfully utilized to fabricate electrostatic ESAM thin films [27,256-270]. Araki et al. have fabricated ESAM thin films even with two oppositely charged porphyrins as building blocks [256].

Some aspects resulted to be very important in the preparation of these composite thin films: control of the dye amount adsorbed, aggregation mode of the dye in the ESAM films and the stability of the polyion/dye thin film depending on the medium properties. Kunitake and co-workers have demonstrated that the stability of the polycation/dye assembly depended on the polycation structure and the aggregation mode of the dye on the positive surface depended on the dye structure being independent on the dye aggregation in solution [258]. Advincula et al. have investigated the molecular assembly process of various multicharged azo dyes with different polycations: PDADMAC, PAH and PEI [264]. Among the other dyes, Direct Red 80 (DR80) has shown significant photoalignment properties [266].

Azobenzene chromophores are frequently utilized as photoreactive components in the multilayer thin films due to their property to photoisomerize from the thermodinamical more stable trans configuration (E-form) to the cis form (Z-form) by UV-light irradiation ($\lambda < 370$ nm). The reverse process takes place by thermal treatment and/or by visible light irradiation ($\lambda > 450$ nm) [43,44,172]. Therefore, the most dyes used as building blocks in the multilayer buildup are azo dyes. An almost linear growth of the absorbance and thickness with the number of the PEL/dye bilayers was observed [264,266,270,274], and this showed a regular increase of such nanoarchitectures in both salt-free aqueous solutions and NaCl containing solutions. However, the details regarding the influence of dye structure and counterion nature on the growth and properties of the polycation/multicharged azo dyes nanostructures are very few [258,260,264]. In this context, we have recently investigated the preparation of LbL thin films incorporating several multicharged azo dyes, DR80, Direct Blue 1 (DB1), Ponceau SS (PSS) and Crocein Scarlet MOO (CSMOO), the polycation being one with N,N-dimethyl-2-hydroxypropyleneammonium chloride units in the backbone (PCA_5) [276,277], adsorbed from aqueous solutions with different ionic strengths, the counterion nature (Cl^-, Br^-, SO_4^{2-}) being also taken under consideration [278-282]. The characteristics of the polycation/azo dye complexes in the solid state could be qualitatively at least correlated with the properties of the ESAM PCA_5/azo dye thin films [283].

The polycation conformation was controlled by the polycation adsorption from both salt-free aqueous solution and 0.2 M electrolyte (NaCl or Na_2SO_4). UV-vis spectroscopy and the electrokinetic measurements were employed to monitor the polycation/azo dye multilayer buildup and stability. A special attention was paid on the dye aggregation mode in the multilayer in dependence on the dye structure and the polycation conformation, comparative with the PCA_5/dye interaction in dilute aqueous solutions [280,281], two azo dyes being compared, CSMOO and PSS. A bathochromic shift from 510 nm at about 516 nm and a definite increase of the maximum at 350 nm were observed with the increase of the layer pair number when CSMOO was used, unlike the complexation of the dye with PCA_5 in dilute aqueous solution [281]. The UV-vis spectral changes from the aqueous solution to the adsorbed layer indicated a different binding mode of CSMOO adsorbed in the multilayer, irrespective of the polycation conformation (PCA_5 adsorbed from either salt-free aqueous solution or 0.2 M Na_2SO_4 aqueous solution).

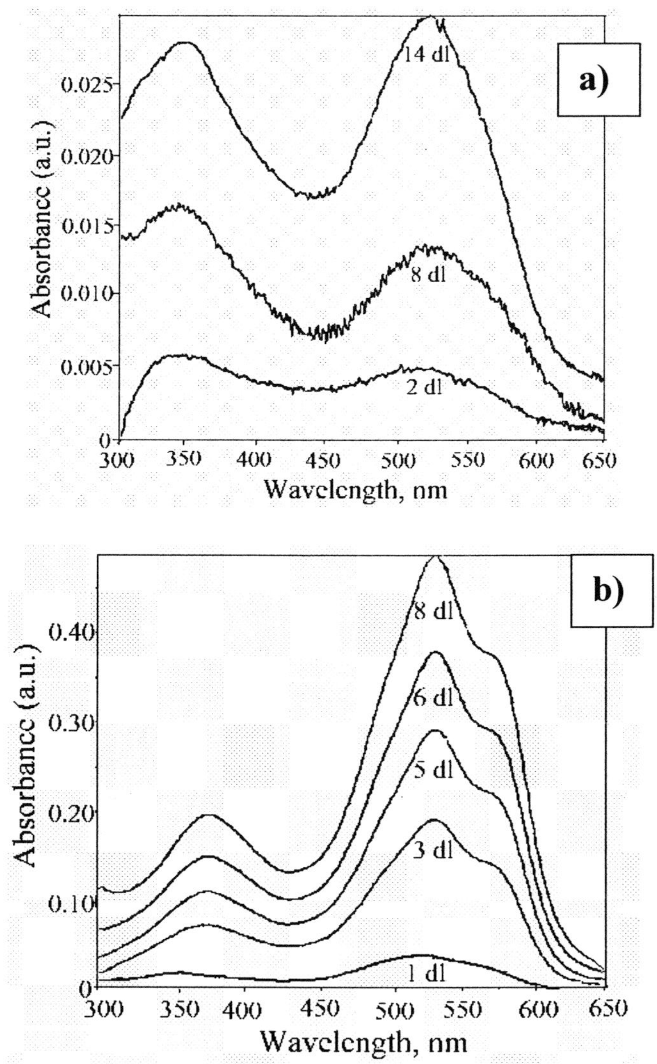

Figure 6. UV-Vis spectra of the $(PCA_5/PSS)_n$ self-assembled multilayers as a function of double layer number: a) salt-free aqueous solutions; b) 0.2 M Na_2SO_4 aqueous solution.

Taking into account the differences between the UV-vis spectra of the $(PCA_5/CSMOO)_n$ multilayers and the UV-vis spectra of this dye in the presence of PCA_5 in dilute aqueous solutions, one can assume that the CSMOO binding mode in the multilayer was somewhat different compared with PCA_5/CSMOO interaction in dilute aqueous solutions. The CSMOO binding in the $(PCA_5/CSMOO)_n$ multilayers could be a combination between the electrostatic and hydrophobic interactions. We suppose that, because of their proximity, both sulfonic groups of the adsorbed CSMOO molecules are involved from the beginning in the interaction with the polycation layer, but the charge reversal of the multilayer, with every layer adsorbed, requires that a further dye amount to be adsorbed. This process could be achieved by the hydrophobic interactions between the azobenzene-containing parts of the dye molecules. The steady increase of the maximum at 350 nm let us to assume that a H-aggregation mode of the dye is possible in this case. This interaction mode could explain the ability of the CSMOO as the last adsorbed layer to bind a new polycation layer mainly by electrostatic interactions.

On the other hand, the polycation adsorption medium had a clear influence on the PSS adsorption mode in the multilayer (Figure 6). Thus, when the polycation was adsorbed from salt-free aqueous solution, the UV-Vis spectra of the $(PCA_5/PSS)_n$ multilayers were similar to those characteristic for the $(PCA_5/CSMOO)_n$ multilayers. On the contrary, the UV-Vis spectra of the $(PCA_5/PSS)_n$ multilayers, with PCA_5 adsorbed from 0.2 M Na_2SO_4 aqueous solution, were similar to those characteristic for PSS binding to PCA_5 in dilute aqueous solution [281]. PSS binding to the PCA_5 in the multilayer appeared to be similar with CSMOO only when the polycation was adsorbed from saltfree aqueous solution, but a J-aggregation occured when the polycation was adsorbed from 0.2 M Na_2SO_4 aqueous solutions. To explain this different behavior, we suppose that only one sulfonic group of PSS was bound on the polycation layer matrix and the other remained free and reversed the charge of the surface, which was thus able to adsorb a new polycation layer. A J-aggregation mode was also observed by Ariga et al. when CR was adsorbed on a positive surface [258].

Not only dyes have been successfully used as building blocks but also other low molar mass organic compounds [264,271-275,], anionic and cationic bipolar amphiphiles being between the first building blocks used by Decher and Hong [271] in the ESAM thin films construction.

5.2. Dyes as Probes for the Structure of ESAM Thin Films

Ionic dyes have been often used to test the diffusion [284], surface charge and driving force of the multilayer growth [135,214,275,285,286], or to check the loading/release properties of the ESAM thin films [202,287]. To test the permeability of ESAM thin films and to elaborate a diffusion model, v. Klitzing and Möhwald embedded a fluorescent dye in the multilayer film at certain distances to the surface [284]. Rodhamine was added in the outer aqueous phase solution as a quencher of fluorescence and the permeability to rodhamine was determined from the data of fluorescence quenching, by variation of the PEL conformation (deposition from 1 M NaCl solution or without any electrolyte) and by sequential deposition of NaPSS with amphiphilic layers of dimethyldioctadecylammonium bromide into the multilayer film.

The relative charge amount on the film surface is often tested with positively or negatively charged dyes [135,214,275,285]. To determine the extent of chain interpenetration in the construction of ESAMs thin films with weak PELs as building blocks (PAH and PAA) as a function of the pH of the deposition solution, Rubner et al. used the surface dying technique which consits in adsorbing a positively charged dye MB onto the SAM film and measuring the amount of dye adsorbed by visible spectroscopy [135]. The amount of MB adsorption was determined mainly by the availability of free carboxylate groups not involved in forming contact ion pairs with cationic PAH when the outermost layer was PAA. Thus, it was observed that the amount of MB adsorbed onto the $(PAH/PAA)_n$ film decreased as the pH of PAA deposition solution increased, this indicating that the additional carboxylate groups formed by increasing pH were not available as binding sites for MB, being involved in forming contact ion pairs with the previously adsorbed PAH chains [135]. Therefore, the number of free carboxylic groups available for ionization to free carboxylate groups during MB adsorption decreased when PAA deposition pH increased, and almost no MB adsorption was detected when both PAH and PAA have been adsorbed at pH 6.5 or higher. A linear increase of the MB adsorbed amount onto the DS/CHI assemblies ended with DS as an anionic surface layer as a function of the NaCl concentration of the deposition solutions was found by Serizawa et al. [214]. Almost no dye was adsorbed when CHI capped the multilayer film and this indicated that MB was electrostatically adsorbed onto the negatively charged surface. The absence of MO adsorption when CHI was the last layer deposited indicated that only a small fraction of positive charges from CHI remained on the film surface.

To probe the binding stoichiometry between oppositely charged PELs forming ESAM thin films, Caruso et al. followed the binding of two fluorescent probes, pyrenetetrasulfonic acid and 6-carboxyfluorescein to $(PAH/PSS)_n$ and $(PDADMAC/PSS)_n$ multilayers formed on PS latex particles [285]. The dye binding was observed only when the outermost layer was positively charged, and the amount of dye linearly increased with the increase of the PEL layer number up to 11 layers for the PAH/PSS film. It was thus demonstrated that a minimum of about 10-30 % of the cationic sites of the polycations in the upper region of the ESAM film were not electrostatically used in the multilayer formation with opposite charges of NaPSS [285]. The binding properties towards MO of the multilayer films formed from PAH and sulfonated α-cyclodextrin (s-α-CyD) and β-cyclodextrin (s-β-CyD) have been investigated by Sato et al. [275]. MO was adsorbed to the film by forming an inclusion complex in the (PAH/s-α-CyD) film, but formed only aggregates onto the surface of the (PAH/s-β-CyD) film.

Loading and release of the dyes as models for drug delivery applications have been investigated by Burke and Barrett [202] in PAH/HA thin film by using Chromotrope 2R and Indoine Blue as probes, and Chung and Rubner [287] in PAH/PAA multilayer films with MB as an indicator. It was found that the incorporation and release of small molecules depended on the degree of swelling of the film, i.e., the hydrophilic/hydrophobic balance, the ability of the dye molecule to aggregate into the film, as well as the attractive/repulsive interactions occurring between the probe dye molecules and the acid/base functional groups into the multilayer film formed from weak PELs [202].

6. ESAMs Based on Colloidal Particles as Building Blocks

Numerous colloidal particles both organic and inorganic have been used in the ESAM films construction [288-291,292 and references therein,293-333]. Some recently reported nanocomposites built with colloidal particles as building blocks are summarized in Table 2.

Molecular dynamic simulation on the structure and mechanism of the growth of multilayer films from PELs and nanoparticles has been performed and revealed that the assembly process is electrostatically driven and proceeds through surface overcharging after each deposited layer [334,335].

Table 2. ESAMs nanoarchitectures fabricated with colloidal particles as building blocks

NanoParticle type	Comments	Refs.
	Organic colloidal particles	
PS nanospheres	• Negatively charged PS nanospheres were adsorbed onto positively charged ESAM thin films, the amount of nanoparticles adsorbed being controlled by the deposition conditions of PELs (PAH, NaPSS), mainly the ionic strength, and by the size and concentration of PS nanospheres. A monolayer adsorption per adsorption step was indicated by SEM. The possibility of regulated multilayering of nanospheres has been also studied [291]. • Hammond et al. have employed a polymer-on-polymer stamping method to create alternate regions of (+/-) charge on surfaces [293,295,296]. The patterned PEL surfaces were used as templates to direct negatively charged PS particles onto 2D colloidal arrays. These topographically structured surfaces can be used as templates for selective and non-selective metal plating, cell adhesion, and QDs deposition. • A snowman-like array of charged PS colloids with antireflecting properties was designed by Koo et al. by adsorption of negatively charged monodisperse PS nanospheres onto $(PAH/NaPSS)_2$-PAH film [294]. • SAM films were prepared by LbL deposition from PEI, PS latex particles and PAH, subsequently hydrophobized by immersing in Nafion solution [297]. The level of hydrophobization depended on the concentration of NaCl in Nafion solution and the number of PEI/Nafion bilayers.	[291, 293-298]
Charged Paraffin particles	• Glinel et al. reported a new route to improve the stability of PEL multilayers with respect to water diffusion by the deposition of wax particles onto PEL SAMs, this approach being recommended for fabrication of devices requiring protection against environmental moisture, to reduce the permeation of drugs, peptides, or ions through the film.	[299]
Prussian blue nanoparticles	• Zhao et al. constructed an ESAM glucose biosensor by integrating Prussian blue nanoparticles, protected by PDADMAC (P-PB), as electron mediators. P-PB and glucose oxidase were consecutively adsorbed onto NaPSS-trerminated bilayers. The P-PB nanoparticles can catalyse the electroreduction of H_2O_2, performances of multilayer films being tailored by the number of bilayers.	[300]
Dendrimer	• Liu et al. utilized ESAM thin films, prepared from positively charged third-generation poly(amido amine) dendrimer and negatively charged PELs (NaPSS and PAA), as nanoreactors for the formation of silver nanoparticles. Silver ions preorganized in LbL film were subsequently reduced to prepare silver nanoparticles having a strong negative redox potential.	[301]
Cellulose nanocrystals	Composite SAM films containing cellulose nanocrystals as a reinforcing material, with a high potential for biomedical applications, were prepared by Kotov et al. using a LbL technique. A linear increase of the multilayer thickness with the layer number was reported when PDADMAC was used as polycation.	[302]
	Inorganic colloidal particles	

Table 2. (Continued)

Gold nanoparticles	• Kim et al. alternately deposited PDADMAC and NaPSS onto functionalized multiwall carbon nanotubes (fMWCNTs). Complex formation of positively charged gold nanoparticles and the negatively charged fMWCNTs was achieved with and without multilayer coatings. The density of positively charged gold nanoparticles self-assembled on raw fMWCNTs was lower than those anchored on PDADMAC/NaPSS bilayer coated fMWCNTs.	[125]
Gold nanoparticles	• Two hydrogen bonded assembly routes were employed [303]: - PVPy + Au nanoparticles with carboxylic groups tailored surfaces; - PAA + Au nanoparticles with pyridine groups tailored surfaces. A linear increase of the film absorbance with the number of adsorbed layers was found. • A highly uniform, extremely robust and long-living free-standing membranes with a central layer of gold nanoparticles embedded between two (PAH/NaPSS) multilayers, i.e., $(PAH/NaPSS)_nPAH/Au/(PAH/NaPSS)_nPAH$ was fabricated by Jiang et al. by using a spin-assembly deposition method, which combined the spin coated and LbL technique [304]. The film thickness linearly increased with the number of bilayers showing a well-ordered nanostructure of the film. • Nanocomposite films $[Au/(PAH/PSS)_nPAH]_m$ with a well-organized microstructure and uniform surface morphology, with a linear increase of the multilayer thickness with the number of bilayers, were fabricated by Jiang et al. by using spin-assembly method [305]. • Positively charged gold nanorods (Au NRs modified with hexadecyltrimethylammonium bromide) and NaPSS have been alternately deposited onto indium-doped tin oxide (ITO) with a previously deposited layer of linear PEI, via electrostatic interactions, to prepare an active surface-enhanced Raman spectroscopy (SERS) substrate with applications in nanodevices and single molecule detection [306].	[303-306]
Silver nanoparticles	• Two approaches for preparation active SERS substrates were reported by Li et al.: – AgI and PDADMAC have been sequentially deposited on a quartz substrate followed by the reduction of Ag^+ to Ag nanoparticles, the multilayer working like a nanoreactor for this step; – preformed Ag nanoparticles and PDADMAC were alternately deposited on a quartz substrate.	[307]

Table 2. (Continued)

Europium complex nanoparticles	• For technological applications, the luminescent europium complexes have been incorporated into a stable matrix such as silica, polymers, organic-inorganic materials, as well as nanocomposite LbL thin films, to get unique optical properties with potential applications for luminescent devices. Cui et al. reported a linear growth of the multilayer nanocomposite film from the complex nanoparticles synthesized with trivalent europium, dibenzoylmethane as the ligand and PEI as the stabilizing agent (PED), on the one side, and NaPSS as polyanion, on planar substrates and on silica microspheres, the final multilayer being $(PSS/PED)_n$.	[308]
Titania precursor, nanotubes, nanosheets	• A negatively charged water soluble titania precursor (TALH) was used to form well-defined ESAMs thin films by electrostatically interaction with different polycations (PAH, PDADMAC, CHI); a uniform amount of TALH was deposited with each alternate deposition cycle of PEL and TALH. The electrostatic binding of TALH to polycations was confirmed using fluorescence spectroscopy [286]. • Ma et al. fabricated LbL multilayer thin films on various substrates by electrostatically deposition of titania nanotubes and PDADMAC. Single crystalline Au and Ag nanoparticles with narrow size distribution confined in the hollow cavity or deposited on the nanotubes were used as building blocks to regularly build up LbL multilayers with a linear increase of the absorbance as a function of the bilayer number. Heteronanocomposites have been successfully fabricated by alternately assembling layer pairs of PDADMAC/titania nanotubes and PDADMAC/titania nanosheets. Such nanocompostes have been explored for various applications in photocatalysis, solar energy conversion and self-cleaning devices [310].	[286,309-311]
Nanocomposites including silica nanoparticles as building blocks	• Electrostatically assembled multilayer films formed by the interaction between anionic silica nanoparticles as building blocks and a copolymer of AA and (3-methacrylamido propyl)-trimethylammonium chloride were prepared by Sennerfors et al. [312]. The influence of ionic strength on the ESAM buildup, structure and stability was emphasized by AFM and XPS. • An interesting type of building block was prepared by Liu et al. [313] as follows: positively charged proteins and NaPSS were alternately adsorbed onto SiO_2 nanoparticles forming core-shell SiO_2 – $(protein/NaPSS)_m$ nanoclusters; negatively charged nanoclusters have been then LbL adsorbed on various substrates by electrostatic interactions with PEI forming at the end $\{[SiO_2 - (protein/NasPSS)_m]/PEI\}_n$ films, the advantages of such nanostructured thin films being a larger fraction of electroactive protein and higher specific biocatalitic activity. To enhance the storability of loaded molecules in nanocapsules, oppositely charged PELs and silica nanoparticles were used to generate layered coatings onto nanocapsules [315].	[312-316]

Table 2. (Continued)

(Exfoliated) clay platelets	• Composite ESAM films were fabricated by Laschewsky et al. by using an inorganic polyanion, namely a synthetic exfoliated clay (hectorite), and a large variety of polycations, alternately deposited on planar substrates [317,318]. To enhace their stability, the hybrid multilayers were crosslinked by UV light, either at the end of the multilayer buildup or after each deposition step, by using a reactive polycation, the photoreaction reducing the film roughness and promoting a more regular growth [318].	[317-319]
ESAMs as nanoreactors for nanoparticles	• Catalytic properties of Pd nanoparticles embedded in SAM films have been studied [320,321]. It was reported a size-based selectivity in hydrogenation of small alcohols correlated with the different rates of transport of the reactant molecules to catalytic sites on the Pd nanoparticles. • Precursor ESAM films were constructed from NaPSS and $Fe(OH)_3$ colloid onto quartz plate, followed by calcinations in Ar/H_2 at 750 °C to produce iron nanoparticles for aligned the growth of multi-walled carbon nanotubes with a controlled structure and morphology [322]. • Zhang et al. prepared core-shell latex spheres with CdS/PEL composite SAM films with applications in biological fluorescence assay areas [324]. • PEL modified ZnS-capped CdSe quantum dots (QDs) were LbL deposited on patterned surfaces to produce multi-QD self-assembled heterostructures [325]. The authors demonstrated the feasibility of this technique for manipulating QD surface chemistry and directed spatial organization for their integration in potential devices, probes and sensors.	[320-327]
CNTs	• Single walled carbon nanotubes (SWCNTs) have been successfully utilized as building blocks in the fabrication of various hybrid SAM films [326-329]. A linear growth of the film absorbance as a function of the number of deposition cycles with PDADMAC and PEI as polycations was reported by Rouse et al. [326,327]. A stepwise formation of weak PEL/SWCNT films was reported also when PVPy was used as counterpart, the ability to modify these films via a sol-gel chemistry to produce hybrid nanotube/silica/polymer thin films being demonstrated [327]. • Negatively and positively charged SWCNTs were alternately deposited with polycations (PDADMAC or PPAH), and NaPSS, respectively, to prepare hybrid ESAM thin films, with an almost linear growth of the multilayer [329]. The electrostatic nature of the driving force in the multilayer buildup was supported by the intrinsic compensation of charges in the multilayers. The influence of the ionic strength on the film growth has been also studied, a clear acceleration being found for the SWCNT/PEL multilayers, whereas the SWCNT/ SWCNT multilayers were more inert to this factor.	[326-333]

Table 2. (Continued)

	• MWCNTs have been also successfully utilized as building blocks in the construction of hybrid SAM films either on planar or spherical colloids [330-333], a linear increase of the absorbance with the increase of bilayer number being reported [330,333]. Decomposition of the colloidal template led to the formation of hollow CNT spheres [331]. Optically homogeneous thin films with controlled thickness and regular nanostructure, needed for the fabrication of chemical sensors, switchers, nanoelectronic devices, were built by LbL technique. The good stability, the electrocatalytic activity toward O2 reduction, the good conductivity and the high surface area recommend the fMWCNTs/PEL multilayer thin films for various applications [330,333].	

7. MICROCAPSULES

The numerous applications of nanocapsules in biology, chemistry and material science greately accelerated last years owing to their ability to accommodate a large variety of guest molecules [13,315,336-352]. Almost all the types of building blocks discussed above have been used in the fabrication of micro- and nanocapsules, deposited on different colloidal particles such as PS, MF and inorganic salts as colloidal particles, followed by the dissolution/decomposition of the template core. Composite capsules with special functionalities have been thus designed. However, there is not enough room in this survey to discuss all the important contributions in the filed. Therefore, the reader can find the most valuable information on this topic in the very comprehensive reviews dedicated to the intelligent micro- and nanocapsules recently given by Sukhorukov et al. [13], and to the PEL multilayer capsules as vehicles with tunable permeability, given by Antipov and Sukhorukov [351].

Small molecules such as the decomposition products of the core can penetrate through the shell wall whereas the high molar mass compounds (> 5000 Da) are normally excluded or trapped [336,337,339]. However, it was later showed that it is possible to tune the shell capsule permeability, i.e., to switch from "closed" to "open", by using adequate organic solvents [13,341,350]. Actually, the shell permeability could be tuned by various factors such as the ionic strength, temperature annealing, pH, etc. [351]. One of the very intensively studied practical properties of the ESAM capsules is their mechanical stability, which has been demonstrated to be strongly dependent on both the size of the core and the structure of the polycation used as building block, a higher elasticity being found in the case of PAH/NaPSS system compared with PDADMAC/NaPSS [see more details in ref. 338].

Some information concerning the construction and applications of micro- and nanocapsules, mostly those recently reported, will be briefly reviewed in this section, taking the type of the building blocks utilized in their fabrication as a criterion.

The first and the most utilized PELs as building blocks in the fabrication of the core/shell microcapsules is the polyion pair PAH/NaPSS [336-339,341,342, 347-350,352]. Mechanical properties of the "filled" microcapsules were investigated by Lulevich et al. by using an AFM-related force-measuring device [341]. The "filled" capsules were prepared from the hollow capsules (after removing the MF utilized as a template core) by changing the shell permeability for a polymer with a high molar mass (fluorescein isothiocyanate-dextran, $M_w \sim$

500 kDa) by opening the shell adding an organic solvent (acetone) [341]. The treatment with organic solvent required to induce transition from "closed" to "open" state of the capsule shell led to a softening of the "filled" capsules. The effect of pH and ionic strength on the stiffness of PEL multilayer microcapsules from PAH/NaPSS were investigated by Vinogradova et al. both as "hollow" [349,350] and "filled" capsules [350]. The dramatic softening of the "hollow" capsules at high pH was explained by the decrease of the charge density of PAH, leading to the corresponding reduction of the number of ionic crosslinks [349].

Polyaniline (PANI) was used either as building block in the fabrication of conductive core-shell particles by the electrostatic LbL deposition of PANI and NaPSS onto the PS colloidal particles [343] or as a thin shell adsorbed onto the PDADMAC/NaPSS multilayers deposited on MF particles used as templates [344]. The regular growth of PANI/NaPSS multilayers was confirmed by different methods (electrophoresis, FTIR, and TEM). Hollow PANI/NaPSS shells were produced by the dissolution of PS core [343].

Hydrogen bonded multilayers (HSAMs) have been also used in the preparation of hollow capsules by Sukhishvili et al. [345] who have employed PVPON, PEO and PMAA as building blocks for this purpose and cadmium carbonate as a template. HSAM capsules have been stabilized by covalently crosslinking the carboxylic groups by the carbodiimide chemistry.

The potential application of microspheres prepared by the LbL deposition of PELs on MF particles as carriers for enzymes has been explored and recently reported by Balabushevich et al. [346]. The microspheres have been fabricated by stepwise adsorption of two natural PELs, DS and protamine (PRM), on MF particles and subsequent decomposition of the nuclei under pH 1.7. To demonstrate the potential application of the hollow microspheres as carriers for enzymes, the system GOD/peroxidase (POD) was used as a model. The hollow microcapsules showed a high capacity of loading the bienzyme system, this system being suitable for the kinetic glucose content assay and more stable than the GOD/POD system in aqueous solution.

A new approach in the fabrication of core/shell microcapsules with PEL/silica shells was recently reported by Sun et al. [315]. According to this approach, silica has been grown directly on the PEL–modified surfaces by the Stöber method. This coating strategy has some advantages, as follows: it can be applied to all surfaces that can be modified with PEL; the presence of silica in the layered structure make the microcapsules mechanically robust increasing also drastically the storability of loaded nanocapsules; the coating does not impede the release of loaded molecules from the capsules.

8. Mechanical Properties of SAMs

Mechanical properties of the multilayer assemblies and the adhesion strength of bonds between the multilayers and the charged substrate are properties which concern the researchers interested in polymer surface modification by the LbL adsorption technique, microcapsules, or in the preparation of hybrid materials for applications like friction reduction coatings, sensors, electronic devices, composite membranes, etc. However, a small number of data are available for the mechanical properties of SAMs owing to the experimental difficulties associated with these measurements. From the beginning, the reader is guided to the very comprehensive review on the viscoelastic properties of PEL multilayer thin films recently published by Walhout [120]. Viscoelasticity of the ESAMs type materials have been discussed in terms of two key parameters, Young's modulus and the glass transition temperature, as a function of the main parameters affecting the viscoelasticity of the electrostatically crosslinked thin films, such as: ionic strength and counterion nature, identity of the oppositely charged counterparts, and pH of the dipping solution. The density of ionic crosslinking, which can be reversibly changed by pH in the case of weak PELs, and by ionic strength in the case of strong PELs, allow to control the mechanical properties of LbL assembled thin films [120].

Some very recently published results concerning the mechanical properties of the LbL assembled multilayers, constructed with a variety of building blocks, such as: strong PELs [121-124], weak PELs [163,165,168,183,187,207], nanoparticles [304,305,326,327,352], will be briefly discussed in the followings. Ferry et al. developed a new technique for the study of mechanical behavior of the free-standing nanocomposite films on a pendant drop using LbL adsorption of PELs onto a charged insoluble monolayer template [121], constructed by a novel technique [31]. The Young's modulus measured for these materials was consistent with a rubbery polymeric material. The multilayer stretched semi-reversibly upon large deformation with an increasing dependence of the film surface elastic modulus on film thickness and template charge density. The thicker films (> 20 nm) displayed glassy mechanical behavior. Gao et al. have studied the deformability of the PDADMAC/NaPSS multilayers deposited on mica [122]. When the LbL multilayer was put in contact with a PDMS stamp, the authors observed a compression of the film at the contact regions. The compression extent varied from several to hundreds of nanometers following a positive correlation with the multilayer thickness, controlled by the NaCl concentration, the deformability being affected by the desiccation of the multilayers. A lower wettability, lower permeability and smoother surface have been found for the uniformely compressed multilayers compared with uncompressed films.

A micromechanical analysis of the electrostatically crosslinked (PDADMAC/NaPSS) multilayers was performed as a function of the salt concentration which allows to control the density of ion pairs into the bulk of multilayers, and hence their mechanical behavior [124]. Viscoelastic behavior of PEL multilayers presented a linear or nonlinear behavior as a function of NaCl concentration, i.e., a linear increase of stress versus strain, for concentrations of NaCl up to 0.4 M, and a nonlinear increase for concentrations ≥ 0.8 M. The ruberry nature of the HSAM films under ambient humidity was confirmed by tensile testing. The mechanical characterization of these films indicated that HSAMs assemblies were less stiff than the ESAMs counterparts, when both are compared under hydrated conditions.

The mechanical properties of the multilayer films can be improved by including inorganic fillers such as: carbon nanotubes [204,199,204,332,333,353], metal or organic nanoparticles [294,302,304,305], or by crosslinking of the PEL counterparts, with transformation of ion pairs and hydrogen bondings in covalent bonds [72,150,165,167,189,208]. Richert et al. studied by AFM the mechanical properties of PLL and sodium hyaluronate (HA) multilayers deposited on glass slides. The SAMs films were crosslinked with a water soluble carbodiimide (EDC) [165]. The native films were nonadhesive whereas the crosslinked ones were highly adhesive. The Young's modulus of native films was dependent on the thickness of the films and increased with the number of crosslinking steps.

Pavoor et al. studied the elastic modulus and hardness of PAH/PAA multilayers deposited on glass by quasi-static nanoindentation, both in dry and wet state [168]. Moduli of the dry films were in the range of 10 GPa, while the same films immersed in water showed modulus value about two orders of magnitude lower. Properties of the dry films were pH dependent and independent to post-assembly crosslinking. The highest hardness was found when both PELs were deposited at pH 6.5, the lowest values of the modulus was associated with the more loopy structure of the film (deposition pH for PAH of 7.5 and for PAA at 3.5). The modulus and the hardness of the hydrated films were affected by covalent crosslinks induced by post-assembly thermal treatment, the smaller degree of swelling resulted in larger values of elastic modulus and hardness. Notley et al. reported on the viscoelastic properties of the adsorbed multilayers based on PAH and PAA by QCM-D (QCM with energy dissipation measurement) [186]. This method gave novel information inside the multilayers and viscoelastic poperties of the formed thin films as well as on the kinetic of the deposition process. The authors determined relative stiffness of an adsorbed layer by observing the change in energy dissipation. The dissipation measurement showed a stiffning of the adsorbed multilayer with PAA as outermost layer. A more compact and rigid film was formed when both PELs were adsorbed in a highly charged state. It was observed also that the adhesion between multilayer covered silica surfaces was less when the more rigid PAA was the outermost layer. The higher adhesion was explained by the more conformable and looser structure of the multilayer when PAH was on the outer layer.

Olek et al. reported that mechanical properties of the hybrid films containing MWCNT exceeded many CNT composites made by mixing or in situ polymerization [204]. Composites based on "bamboo" type CNT showed tensile strength and Young modulus higher than those found for the composites made with hollow MWCNTs. This approach opens new possibilities for the fabrication of nanostructured materials for nanoscale electronic devices with high strength requirements [204].

Robust gold nanoparticle-PEL multilayer free-standing films of the general formula $(PAH/NaPSS)_nPAH/Au/(PAH/NaPSS)_nPAH$ were time-efficiently fabricated by Tsukruk et al. by using a spin-assembly method [304,305]. The high uniformity and integrity of the free-standing films with a thickness less than 70 nm was demonstrated by mechanical tests which could be repeated several times.

Ko et al. recently reported the successful fabrication of CNT microscopic arrays encapsulated into freely suspended ESAM membranes using spin-assisted LbL assembly and microcontact printing via sacrificial polymer patterning (PS) [353]. In this nanostructure, the CNT arrays have been sandwiched between two ESAM films, the overall microstructure being described as $(PAH/NaPSS)_9PAH/CNT/(PAH/NaPSS)_9PAH$. The effect of the CNT encapsulation on the elastic properties of the freely suspended films has been addressed in this article for the first time, the method used being interferometry. The authors found elastic modulus very high compared with polymer composites with similar matrixes and thus confirmed that the filler toughning mechanism effectively enhanced the elastic properties of the patterned nanomembranes [353].

9. Characterization Methods of SAMs Thin Films

The growth and structure of the SAMs thin films have been examined by several techniques which have been summarized in Table 3.

Table 3. The most often utilized methods for the characterization of the SAMs thin films

Characterization methods	What was measured	References
Atomic force microscopy (AFM)	Surface roughness, film thickness	[17], [23], [25], [29], [32], [33], [37], [40], [47], [51], [53-55], [57], [62,63], [83], [87], [91], [103], [106,107], [109], [116], [122], [140], [144], [156,157], [163], [164,165], [169], [173], [175], [180], [182-186], [188,199], [193], [196], [198,199], [202-204], [206,207], [209], [220-222], [227], [231], [237-239], [243], [247,248], [250,251], [254], 264-266], [269], [272], [282], [288,289], [293], [295], [298-300], [302], [307], [310], [312], [319], [322], [325], [328,329], [331], [334], [341], [349,350]
Confocal laser scanning microscopy (CLSM)	Fluorescence	[86], [91], [126], [152], [164], [174,175], [211], [216], [220], [233], [338], [346], [349]
Contact angle measurements	Contact angle	[20], [22], [80,81], [90-93], [106], [122], [135], [177], [201], [210], [297], [299]
Cyclic voltammetry	Current intensity	[18], [21], [28], [30], [34-36], [40], [64], [74], [93,94], [118], [137], [142], [149], [156], [183], [192], [221], [235], [244], [300,301], [313,314], [323], [330], [332], [335], [343]
Diffusion dialysis	Molecule concentrations	[101], [103], [148,149], [176]
Electrokinetic measurements	ζ-potential	[16], [26], [27], [57], [83], [99], [117], [140], [146,147], [158], [164], [201,202], [230], [234], [236,237], [240], [279-282], [311], [313], [316], [324], [343], [348]
Ellipsometry	Film thickness	[17,18], [22], [26], [28], [46], [52], [54], [57], [59], [61], [64], [74], [90], [99], [102,103], [123], [139], [152], [155], [172], [175], [177], [178], [182], [187], [193], [201,202], [204], [206], [208], [210], [218,219], [221], [231], [251], [254,255], [261], [264], [266], [273], [279], [282], [287], [299], [302], [312], [317,318]
Fluorescence spectrometry	Fluorescence	[109,110], [114], [117], [146], [152], [160], [197,198], [230], [236], [241], [253], [263], [268], [284], [295], [308], [325], [337], [339], [340,341], [345], [348]
Fourier transform infrared spectroscopy (FTIR)	Absorbance	[28], [30], [38], [40], [51], [54], [67], [69], [72], [104], [106], [124], [127], [136,137], [139], [143,144], [146], [148-152], [155,156], [159], [162], [177], [180-182], [188], [190], [194], [196,197], [199], [201], [203], [205], [208], [212,213], [215], [222], [227], [229], [248], [250,251], [253-255], [270], [303], [343], [344]

Table 3. (Continued

Characterization methods	What was measured	References
Impedance and dielectric measurements	Impedance, dielectric constant	[78], [137]
In situ surface tension measurements	Surface tension	[31], [121]
Ion permeation measurements	Conductivity	[95], [217]
Nanofiltration	Ion content	[64], [98,99], [101-104]
Neutron reflectometry	Neutron reflectivity	[58], [71], [191], [274], [299]
Nuclear magnetic resonance (NMR)	Chemical shift	[34], [49], [113], [127], [318]
Optical waveguide light mode spectroscopy (OWLS)	Film thickness, PEL adsorbed amount	[79], [83], [85], [140], [175], [196], [213], [220], [247], [250], [251]
Pervaporation	Flux, water content	[96,97], [105], [147], [194]
Polyelectrolyte titration (PET)	Potential	[80], [157,158], [180], [202]
Quartz crystal microbalance (QCM)	Frequency change	[19], [25], [29], [36], [51], [53,54], [77], [85,86], [140], [155-157], [162,163], [179,180], [183], [186], [188], [195,196], [206], [213], [218], [220], [224], [226,227], [235], [237], [242], [248], [252], [258], [291], [313,314], [332], [344]
Scanning angle reflectometry (SAR)	Optical thickness, mean refractive index	[27], [246], [249]
Scanning electron microscopy (SEM)	3D images	[25], [46], [80], [93,94], [104], [114], [117], [126,127], [139], [147], [152], [176], [178], [189], [196], [204,205], [211], [224], [227], [242], [291], [293,294], [296,297], [306], [311], [316], [318,319], [322], [324], [328,329], [332-336], [344], [348], [350], [352]
Small angle X-ray reflectivity (SAXR)	Reflected intensity	[15], [24], [49,50], [52], [54], [82], [228], [257], [260], [268], [271], [289], [317,318]
Surface force apparatus (SFA)	Force	[24], [68], [73], [173]
Surface plasmon resonance spectroscopy (SPR)	Reflectivity	[19], [29], [57], [87], [157], [195], [197], [226], [288,289], [331], [350]
Transmision electron microscopy (TEM)	2D images	[146], [153], [166], [167], [204], [205], [217], [230], [238], [240], [242], [301], [306], [308], [310], [316], [320], [321], [322], [324], [325], [333], [343], [344], [348], [352]

Table 3. (Continued

Characterization methods	What was measured	References
UV-vis spectroscopy	Absorbance	[14], [18], [19], [21], [32], [33], [34], [37], [38], [39], [40], [46], [47], [49], [51], [54], [55], [56], [57], [63], [127], [136], [152], [153], [155], [167], [172], [180], [188], [202], [203], [208], [209], [224], [234], [237-240], [242], [244], [251], [255], [256-261], [263-266], [268], [270-273], [279-282], [286,287], [296], [300-303], [306-308], [310], [313], [314], [316], [318], [322-325], [328], [329], [331], [332], [335], [347], [348]
X-ray diffraction	Fringes	[38], [136], [144], [209], [241], [289]
X-ray photoelectron spectroscopy (XPS)	Elemental content, film thickness	[18-20], [40], [57], [144], [166], [171], [231], [244], [256], [312], [314], [316,317], [321-323], [328], [331]

10. Conclusion

The differences in the multilayer growth with strong PELs as building blocks reveал led that not only the charge density is important but also the whole chemical structure of the oppositely charged PELs, mainly the nonionic comonomer, strongly influences the multilayer growth and morphology of ESAMs thin films. Concerning the influence of the ionic strength and salt nature, for a given pair of strongly dissociated polycations and polyanions, the salt concentration in the deposition solution has a strong influence on the thickness of each polymer layer, which is approximately proportional to salt concentration, but the counterion/polyion interaction has also a strong influence on the multilayer growth and morphology. The solvent quality could be also used to tune the multilayer properties like thickness, permeability, and porosity, because changing the dielectric constant of a mixed solvent, conformational changes of the PEL chains occur. Besides the well established experimental variables with a strong impact on the buildup of ESAMs films, it was recently demonstrated that temperature is a significant factor of influence, the effect of the temperature increase being similar with that of the increase of ionic strength and the decrease of the solvent dielectric constant.

The driving force in the multilayer construction with weak PELs as building blocks can be either electrostatic, at a pH where the PELs are fully charged, or hydrogen bondings. The SAMs stabilized by hydrogen bonding interactions known that “hydrogen-bonded multilayers” were called in this review HSAMs, to differentiate them from ESAMs resulted by electrostatic interactions. Normaly, there is a fine balance between electrostatics and hydrogen bondings, their role in the SAM film growth and stabilization being determined by the pH, ionic strength and local microenvironment. The main factors controlling the level of interlayer interpenetration, the film composition and thickness of both the internal bilayers and the outermost layer of the SAMs thin films made with weak PELs as building blocks, and also the possibility to control the surface wettability have been discussed in the paper.

Due to their property to fast respond to the changes in the environment such as pH, ionic strength, solvent quality, SAMs films with different properties and applications have been constructed with weak PELs as building blocks. Some of the numerous applications of such nanostructured materials are summarized as follows: separations and transport, responsive systems for controlled release, free-standing polymer membranes, corrosion resistant coatings, nanoreactors for nanoparticle synthesis, substrates for cells spreading and growth, surface patterning, sensors, catalysis, functionalization of hydrophobic surfaces, etc. The stability of exponentially preassembled polypeptide multilayers, as a function of various parameters such as pH, ionic strength, and temperature, bas been intensively investigated because their applications for biomedical purposes strongly depend on the film stability at physiological conditions.

As many researchers observed, electrostatic interactions are the main driving force in the the construction of multilayer thin films with proteins and colloidal particles, but hydrogen bonds and hydrophobic interactions play also an important role.

The authors apologize if some key papers in the field of SAMs thin films have not been discussed in this review because it was not enough room to discuss all the valuable contributions in this very attractive field. Therefore, only a fraction of the relevant works has been referenced in the paper.

ACKNOWLEDGEMENT

The financial support of this research from the Project CEEX-MATNANTECH 40/2005 is gratefully acknowledged.

REFERENCES

[1] Iler, R. K. *J. Colloid Interface Sci.* 1966, *21*, 569-594.

[2] Decher, G.; Hong, J.-D. *Macromol. Chem. Macromol. Symp.* 1991, *46*, 321-327.

[3] Decher, G.; Hong, J.-D.; Schmitt, J. *Thin Solid Films* 1992, *210/211*, 831-835.

[4] Lvov, Y.; Decher, G.; Möhwald, H. *Langmuir* 1993, *9*, 481-486.

[5] Lvov, Y.; Decher, G.; Sukhorukov, G. *Macromolecules* 1993, *26*, 5396-5399.

[6] Decher, G. *Science* 1997, *277*, 1232-1237.

[7] Bertrand, P.; Jonas, A.; Laschewsky, A.; Legras, R. *Macromol. Rapid Commun.* 2000, *21*, 319-348.

[8] Hammond, P. T. *Curr. Opin. Colloid Interface Sci.* 2000, *4*, 430-442.

[9] Tripathy, S.; Kumar, J.; Nalwa, H. S.; Eds.; *Handbook of Polyelectrolytes and Their Applications*; American Scientific Publishers: Los Angeles, 2002; Vol. 1.

[10] Decher, G.; Schlenoff, J. B. *Multilayer Thin Films*; Wiley-VCH: Weinheim, 2003.

[11] Schönhoff, M. *Curr. Opin. Colloid Interface Sci.* 2003, *8*, 86-95.

[12] Hammond, P. T. *Adv. Mater.* 2004, *16*, 1271-1293.

[13] Sukhorukov, G.; Fery, A.; Möhwald, H. *Prog. Polym. Sci.* 2005, *30*, 885-897.

[14] Ferreira, M.; Rubner, M. F. *Macromolecules* 1995, *28*, 7104-7114.

[15] Sukhorukov, G. B.; Schmitt, J.; Decher, G. *Ber. Bunsenges. Phys. Chem.* 1996, *100*, 948-953.

[16] Hoogeveen, N. G.; Cohen Stuart, M. A.; Fleer, G. J.; Böhmer, M. R. *Langmuir* 1996, *12*, 3675-3681.

[17] Clark, S. L.; Montague, M. F.; Hammond, P. T. *Macromolecules* 1997, *30*, 7237-7244.
[18] Laurent, D.; Schlenoff, J. B. *Langmuir* 1997, *13*, 1552-1557.
[19] Caruso, F.; Niikura, K.; Furlong, D. N.; Okahata, Y. *Langmuir* 1997, *13*, 3422-3426.
[20] Phuvanartnuruks, V.; McCarthy, T. J. *Macromolecules* 1998, *31*, 1906-1914.
[21] Schlenoff, J. B. Ly, H.; Li, M. *J. Am. Chem. Soc.* 1998, *120*, 7626-7634.
[22] Dubas, S. T.; Schlenoff, J. B. *Macromolecules* 1999, *32*, 8153-8160.
[23] Clark, S. L.; Hammond, P. T. *Adv. Mater.* 1998, *10*, 1515-1519.
[24] Lowack, K.; Helm, C. A. *Macromolecules*, 1998, *31*, 823-833.
[25] Lvov, Y.; Ariga, K.; Onda, M.; Ichinose, I.; Kunitake, K. *Colloids Surf. A* 1999, *146*, 337-346.
[26] Schwarz, S.; Eichhorn, K. -J.; Wischerhoff, E.; Laschewsky, A. *Colloids Surf. A* 1999, *159*, 491-501.
[27] Ladam, G.; Schaad, P.; Voegel, J. C.; Schaaf, P.; Decher, G.; Cuisinier, F. *Langmuir* 2000, *16*, 1249-1255.
[28] Harris, J. J.; Bruening, M. L. *Langmuir* 2000, *16*, 2006-2013.
[29] Poptoshev, E.; Schoeler, B.; Caruso, F. *Langmuir* 2004, *20*, 829-834.
[30] Schlenoff, J. B.; Dubas, S. T.; Farhat, T. R. *Langmuir* 2000, *16*, 9968-9969.
[31] Ferry, J. K.; Dong, W.-F.; Miller, R. *J. Phys. Chem. B* 2005, *109*, 14764-14768.
[32] Kim, H.-J.; Lee, K.; Kumar, S.; Kim, J. *Langmuir* 2005, *21*, 8532-8538.
[33] Pinto, M. R.; Kristal, B. M.; Schanze, K. S. *Langmuir* 2003, *19*, 6523-6533.
[34] Mruk, R.; Prehl, S.; Zentel, R. *Macromol. Rapid Commun.* 2003, *24*, 1014-1018.
[35] Man, K. Y. K.; Wong, H. L.; Chan, W. K.; Kwong, C. Y.; Djurišić, A. B. *Chem. Mater.* 2004, *16*, 365-367.
[36] Liu, A.; Anzai, J. *Langmuir* 2003, *19*, 4043-4046.
[37] Chen, J.; Huang, L.; Ying, L.; Luo, G.; Zhao, X.; Cao, W. *Langmuir* 1999, *15*, 7208-7212.
[38] Sun, J.; Wu, T.; Liu, F.; Wang, Z.; Zhang, X.; Shen, J. *Langmuir* 2000, *16*, 4620-4624.
[39] Cao, T.; Wei, L.; Yang, S.; Zhang, M.; Huang, C.; Cao, W. *Langmuir* 2002, *18*, 750-753.
[40] Feng, Y.; Peng, J.; Han, Z.; Ma, H. *J. Colloid Interface Sci.* 2005, *286*, 589-595.
[41] Dante, S.; Advincula, R.; Frank, C. W.; Stroeve, P. *Langmuir* 1999, *15*, 193-201.
[42] Park, M.-K.; Advincula, R. C. *Langmuir* 2002, *18*, 4532-4535.
[43] Suzuki, I.; Ishizaki, T.; Tomonori, H.; Anzai, Jun-ichi *Macromolecules* 2002, *35*, 577-580.
[44] Suzuki, I.; Ishizaki, T.; Inoue, H.; Anzai, Jun-ichi *Macromolecules* 2002, *35*, 6470-6474.
[45] Wang, H., He, Y., Tuo, X., Wang, X. *Macromolecules* 2004, *37*, 135-146.
[46] Fleischhaker, F.; Arsenault, A. C.; Kitaev, V.; Peiris, F. C.; von Freymann, G.; Manners, I.; Zentel, R.; Ozin, G. A. *J. Am. Chem. Soc.* 2005, *127*, 9318-9319.
[47] Zapotoczny, S.; Golonka, M.; Nowakowska, M. *Macromol. Rapid Commun.* 2005, *26*, 1049-1054.
[48] Kolarik, L.; Furlong, D. N.; Joy, H.; Struijk, C.; Rowe, R. *Langmuir* 1999, *15*, 8265-8275.
[49] Fischer, P.; Laschewsky, A.; Wischerhoff, E.; Arys, X.; Jonas, A.; Legras, R. *Macromol. Symp.* 1999, *137*, 1-24.
[50] Steitz, R.; Jaeger, W.; Klitzing, R. V. *Langmuir 17*, 2001, 4471-4474.

[51] Schoeler, B.; Kumaraswamy, G.; Caruso, C. *Macromolecules* 2002, *35*, 889-897.
[52] Glinel, K.; Moussa, A.; Jonas, A. M.; Laschewsky, A. *Langmuir* 2002, *18*, 1408-1412.
[53] Serizawa, T.; Kawanishi, N.; Akashi, M. *Macromolecules* 2003, *36*, 1967-1974.
[54] Schoeler, B.; Sharpe, S.; Hatton, T. A.; Caruso, F. *Langmuir* 2004, *20*, 2730-2738.
[55] Zhang, J.; Fredin, N. J.; Janz, J. F.; Sun, B.; Lynn, D. M. *Langmuir* 2006, *22*, 239-245.
[56] Sun, Q.; Tong, Z.; Wang, C.; Ren, B.; Liu, X.; Zeng, F. *Polymer* 2005, *46*, 4958-4966.
[57] Licea-Claverie, A.; Schwarz, S.; Simon, F.; Urzúa-Sánchez, O.; Nagel, J.; Pleul, D.; Eichhorn, K.-J.; Janke, A. *Colloid Polym. Sci.* 2005, *283*, 826-835.
[58] Lösche, M.; Schmitt, Decher, G.; Bouwman, W. G.; Kjaer, K. *Macromolecules* 1998, *31*, 8893-8906.
[59] Sui, Z.; Salloum, D.; Schlenoff, J. B. *Langmuir* 2003, *19*, 2491-2495.
[60] Fleer, G. J.; Cohen Stuart, M. A.; Scheutjens, J. M. H. M.; Cosgrove, T.; Vincent, B. *Polymers at Interfaces*; Chapman and Hall: London, 1993, pp 346-350.
[61] Schlenoff, J. B.; Dubas, S. T. *Macromolecules* 2001, *34*, 592-598.
[62] Dubas, S. T.; Schlenoff, J. B. *Langmuir* 2001, *17*, 7725-7727.
[63] McAloney, R. A.; Sinyor, M.; Dudnik, V.; Goh, M. C. *Langmuir* 2001, *17*, 6655-6663.
[64] Farhat, T. R.; Schlenoff, J. B. *Langmuir* 2001, *17*, 1184-1192.
[65] Salomäki, M.; Tervasmäki, P.; Areva, S.; Kankare, J. *Langmuir* 2004, *20*, 3679-3683.
[66] Riegler, H.; Essler, F. *Langmuir* 2002, *18*, 6694-6698.
[67] Izumrudov, V.; Kharlampieva, E.; Sukhishvili, S. A. *Macromolecules* 2004, *37*, 8400-8406.
[68] Blomberg, E.; Poptoshev, E.; Claesson, P. M.; Caruso, F. *Langmuir* 2004, *20*, 5432-5438.
[69] Jomma, H. W.; Schlenoff, J. B. *Langmuir* 2005, *21*, 8081-8084.
[70] Lourenço, J. M. C.; Ribeiro, P. A.; Bothelo do Rego, A. M.; Braz Fernandes, F. M.; Mouthino, A. M. C.; Raposo, M. *Langmuir* 2004, *20*, 8103-8109.
[71] Jomma, H. W.; Schlenoff, J. B. *Macromolecules* 2005, *38*, 8473-8480.
[72] Dragan, E. S.; Mihai, M.; Schauer, J.; Ghimici, L. *J. Polym. Sci. Part A Polym. Chem.* 2005, *43*, 4161-4171.
[73] Blomberg, E.; Poptoshev, E.; Caruso, F. *Langmuir* 2006, *22*, 4153-4157.
[74] Silva, T. H.; Garcia-Morales, V.; Moura, C.; Manzanares, J. A.; Silva, F. *Langmuir* 2005, *21*, 7461-7467.
[75] Büscher, K.; Karlheinz, G.; Ahrens, H.; Helm, C. A. *Langmuir* 2002, *18*, 3585-3591.
[76] Tan, H. L.; McMurdo, M. J.; Pan, G.; Van Patten, P. G. *Langmuir* 2003, *19*, 9311-9314.
[77] Salomäki, M.; Vinokurov, I. A.; Kankare, J. *Langmuir* 2005, *21*, 11232-11240.
[78] Durstock, M. F.; Rubner, M. F. *Langmuir* 2001, *17*, 7865-7872.
[79] Ball, V.; Voegel, J.-C.; Schaaf, P. *Langmuir* 2005, *21*, 4129-4137.
[80] Lingström, R.; Wågberg, L.; Larsson, P. T. *J. Colloid Interface Sci.* 2006, *296*, 396-408.
[81] Kolasinska, M.; Warszynski, P. *Appl. Surface Sci.* 2005, *252*, 759-765.
[82] Gopinadhan, M.; Ahrens, H.; Günther, J.-U.; Steitz, R.; Helm, C. A. *Macromolecules* 2005, *38*, 5228-5235.
[83] Lavalle, Ph.; Gergely, C.; Cuisinier, F. J. G.; Decher, G.; Schaaf, P.; Voegel, J. C.; Picart, C. *Macromolecules* 2002, *35*, 4458-4465.
[84] Arys, X.; Laschewsky, A.; Jonas, A. M. *Macromolecules* 2001, 34, 3318-3330.

[85] Hübsch, E.; Ball, V.; Senger, B.; Decher, G.; Voegel, J.-C.; Schaaf, P. *Langmuir* 2004, *20*, 1980-1985.
[86] Garza, J. M.; Schaaf, P.; Muller, S.; Ball, V.; Stoltz, J.-F.; Voegel, J.-C.; Lavalle, P. *Langmuir* 2004, *20*, 7298-7302.
[87] Gong, H.; Garcia-Turiel, J.; Vasilev, K.; Vinogradova, O. I. *Langmuir* 2005, *21*, 7545-7550.
[88] Adamczyk, Z.; Zembala, M.; Warszyński, P., Jachimska, B. *Langmuir* 2004, *20*, 10517-10525.
[89] Köstler, S.; Ribitsch, V.; Stana-Kleinschek, K.; Jakopic, G.; Strnad, S. *Colloids Surf. A* 2005, *270-271*, 107-114.
[90] Wong, J. E.; Rehfeldt, F.; Hänni, P.; Motomu, T.; v. Klitzing, R. *Macromolecules* 2004, *37*, 7285-7289.
[91] Wang, B.; Feng, J.; Gao, C. *Colloids Surf. A* 2005, *259*, 1-5.
[92] Köstler, S.; Delgado, A. V.; Ribitsch, V. *J. Colloid Interface Sci.* 2005, *286*, 339-348.
[93] Zhao, N.; Shi, F.; Wang, Z.; Zhang, X. *Langmuir* 2005, *21*, 4713-4716.
[94] Farhat, T. R.; Schlenoff, J. B. *Electrochem. Solid State Lett.* 2002, *5*, B13-B15.
[95] Krasemann, L.; Tieke, B. *Langmuir* 2000, *16*, 287-290.
[96] Krasemann, L.; Toutianoush, A.; Tieke, B. *J. Membr. Sci.* 2001, *181*, 221-228.
[97] Tieke, B.; Krasemann, L.; Toutianoush, A. *Macromol. Symp.* 2001, *163*, 97-111.
[98] Jin, W.; Toutianoush, A.; Tieke, B. *Langmuir* 2003, *19*, 2550-2553.
[99] Stanton, B. W.; Harris, J. J.; Miller, M. D.; Bruening, M. L. *Langmuir* 2003, *19*, 7038-7042.
[100] Xie, A. F.; Granick, S. *Macromolecules* 2002, *35*, 1805-1813.
[101] Liu, X.; Bruening, M. L. *Chem. Mater.* 2004, *16*, 351-357.
[102] Miller, M. D.; Bruening, M. L. *Langmuir* 2004, *20*, 11545-11551.
[103] Miller, M. D.; Bruening, M. L. *Chem. Mater.* 2005, *17*, 5375-5381.
[104] Malaisamy, R.; Bruening, M. L. *Langmuir* 2005, *21*, 10587-10592.
[105] Budd, P. M.; Ricardo, N. M. P. S.; Jafar, J. J.; Stephenson, B; Hughes, R. *Ind. Eng. Chem. Res.* 2004, *43*, 1863-1867.
[106] Jiang, X.; Zheng, H.; Gourdin, S.; Hammond, P. T. *Langmuir* 2002, *18*, 2607-2615.
[107] Kidambi, S.; Chan, C.; Lee, I. *J. Am. Chem. Soc.* 2004, *126*, 4697-4703.
[108] Lee, I.; Ahn, J. S.; Hendricks, T. R. *Langmuir* 2004, *20*, 2478-2483.
[109] Park, J.; Hammond, P. T. *Adv. Mater.* 2004, *16*, 520-525.
[110] Park, J.; Hammond, P. T. *Macromolecules* 2005, *38*,10542-10550.
[111] Sakaguchi, H.; Serizawa, T.; Akashi, M. *Chem. Lett.* 2003, *32*, 174.
[112] Serizawa, T.; Sakaguchi, H.; Matsusaki. M.; Akashi, M. *J. Polym. Sci. Part A Polym. Chem.* 2005, *43*, 1062-1067.
[113] McCormick, M.; Smith, R. N.; Graf, R.; Barrett, C. J.; Reven, L.; Spiess, H. W. *Macromolecules* 2003, *36*, 3616-3625.
[114] Müller, K.; Quinn, J. F.; Johnston, A. P. R.; Becker, M.; Greiner, A.; Caruso, F. *Chem. Mater.* 2006, *18*, 2397-2403.
[115] Jaber, J. A.; Schlenoff, J. B. *Macromolecules* 2005, *38*, 1300-1306.
[116] Krass, H.; Papastavrou, G.; Kurth, D. G. *Macromol. Symp.* 2004, *210*, 311-319.
[117] Wosnick, J. H.; Liao, J. H.; Swager, T. M. *Macromolecules* 2005, *38*, 9287-9290.
[118] Bucur, C. B.; Schlenoff, J. B. *Anal. Chem.* 2006, *78*, 2360-2365.

[119] Chen, H.; Zeng, G.; Wang, Z.; Zhang, X.; Peng, M.-L.; Wu, L.-Z.; Tung, C.-H. *Chem. Mater.* 2005, *17*, 6679-6685.
[120] Walhout, P. K. In: *Focus on Ionic Polymers*; Dragan, E. S.; Ed.; Research Signpost: Trivandrum, India, 2005, pp 223-259.
[121] Ferri, J. K.; Dong, W.-F.; Miller, R.; Möhwald, H. *Macromolecules* 2006, *39*, 1532-1537.
[122] Gao, C.; Wang, B.; Feng, J.; Shen, J. *Macromolecules* 2004, *37*, 8836-8839.
[123] Nolte, A. J.; Rubner, M. F.; Cohen, R. E. *Macromolecules* 2005, *38*, 5367-5370.
[124] Jaber, J. A.; Schlenoff, J. B. *J. Am. Chem. Soc.* 2006, *128*, 2940-2947.
[125] Kim, B.; Sigmund, W. M. *Langmuir* 2004, *20*, 8239-8242.
[126] Artyukhin, A. B.; Bakajin, O.; Stroeve, P.; Noy, A. *Langmuir* 2004, *20*, 1442-1448.
[127] Kong, H.; Luo, P.; Gao, C.; Yan, D. *Polymer* 2005, *46*, 2472-2485.
[128] Dragan, S.; Ghimici, L. *Polymer* 2001, *42*, 2887-2891.
[129] Ghimici, L.; Dragan, S. *Colloid Polym. Sci.* 2002, *280*, 130-134.
[130] Ghimici, L.; Dragan, E. S. In: *Focus on Ionic Polymers;* Dragan, E. S.; Ed.; Research Signpost: Trivandrum, India, 2005, pp 49-92.
[131] Helfferich, H. Ion Exchange, McGraw-Hill, New York, 1962, Cap. 5, pp. 103.
[132] Dragan, S.; Cristea, M. In: *Recent Research Developments in Polymer Science*; Gayathry, A.; Ed.; Research Signpost: Trivandrum, India, 2003, Vol. 7, pp 149-181.
[133] Manning, G. S. *J. Chem. Phys.* 1969, *51*, 924-933.
[134] Manning, G. S. *Ber. Bunsenges. Phys. Chem.* 1996, *100*, 909-922.
[135] Yoo, D.; Shiratori, S. S.; Rubner, M. F. *Macromolecules* 1998, *31*, 4309-4318.
[136] Wang, L.; Fu, Y.; Wang, Z.; Fan, Y.; Zhang, X. *Langmuir* 1999, *15*, 1360-1363.
[137] Harris, J. J.; DeRose, P. M.; Bruening, M. L. *J. Am. Chem. Soc.* 1999, *121*, 1978-1979.
[138] Dubas, S. T.; Schlenoff, J. B. *Macromolecules* 2001, *34*, 3736-3740.
[139] Mendelsohn, J. D.; Barrett, C. J.; Chan, V. V.; Pal, A. J.; Mayes, A. M.; Rubner, M. F. *Langmuir* 2000, *16*, 5017-5023.
[140] Picart, C.; Lavalle, Ph.; Hubert, P.; Cuisinier, F. J. G.; Decher, G.; Schaaf, P.; Voegel, J. C. *Langmuir* 2001, *17*, 7414-7424.
[141] Pardo-Yissar, V.; Katz, E.; Lioubashevski, O.; Willner, I. *Langmuir* 2001, *17*, 1110-1118.
[142] Fery, A.; Schöler, B.; Cassagneau, T.; Caruso, F. *Langmuir* 2001, *17*, 3779-3783.
[143] Müller, M.; Heinen, S.; Oertel, U.; Lunkwitz, K. *Macromol. Symp.* 2001, *164*, 197-210.
[144] Fu, Y.; Bai, S.; Cui, S.; Qiu, D.; Wang, Z.; Zhang, X. *Macromolecules* 2002, *35*, 9451-9458.
[145] Kovacevic, D.; van der Burgh, S.; Keizer, A.; Cohen Stuart, M. A. *Langmuir* 2002, *18*, 5607-5612.
[146] Kato, N.; Schuetz, P.; Fery, A.; Caruso, F. *Macromolecules* 2002, *35*, 9780-9787.
[147] Meier-Haack, J.; Müller, M. *Macromol. Symp.* 2002, *188*, 91-103.
[148] Dai, J.; Balachandra, A. M.; Lee, J. I.; Bruening, M. L. *Macromolecules* 2002, *35*, 3164-3170.
[149] Balachandra, A. M.; Dai, J.; Bruening, M. L. *Macromolecules* 2002, *35*, 3171-3178.
[150] Yang, S. Y.; Rubner, M. F. *J. Am. Chem. Soc.* 2002, *124*, 2100-2101.
[151] Sukhishvili, S. A.; Granick, S. *Macromolecules* 2002, *35*, 301-310.
[152] Kharlampieva, E.; Kozlovskaya, V.; Tyutina, J.; Sukhishvili, S. A. *Macromolecules* 2005, *38*, 10523-10531.

[153] Wang, T. C.; Rubner, M. F.; Cohen, R. E. *Langmuir* 2002, *18*, 3370-3375.
[154] Sui, Z.; Schlenoff, J. B. *Langmuir* 2003, *19*, 7829-7831.
[155] Cho, J.; Caruso, F. *Macromolecules* 2003, *36*, 2845-2851.
[156] Serizawa, T.; Nakashima, Y.; Akashi, M. *Macromolecules* 2003, *36*, 2072-2078.
[157] Schoeler, B.; Poptoshev, E.; Caruso, F. *Macromolecules* 2003, *36*, 5258-5264.
[158] Burke, S. E.; Barrett, C. J. *Langmuir* 2003, *19*, 3297-3303.
[159] Zhang, H.; Rühe, J. *Macromol. Rapid Commun.* 2003, *24*, 576-579.
[160] Serpe, M. J.; Jones, C. D.; Lyon, L. A. *Langmuir* 2003, *19*, 8759-8764.
[161] Rubner, M. F. In: *Multilayer Thin Films*; Decher, G.; Schlenoff, J. B.; Eds.; Wiley-VCH: Weinheim, 2003, pp 133-154.
[162] Debreczeny, M.; Ball, V.; Boulmedais, F.; Szalontai, B.; Voegel, J.-C.; Schaaf, P. *J. Phys. Chem. B* 2003, *107*, 12734-12739.
[163] Quinn, J. F.; Caruso, F. *Langmuir* 2004, *20*, 20-22.
[164] Richert, L.; Lavalle, P.; Payan, E.; Shu, X. Z.; Prestwich, G. D.; Stoltz, J.-F.; Schaaf, P.; Voegel, J. C.; Picart, C. *Langmuir* 2004, *20*, 448-458.
[165] Richert, L.; Engler, A. J.; Discher, D. E.; Picart, C. *Biomacromolecules* 2004, *5*, 1908-1916.
[166] Wang, T. C.; Rubner, M. F.; Cohen, R. E. *Chem. Mater.* 2003, *15*, 299-304.
[167] Schuetz, P.; Caruso, F. *Chem. Mater.* 2004, *16*, 3066-3073.
[168] Pavoor, P. V.; Bellare, A. ; Strom, A.; Yang, D.; Cohen, R. E. *Macromolecules* 2004, *37*, 4865-4871.
[169] Sui, Z.; Schlenoff, J. B. *Langmuir* 2004, *20*, 6026-6031.
[170] Nolte, A. J.; Rubner, M. F.; Cohen, R. E. *Langmuir* 2004, *20*, 3304-3310.
[171] Greene, G.; Yao, G.; Tannenbaum, R. *Langmuir* 2004, *20*, 2739-2745.
[172] Johal, M. S.; Ozer, B. H.; Casson, J. L.; John, A. St.; Robinson, J. M.; Wang, H.-L. *Langmuir* 2004, *20*, 2792-2796.
[173] Kulcsar, A.; Lavalle, P.; Voegel, J.-C.; Schaaf, P.; Kekicheff, P. *Langmuir* 2004, *20*, 282-286.
[174] Lavalle, P.; Vivet, V.; Jessel, N.; Decher, G.; Voegel, J.-C.; Mesini, P. J.; Schaaf, P. *Macromolecules* 2004, *37*, 1159-1162.
[175] Porcel, C.; Lavalle, P.; Ball, V.; Decher, G.; Senger, B.; Voegel, J.-C.; Schaaf, P. *Langmuir* 2006, *22*, 4376-4383.
[176] Lajimi, R. H.; Abdallah, A. B.; Ferjani, E.; Roudesli, M. S.; Deratani, A. *Desalination* 2004, *163*, 193-202.
[177] Jensen, A. W. ; Desai, N. K.; Maru, B. S.; Mohanty, D. K. *Macromolecules* 2004, *37*, 4196-4200.
[178] Mallwitz, F.; Laschewsky, A. *Adv. Mater.* 2005, *17*, 1296-1299.
[179] Thierry, B.; Kujawa, P.; Tkaczyk, C.; Winnik, F. M.; Bilodeau, L.; Tabrizian, M. *J. Am. Chem. Soc.* 2005, *127*, 1626-1627.
[180] Tjipto, E.; Quinn, J. F.; Caruso, F. *Langmuir* 2005, *21*, 8785-8792.
[181] Choi, J.; Rubner, M. F. *Macromolecules* 2005, *38*, 116-124.
[182] Itano, K.; Choi, J.; Rubner, M. F. *Macromolecules* 2005, *38*, 3450-3460.
[183] Li, Q.; Quinn, J. F.; Caruso, F. *Adv. Mater.* 2005, *17*, 2058-2062.
[184] Berg, M. C.; Zhai, L.; Cohen, R. E.; Rubner, M. F. *Biomacromolecules* 2006, *7*, 357-364.
[185] Feng, Q.; Zeng, G.; Yang, P.; Wang, C.; Cai, J. *Colloids Surf. A* 2005, *257-258*, 85-88.

[186] Notley, S. M.; Eriksson, M.; Wågberg, L. *J. Colloid Interface Sci.* 2005, *292*, 29-37.
[187] Guyomard, A.; Muller, G.; Glinel, K. *Macromolecules* 2005, *38*, 5737-5742.
[188] Takeoka, Y.; Sasada, K.; Nishiwaki, Y.; Rikukawa, M.; Sanui, K. *Colloids Surf. A* 2005, *257-258*, 485-488.
[189] Cheng, K.; Yang, M.-H.; Chiu, W. W. W.; Huang, C.-Y.; Chang, J.; Ying, T.-F.; Yang, Y. *Macromo. .Rapid Commun.* 2005, *26*, 247-264.
[190] Schneider, A.; Francius, G.; Obeid, R.; Schwinté, P.; Hemmerlé, J.; Frisch, B.; Schaaf, P.; Voegel, J.-C.; Senger, B.; Picart, C. *Langmuir* 2006, *22*, 1193-1200.
[191] Tanchak, O. M.; Yager, K. G.; Fritzsche, H.; Harroun, T.; Katsaras, J.; Barrett, C. J. *Langmuir* 2006, *22*, 5137-5143.
[192] Noguchi, T.; Anzai, J. *Langmuir* 2006, *22*, 2870-2875.
[193] Ono, S. S.; Decher, G. *Nano Lett.* 2006, *6*, 592-598.
[194] Lenk, W.; Meier-Haack, J. *Desalination* 2002, *148*, 11-16.
[195] Yap, H. P.; Quinn, J. F.; Ng, S. M.; Cho, J.; Caruso, F. *Langmuir* 2005, *21*, 4328-4333.
[196] Ball, V.; Michel, M.; Boulmedais, F.; Hemmerle, J.; Haikel, Y.; Schaff, P.; Voegel, J. C. *Cryst. Growth Des.* 2006, *6*, 327-334.
[197] Stroock, A. D.; Kane, R. S.; Weck, M.; Metallo, S. J.; Whitesides, G. M. *Langmuir* 2003, *19*, 2466-2472.
[198] Reyes, D. R.; Perruccio, E. M.; Becerra, S. P.; Locascio, L. E.; Gaitan, M. *Langmuir* 2004, *20*, 8805-8811.
[199] Qin, S.; Qin, D.; Ford, W. T.; Herrera, J. E.; Resasco, D. E. *Macromolecules* 2004, *37*, 9963-9967.
[200] Lutkenhaus, J. L.; Hrabak, K. D.; McEnnis, K.; Hammond, P. T. *J. Am. Chem. Soc.* 2005, *127*, 17228-17234.
[201] Burke, S. E.; Barrett, C. J. *Biomacromolecules* 2003, *4*, 1773-1783.
[202] Burke, S. E.; Barrett, C. J. *Macromolecules* 2004, *37*, 5375-5384.
[203] Yang, Z.; Cao, W. *Macromol. Chem. Phys.* 2004, *205*, 241-246.
[204] Olek, M.; Ostrander, J., Jurga, S.; Möhwald, H.; Kotov, N.; Kempa, K.; Giersig, M. *Nano Lett.* 2004, *4*, 1889-1895.
[205] Carrilo, A.; Swartz, J. A.; Gamba, J. M.; Kane, R. S.; Chakrapani, N.; Wei, B.; Ajayan, P. M. *Nano Lett.* 2003, *3*, 1437-1440.
[206] Michel, M.; Vautier, D.; Voegel, J.-C.; Schaaf, P.; Ball, V. *Langmuir* 2004, *20*, 4835-4839.
[207] Collin, D.; Lavalle, P.; Garza, J. M.; Voegel, J.-C.; Schaff, P.; Martinoty, P. *Macromolecules* 2004, *37*, 10195-10198.
[208] Stockton, W. B.; Rubner, M. F. *Macromolecules* 1997, *30*, 2717-2725.
[209] Wang, L. Y.; Wang, Z. Q.; Zhang, X.; Shen, J. C.; Chi, L. F.; Fuchs, H. *Macromol. Rapid Commun.* 1997, *18*, 509-514.
[210] Elbert, D. L.; Herbert, C. B.; Hubbell, J. A. *Langmuir* 1999, *15*, 5355-5362.
[211] Kozlovskaya, V.; Sukhishvili, S. A. *Macromolecules* 2006, *39*, 5569-5572.
[212] Boulmedais, F.; Schwinte, P.; Gergerly, C.; Voegel, J.-C.; Schaaf, P. *Langmuir* 2002, *18*, 4523-4525.
[213] Boulmedais, F.; Ball, V.; Schwinte, P.; Frisch, B.; Schaaf, P.; Voegel, J.-C. *Langmuir* 2003, *19*, 440-445.
[214] Serizawa, T.; Yamaguchi, M.; Akashi, M. *Biomacromolecules* 2002, *3*, 724-731.

[215] Benkirane-Jessel, N.; Schwinté, P.; Falvey, P.; Darcy, R.; Haïkel, Y.; Schaaf, P.; Voegel, J.-C.; Ogier, J. *Adv. Funct. Mater.* 2004, *14*, 174-182.
[216] Etienne, O.; Gasnier, C.; Taddei, C.; Voegel, J.-C.; Aunis, D.; Schaaf, P.; Metz-Boutigue, M.-H.; Bolcato-Bellemin, A.-L.; Egles, C. *Biomaterials* 2005, *26*, 6704-6712.
[217] Hollman, A. M.; Bhattacharyya, D. *Langmuir* 2004, *20*, 5418-5424.
[218] Halthur, T. J.; Elofsson, U. M. *Langmuir* 2004, *20*, 1739-1745.
[219] Halthur, T. J.; Claesson, P. M.; Elofsson, U. M. *J. Am.Chem. Soc.* 2004, *126*, 17009-17015.
[220] Tezcaner, A.; Hicks, D.; Boulmedais, F.; Sahel, J.; Schaaf, P.; Voegel, J.-C.; Lavalle, P. *Biomacromolecules* 2006, *7*, 86-94.
[221] Michel, M.; Izquierdo, A.; Decher, G.; Voegel, J. C.; Schaaf, P.; Ball, V. *Langmuir* 2005, *21*, 7854-7859.
[222] Michel, M.; Arntz, Y.; Fleith, G.; Toquant, J.; Haikel, Y.; Voegel, J.-C.; Schaaf, P.; Ball, V. *Langmuir* 2006, *22*, 2358-2364.
[223] Kidambi, S.; Lee, I.; Chan, C. *J. Am. Chem. Soc.* 2004, *126*, 16286-16287.
[224] Cooper, C. L.; Dubin, P. L.; Kayitmazer, A. B.; Turksen, S. *Curr. Opin. Colloid Interface Sci.* 2005, *10*, 52-78.
[225] Lvov, Y.; Ariga, K.; Ichinose, I.; Kunitake, T. J. Am. Chem. Soc. 1995, 117, 6117-6123.
[226] Caruso, F.; Niikura, K.; Furlong, D. N.; Okahata, Y. *Langmuir* 1997, *13*, 3427-3433.
[227] Caruso, F.; Furlong, D. N.; Ariga, K.; Ichinose, I. ; Kunitake, T. *Langmuir* 1998, *14*, 4559-4565.
[228] Cassier, T.; Lowack, K.; Decher, G. *Supramolecular Science* 1998, *5*, 309-315.
[229] Brynda, E.; Houska, M. *Macromol. Rapid Commun.* 1998, *19*, 173-176.
[230] Caruso, F.; Möhwald, H. *J. Am. Chem. Soc.* 1999, *121*, 6039-6046.
[231] Werner, C.; Eichhorn, K.-J.; Grundke, K.; Simon, F.; Grählert, W.; Jacobasch, H.-J. *Colloids Surf. A* 1999, *156*, 3-17.
[232] Ladam, G.; Gergely, C.; Senger, B.; Decher, G.; Voegel, J.-C.; Schaaf, P.; Cuisinier, F. J. G. *Biomacromolecules* 2000, *1*, 674-687.
[233] Schüler, C.; Caruso, F. *Macromol. Rapid Commun.* 2000, *21*, 750-753.
[234] Jessel, N.; Atalar, F.; Lavalle, P.; Mtterer, J.; Decher, G.; Schaaf, P.; Voegel, J.-C.; Ogier, J. *Adv. Mater.* 2003, *15*, 692-695.
[235] Lojou, E.; Bianco, L. *Langmuir* 2004, *20*, 748-755.
[236] Hiller, S.; Leporatti, S.; Schnäckel, A.; Typlt, E.; Donath, E. *Biomacromolecules* 2004, *5*, 1580-1587.
[237] Patel, D. S.; Aithal, R. K.; Krishna, G.; Lvov, Y. M.; Tien, M.; Kuila, D. *Colloids Surf. B* 2005, *43*, 13-19.
[238] Ren, K.; Ji, J.; Shen, J. *Macromol. Rapid Commun.* 2005, *26*, 1633-1638.
[239] Johnston, A. P. R.; Read, E. S.; Caruso, F. *Nano Lett.* 2005, *5*, 953-956.
[240] Liang, Z.; Wang, C.; Tong, Z.; Ye, W.; Ye, S. *React. Funct. Polym.* 2005, *63*, 85-94.
[241] Desigaux, L.; Belkacem, M. B.; Richard, P.; Cellier, J.; Leone, P.; Cario, L.; Leroux, F.; Taviot-Guého, C.; Pitard, B. *Nano Lett.* 2006, *6*, 199-204.
[242] Yu, A.; Liang, Z.; Caruso, F. *Chem. Mater.* 2005, *17*, 171-175.
[243] Yang, S.; Li, Y.; Jiang, X.; Chen, Z.; Lin, X. *Sens. Actuators B* 2006, *114*, 774-780.
[244] Recksiedler, C. L.; Deore, B. A.; Freund, M. S. *Langmuir* 2006, *22*, 2811-2815.

[245] Ichinose, I.; Kuroiwa, K.; Lvov, Y.; Kunitake, T. In: *Multilayer Thin Films*; Decher, G.; Schlenoff, J. B.; Eds.; Wiley-VCH: Weinheim, 2003, pp 155-175.

[246] Ladam, G.; Gergely, C.; Senger, B.; Decher, G.; Voegel, J. C.; Schaaf, P.; Cuisinier, F. *Biomacromolecules* 2000, *1*, 674-687.

[247] Ngankam, A. P.; Mao, G.; Van Tassel, P. R. *Langmuir* 2004, *20*, 3362-3370.

[248] Liu, Y.; Lu, H.; Zhong, W.; Song, P.; Kong, J.; Yang, P.; Girault, H. H.; Liu, B. *Anal. Chem.* 2006, *78*, 801-808.

[249] Ladam, G.; Schaaf, P.; Cuisinier, F.; Decher, G.; Voegel, J. C. *Langmuir* 2001, *17*, 878-882.

[250] Gergerly, C.; Bahi, S.; Szalontai, B.; Flores, H.; Schaaf, P.; Voegel, J. C.; Cuisinier, F. J. G. *Langmuir* 2004, *20*, 5575-5582.

[251] Salloum, S. D.; Schlenoff, J. B. *Biomacromolecules* 2004, *5*, 1089-1096.

[252] Weidinger, I. M. Murgida, D. H.; Dong, W.; Möhwald, H.; Hildebrandt, P. *J. Phys. Chem. B* 2006, *110*, 522-529.

[253] Dai, J.; Baker, G. L.; Bruening, M. L. *Anal. Chem.* 2006, *78*, 135-140.

[254] Müller, M.; Kessler, B.; Houbenov, N.; Bohata, K.; Pientka, Z.; Brynda, E. *Biomacromolecules* 2006, *7*, 1285-1294.

[255] Cooper, Th. M.; Campbell, A. L.; Crane, R. L. *Langmuir* 1995, *11*, 2713-2718.

[256] Araki, K.; Wagner, M. J.; Wrighton, M. S. *Langmuir* 1996, *12*, 5393-5398.

[257] Laschewsky, A.; Mayer, B.; Wischerhoff, E. *Ber. Bunsenges. Phys. Chem.* 1996, *100*, 1033-1038.

[258] Ariga, K.; Lvov, Y.; Kunitake, T. *J. Am. Chem. Soc.* 1997, *119*, 2224-2231.

[259] Yoo, D.; Wu, A.; Lee, J.; Rubner, M. F. *Synth. Met.* 1997, *85*, 1425-1426.

[260] Linford, M. R.; Auch, M.; Möhwald, H. *J. Am. Chem. Soc.* 1998, *120*, 178-182.

[261] Koetse, M.; Laschewsky, A.; Verbiest, T. *Mater. Sci. Eng. C* 1999, *10*, 107-113.

[262] Kiyoi, T.; Inoue, Y.; Ohmoto, H.; Yoshida, M.; Kiso, M.; Kondo, H.; Menzel, H.; Ruther, M.; Stumpe, J.; Fischer, T. *Supramolecular Sci.* 1998, *5*, 49-59.

[263] Lenahan, K. M.; Wang, Y.-X.; Liu, Y.; Claus, R. O.; Heflin, J. R.; Marciu, D.; Figura, C. *Adv. Mater.* 1998, *10*, 853-855.

[264] Advincula, R. C.; Fells, E.; Park, M. *Chem. Mater.* 2001, *13*, 2870-2878.

[265] Das, S.; Pal, A. J. *Langnuir* 2002, *18*, 458-461.

[266] Advincula, R.; Park, M.-K.; Baba, A.; Kaneko, F. *Langmuir* 2003, *19*, 654-665.

[267] Nicol, E.; Habib-Jiwan, J.-L.; Jonas, A. M. *Langmuir* 2003, *19*, 6178-6186.

[268] Tieke, B.; Pyrasch, M.; Toutianoush, A. In: *Multilayer Thin Films*; Decher, G.; Schlenoff, J. B.; Eds.; Wiley-VCH: Weinheim, 2003, pp 427-460.

[269] Zucolotto, V.; He, J.-A.; Constantino, C. J. L.; Barbosa Neto, N. M.; Rodrigues Jr., J. J.; Mendonça, C. R.; Zilio, S. C.; Li, L.; Aroca, R. F.; Oliviera Jr., O. N.; Kumar, J. *Polymer* 2003, *44*, 6129-6133.

[270] Zhaohui, Y.; Weixiao, C. *J. Polym. Sci. Part A Polym. Chem.* 2003, *41*, 3103-3108.

[271] Decher, G.; Hong, J. D. *Ber. Bunsenges. Phys. Chem.* 1991, *95*, 1430-1434.

[272] Toutianoush, A.; Saremi, F.; Tieke, B. *Mater. Sci. Eng. C* 1999, *8-9*, 343-352.

[273] Hong, J.-D.; Park, E.-S.; Park, A.-L. *Langmuir* 1999, *15*, 6515-6521.

[274] Delajon, C.; Gutberlet, T.; Steitz, R.; Möhwald, H.; Krastev, R. *Langmuir* 2005, *21*, 8509-8514.

[275] Sato, K.; Suzuki, I.; Anzai, J. *Langmuir* 2003, *19*, 7406-7412.

[276] Dragan, S., Ghimici, L. *Angew. Makromol. Chem.* 1991, *192*, 199-211.

[277] Ghimici, L.; Dragan, S.; Popescu, F. *J. Polym. Sci. Part B Polym. Phys*. 1997, *35,* 2571-2581.

[278] Dragan, S.; Ghimici, L.; Cristea, M.; Airinei, A. *Acta Polym.* 1999, *50*, 260-266.

[279] Dragan, S.; Schwarz, S.; Eichhorn, K.-J.; Lunkwitz, K. *Colloids Surf. A* 2001, *195*, 243-251.

[280] Dragan, S.; Schwarz, S. *Macromol. Symp.* 2002, *181*, 155-166.

[281] Dragan, S.; Schwarz, S. *Prog. Colloid Polym. Sci.* 2003, *122*, 8-15.

[282] Dragan, E. S.; Schwarz, S.; Eichhorn, K.-J. Manuscript in preparation.

[283] Dragan, S.; Timpu, D. *J. Polym. Sci. Part A Polym. Chem.* 2003, *41*, 264-272.

[284] v. Klitzing, R.; Möhwald, H. *Macromolecules* 1999**,** *29*, 6901-6906.

[285] Caruso, F.; Lichtenfeld, H.; Donath, E.; Möhwald, H. *Macromolecules* 1999**,** *32*, 2317-2328.

[286] Shi, X.; Cassagneau, T.; Caruso, F. *Langmuir* 2002, *18*, 904-910.

[287] Chung, A. J.; Rubner, M. F. *Langnuir* 2002, *18*, 1176-1183.

[288] Fendler, J. H. *Chem. Mater.* 1996, *8*, 1616-1624.

[289] Kotov, N. A.; Haraszti, T.; Turi, L.; Zavala, G.; Geer, R. E.; Dékány, I.; Fendler, J. H. *J. Am. Chem. Soc.* 1997, *119*, 6821-6832.

[290] Tsukruk, V. V. *Prog. Polym. Sci.* 1997, *22*, 247-311.

[291] Serizawa, T.; Takeshita, H.; Akashi, M. *Langmuir* 1998, *14*, 4088-4094.

[292] Kotov, N. A. In: *Multilayer Thin Films*; Decher, G.; Schlenoff, J. B.; Eds.; Wiley-VCH: Weinheim, 2003, pp 207-243.

[293] Zheng, H.; Rubner, M. F.; Hammond, P. T. *Langmuir* 2002, *18*, 4505-4510.

[294] Koo, H. Y.; Yi, D. K.; Yoo, S. J.; Kim, D.-Y. *Adv. Mater.* 2004, *16*, 274-277.

[295] Park, J.; Fouche, L. D.; Hammond, P. T. *Adv. Mater.* 2005, *17*, 2575-2579.

[296] Ahn, J. S.; Hammond, P. T.; Rubner, M. F.; Lee, I. *Colloids Surf. A* 2005, *259*, 45-53.

[297] Sangribsub, S.; Tangboriboonrat, P.; Pith, T.; Decher, G. *Polym. Bull.* 2005, *53*, 425-434.

[298] Nyamjav, D.; Ivanisevic, A. *Talanta* 2005, *67*, 503-505.

[299] Glinel, K.; Prevot, M.; Krustev, R.; Sukhorukov, G. B.; Jonas, A. M.; Möhwald, H. *Langmuir* 2004, *20*, 4898-4902.

[300] Zhao, W.; Xu, J.-J.; Shi, C.-G.; Chen, H.-Y. *Langmuir* 2005, *21*, 9630-9634.

[301] Liu, Z.; Wang, X.; Wu, H.; Li, C. *J. Colloid Interface Sci.* 2005, *287*, 604-611.

[302] Podsiadlo, P.; Choi, S.-Y.; Shim, B.; Lee, J.; Cuddihy, M.; Kotov, N. A. *Biomacromolecules* 2005, *6*, 2914-2918.

[303] Hao, E.; Lian, T. *Chem. Mater.* 2000, *12*, 3392-3396.

[304] Jiang, C.; Markutsya, S.; Tsukruk, V. V. *Adv. Mater.* 2004, *16*, 157-161.

[305] Jiang, C.; Markutsya, S.; Tsukruk, V. V. *Langmuir* 2004, *20*, 882-890.

[306] Hu, X.; Cheng, W.; Wang, T.; Wang, Y.; Wang, E.; Dong, S. *J. Phys. Chem. B* 2005, *109*, 19385-19389.

[307] Li, X.; Xu, W.; Zhang, J.; Jia, H.; Yang, B.; Zhao, B.; Li, B.; Ozaki, Y. *Langmuir* 2004, *20*, 1298-1304.

[308] Cui, F.; Zhang, J.; Cui, T.; Han, K.; Xie, B.; Lin, Q.; Yang, B. *Colloids Surf. A* 2006, *278*, 39-45.

[309] Sasaki, T.; Ebina, Y.; Watanabe, M.; Decher, G. *Chem. Commun.* 2000, 2163-2164.

[310] Ma, R.; Sasaki, T.; Bando, Y. *J. Am. Chem. Soc.* 2004, *126*, 10382-10388.

[311] Nakamura, H.; Ishii, M.; Tsukigase, A.; Harada, M.; Nakano, H. *Langmuir* 2006, *22*, 1268-1272.
[312] Sennerfors, T.; Bogdanovic, G.; Tiberg, F. *Langmuir* 2002, *18*, 6410-6415.
[313] Liu, H.; Rusling, J. F.; Hu, N. *Langmuir* 2004, *20*, 10700-10705.
[314] He, P.; Hu, N.; Rusling, J. F. *Langmuir* 2004, *20*, 722-729.
[315] Sun, B.; Mutch, S. A.; Lorenz, R. M.; Chiu, D. T. *Langmuir* 2005, *21*, 10763-10769.
[316] Yuan, J.; Zhou, S.; You, B.; Wu, L. *Chem. Mater.* 2005, *17*, 3587-3594.
[317] Glinel, K.; Laschewsky, A.; Jonas, A. M. *Macromolecules* 2001, *34*, 5267-5274.
[318] Villaume, P. Y.; Jonas, A. M.; Laschewsky, A. *Macromolecules* 2002, *35*, 5004-5012.
[319] Hua, F.; Cui, T.; Lvov, Y. M. *Nano Lett.* 2004, *4*, 823-825.
[320] Kidambi, S.; Dai, J.; Bruening, M. L. *J. Am. Chem. Soc.* 2004, *126*, 2658-2659.
[321] Kidambi, S.; Bruening, M. L. *Chem. Mater.* 2005, *17*, 301-307.
[322] Liu, J.; Li, X.; Schrand, A.; Ohashi, T.; Dai, L. *Chem. Mater.* 2005, *17*, 6599-6604.
[323] Nakayama, M.; Tagashira, H. *Langmuir* 2006, *22*, 3864-3869.
[324] Zhang, S.; Zhu, Y.; Yang, Li, C. *Colloids Surf. A* 2005, *264*, 215-218.
[325] Jaffar, S.; Nam, K. T.; Khademhosseini, A.; Xing, J.; Langer, R. S.; Belcher, A. M. *Nano Lett.* 2004, *4*, 1421-1425.
[326] Rouse, J. H.; Lillehei, P. T. *Nano Lett.* 2003, *3*, 59-62.
[327] Rouse, J. H.; Lillehei, P. T.; Sanderson, J.; Siochi, E. J. *Chem. Mater.* 2004, *16*, 3904-3910.
[328] Liu, J.; Tian, S.; Knoll, W. *Langmuir* 2005, *21*, 5596-5599.
[329] Paloniemi, H.; Lukkarinen, M.; Ääritalo, T.; Areva, S.; Leiro, J.; Heinonen, M; Haapakka, K.; Lukkari, J. *Langmuir* 2006, *22*, 74-83.
[330] Zhang, M.; Yan, Y.; Gong, K.; Mao, L.; Guo, Z.; Chen, Y. *Langmuir* 2004, *20*, 8781-8785.
[331] Corre-Duarte, M. A.; Kosiorek, A.; Kandulski, W.; Giersig, M.; Liz-Marzan, L. M. *Chem. Mater.* 2005, *17*, 3268-3272.
[332] Firkowska, I.; Olek, M.; Pazos-Peréz, N.; Rojas-Chapana, J.; Giersig, M. *Langmuir* 2006, *22*, 5427-5434.
[333] Xu, Z.; Gao, N.; Dong, S. *Talanta* 2006, *68*, 753-758.
[334] Abu-Sharkh, B. *Langmuir* 2006, *22*, 3028-3034.
[335] Jeon, J.; Panchagnula, V.; Pan, J.; Dobrynin, A. V. *Langmuir* 2006, *22*, 4629-4637.
[336] Sukhorukov, G. B.; Dähne, L.; Hartmann, J.; Donath, E.; Möhwald, H. *Adv. Mat.* 2000, *12*, 112-115.
[337] Antipov, A. A.; Sukhorukov, G. B.; Donath, E.; Möhwald, H. *J. Phys. Chem. B* 2001, *105*, 2281-2284.
[338] Gao, C.; Leporetti, S.; Moya, S.; Donath, E.; Möhwald, H. *Langmuir* 2001, *17*, 3491-3495.
[339] Shi, X.; Caruso, F. *Langmuir* 2001, *17*, 2036-2042.
[340] Antipov, A. A.; Sukhorukov, G. B.; Möhwald, H. *Langmuir* 2003, *19*, 2444-2448.
[341] Lulevich, V. V.; Radtchenko, I. L.; Sukhorukov, G. B.; Vinogradova, O. I. *Macromolecules* 2003, *36*, 2832-2837.
[342] Möhwald, H.; Donath, E.; Sukhorukov, G. B. In: *Multilayer Thin Films*; Decher, G.; Schlenoff, J. B.; Eds.; Wiley-VCH: Weinheim, 2003, pp 363-392.
[343] Park, M. K.; Onishi, K.; Locklin, J.; Caruso, F.; Advincula R. C. *Langmuir* 2003, *19*, 8550-8554.

[344] Shi, X.; Briseno, A. L.; Sanedrin, R. J.; Zhou, F. *Macromolecules* 2003, *36*, 4093-4098.
[345] Kozlovskaya, V.; Ok, S.; Sousa, A.; Libera, M.; Sukhishvili, S. A. *Macromolecules* 2003, *36*, 8590-8592.
[346] Balabushevich, N. G.; Sukhorukov, G. B.; Larionova, N. I. *Macromol. Rapid Commun.* 2005, *26*, 1168-1172.
[347] Yang, X.; Han, X.; Zhu, Y. *Colloids Surf. A* 2005, *264*, 49-54.
[348] Zhang, S.; Zhu, Y.; Yang, X.; Li, C. *Colloids Surf. A* 2005, *264*, 215-218.
[349] Lulevich, V. V.; Vinogradova, O. I. *Langmuir* 2004, *20*, 2874-2878.
[350] Lebedeva, O. V.; Kim, B.-S.; Vasilev, K.; Vinogradova, O. I. *J. Colloid Interface Sci.* 2005, *284*, 455-462.
[351] Antipov, A. A.; Sukhorukov, G. B. *Adv. Colloid Interf. Sci.* 2004, *111*, 49-61.
[352] Shchukin, D. G., Ustinovich, E. A.; Sukhorukov, G. B.; Möhwald, H.; Sviridov, D. V. *Adv. Mater.* 2005, *17*, 468-472.
[353] Ko, H.; Jiang, C.; Shulha, H.; Tsukruk. V. V. *Chem. Mater.* 2005, *17*, 2490-2493.

INDEX

A

B

C

D

E

F

G

H

I

J

K

L

M

N

O

P

Q

R

S

T

Y

Z